Atlantis Studies in Dynamical Systems

Volume 5

The “Atlantis Studies in Dynamical Systems” publishes monographs in the area of dynamical systems, written by leading experts in the field and useful for both students and researchers. Books with a theoretical nature will be published alongside books emphasizing applications.

More information about this series at http://www.atlantis-press.com

Svetlin G. Georgiev

Integral Equations on Time Scales

ATLANTIS PRESS

Svetlin G. Georgiev
Department of Differential Equations
Sofia University
Sofia
Bulgaria

Atlantis Studies in Dynamical Systems
ISBN 978-94-6239-227-4 ISBN 978-94-6239-228-1 (eBook)
DOI 10.2991/978-94-6239-228-1

Library of Congress Control Number: 2016951703

Printed on acid-free paper

Preface

Many problems arising in applied mathematics or mathematical physics, can be formulated in two ways namely as differential equations and as integral equations. In the differential equation approach, the boundary conditions have to be imposed externally, whereas in the case of integral equations, the boundary conditions are incorporated within the formulation, and this confers a valuable advantage to the latter method. Moreover, the integral equation approach leads quite naturally to the solution of the problem as an infinite series, known as the Neumann expansion, the Adomian decomposition method, and the series solution method in which the successive terms arise from the application of an iterative procedure. The proof of the convergence of this series under appropriate conditions presents an interesting exercise in an elementary analysis.

This book encompasses recent developments of integral equations on time scales. For many population models biological reasons suggest using their difference analogues. For instance, North American big game populations have discrete birth pulses, not continuous births as is assumed by differential equations. Mathematical reasons also suggest using difference equations—they are easier to construct and solve in a computer spreadsheet. North American large mammal populations do not have continuous population growth, but rather discrete birth pulses, so the differential equation form of the logistic equation will not be convenient. Age-structured models add complexity to a population model, but make the model more realistic, in that essential features of the population growth process are captured by the model. They are used difference equations to define the population model because discrete age classes require difference equations for simple solutions. The discrete models can be investigated using integral equations in the case when the time scale is the set of the natural numbers. A powerful method introduced by Poincaré for examining the motion of dynamical systems is that of a Poincaré section. This method can be investigated using integral equations on the set of the natural numbers. The total charge on the capacitor can be investigated with an integral equation on the set of the harmonic numbers.

This book contains elegant analytical and numerical methods. This book is intended for the use in the field of integral equations and dynamic calculus on time scales. It is also suitable for graduate courses in the above fields. This book contains nine chapters. The chapters in this book are pedagogically organized. This book is specially designed for those who wish to understand integral equations on time scales without having extensive mathematical background.

The basic definitions of forward and backward jump operators are due to Hilger. In Chap. 1 are given examples of jump operators on some time scales. The graininess function, which is the distance from a point to the closed point on the right, is introduced in this chapter. In this chapter, the definitions for delta derivative and delta integral are given and some of their properties are deducted. The basic results in this chapter can be found in [2]. Chapter 2 introduces the classification of integral equations on time scales and necessary techniques to convert dynamic equations to integral equations on time scales. Chapter 3 deals with the generalized Volterra integral equations and the relevant solution techniques. Chapter 4 is concerned with the generalized Volterra integro-differential equations and also solution techniques. Generalized Fredholm integral equations are investigated in Chap. 5. Chapter 6 is devoted on Hilbert–Schmidt theory of generalized integral equations with symmetric kernels. The Laplace transform method is introduced in Chap. 7. Chapter 8 deals with the series solution method. Nonlinear integral equations on time scales are introduced in Chap. 9.

The aim of this book was to present a clear and well-organized treatment of the concept behind the development of mathematics and solution techniques. The text material of this book is presented in highly readable, mathematically solid format. Many practical problems are illustrated displaying a wide variety of solution techniques. Nonlinear integral equations on time scales and some of their applications in the theory of population models, biology, chemistry, and electrical engineering will be discussed in a forthcoming book "Nonlinear Integral Equations on Time Scales and Applications."

The author welcomes any suggestions for the improvement of the text.

Paris, France Svetlin G. Georgiev
June 2016

Contents

Chapter 1
Elements of the Time Scale Calculus

This chapter is devoted to a brief exposition of the time scale calculus that provide the framework for the study of integral equations on time scales. Time scale calculus is very interesting in itself, this challenging subject has been developing very rapidly in the last decades. A detailed discussion of the time scale calculus is beyond the scope of this book, for this reason the author confine to outlining a minimal set of properties needed in the further proceeding. The presentation in this chapter follows the book [2]. A deep and thorough insight into the time scale calculus, as well as the discussion of the available bibliography on this issue, can be found in the book [2].

1.1 Forward and Backward Jump Operators, Graininess Function

Definition 1 A time scale is an arbitrary nonempty closed subset of the real numbers.

We will denote a time scale by the symbol $\mathscr{T}$.

We suppose that a time scale $\mathscr{T}$ has the topology that inherits from the real numbers with the standard topology.

Example 1 $[1, 2]$, $\mathscr{R}$, $\mathscr{N}$ are time scales.

Example 2 $[a, b)$, $(a, b]$, (a, b) are not time scales.

Definition 2 For $t \in \mathscr{T}$ we define the *forward jump operator* $\sigma : \mathscr{T} \longmapsto \mathscr{T}$ as follows

$$\sigma(t) = \inf\{s \in \mathscr{T} : s > t\}.$$

We note that $\sigma(t) \geq t$ for any $t \in \mathscr{T}$.

S.G. Georgiev, *Integral Equations on Time Scales*, Atlantis Studies in Dynamical Systems 5, DOI 10.2991/978-94-6239-228-1_1

Definition 3 For $t \in \mathscr{T}$ we define the *backward jump operator* $\rho : \mathscr{T} \longmapsto \mathscr{T}$ by

$$\rho(t) = \sup\{s \in \mathscr{T} : s < t\}.$$

We note that $\rho(t) \le t$ for any $t \in \mathscr{T}$.

Definition 4 We set

$$\inf \emptyset = \sup \mathscr{T}, \quad \sup \emptyset = \inf \mathscr{T}.$$

Definition 5 For $t \in \mathscr{T}$ we have the following cases.

1. If $\sigma(t) > t$, then we say that t is *right-scattered.*
2. If $t < \sup \mathscr{T}$ and $\sigma(t) = t$, then we say that t is *right-dense.*
3. If $\rho(t) < t$, then we say that t is *left-scattered.*
4. If $t > \inf \mathscr{T}$ and $\rho(t) = t$, then we say that t is *left-dense.*
5. If t is left-scattered and right-scattered at the same time, then we say that t is *isolated.*
6. If t is left-dense and right-dense at the same time, then we say that t is *dense.*

Example 3 Let $\mathscr{T} = \{\sqrt{2n+1} : n \in \mathscr{N}\}$. If $t = \sqrt{2n+1}$ for some $n \in \mathscr{N}$, then $n = \dfrac{t^2-1}{2}$ and

$$\sigma(t) = \inf\{l \in \mathscr{N} : \sqrt{2l+1} > \sqrt{2n+1}\} = \sqrt{2n+3} = \sqrt{t^2+2} \quad for \quad n \in \mathscr{N},$$
$$\rho(t) = \sup\{l \in \mathscr{N} : \sqrt{2l+1} < \sqrt{2n+1}\} = \sqrt{2n-1} = \sqrt{t^2-2} \quad for \quad n \in \mathscr{N}, n \ge 2.$$

For $n = 1$ we have

$$\rho(\sqrt{3}) = \sup \emptyset = \inf \mathscr{T} = \sqrt{3}.$$

Since

$$\sqrt{t^2-2} < t < \sqrt{t^2+2} \qquad \text{for} \qquad n \ge 2,$$

we conclude that every point $\sqrt{2n+1}$, $n \in \mathscr{N}$, $n \ge 2$, is right-scattered and left-scattered, i.e., every point $\sqrt{2n+1}$, $n \in \mathscr{N}$, $n \ge 2$, is isolated.

Because

$$\sqrt{3} = \rho(\sqrt{3}) < \sigma(\sqrt{3}) = \sqrt{5},$$

we have that the point $\sqrt{3}$ is right-scattered.

Example 4 Let $\mathscr{T} = \left\{\dfrac{1}{2n} : n \in \mathscr{N}\right\} \cup \{0\}$ and $t \in \mathscr{T}$ be arbitrarily chosen.

1. $t = \dfrac{1}{2}$. Then

$$\sigma\left(\frac{1}{2}\right) = \inf\left\{\frac{1}{2l}, 0 : \frac{1}{2l}, 0 > \frac{1}{2}, l \in \mathscr{N}\right\} = \inf \emptyset = \sup \mathscr{T} = \frac{1}{2},$$
$$\rho\left(\frac{1}{2}\right) = \sup\left\{\frac{1}{2l}, 0 : \frac{1}{2l}, 0 < \frac{1}{2}, l \in \mathscr{N}\right\} = 0 < \frac{1}{2},$$

i.e., $\frac{1}{2}$ is left-scattered.

2. $t = \frac{1}{2n}$, $n \in \mathscr{N}$, $n \geq 2$. Then

$$\sigma\left(\frac{1}{2n}\right) = \inf\left\{\frac{1}{2l} : \frac{1}{2l} > \frac{1}{2n}, l \in \mathscr{N}\right\} = \frac{1}{2(n-1)} > \frac{1}{2n},$$
$$\rho\left(\frac{1}{2n}\right) = \sup\left\{\frac{1}{2l}, 0 : \frac{1}{2l}, 0 < \frac{1}{2n}, l \in \mathscr{N}\right\} = \frac{1}{2(n+1)} < \frac{1}{2n}.$$

Therefore all points $\frac{1}{2n}$, $n \in \mathscr{N}$, $n \geq 2$, are right-scattered and left-scattered, i.e., all points $\frac{1}{2n}$, $n \in \mathscr{N}$, $n \geq 2$, are isolated.

3. $t = 0$. Then

$$\sigma(0) = \inf\{s \in \mathscr{T} : s > 0\} = 0,$$
$$\rho(0) = \sup\{s \in \mathscr{T} : s < 0\} = \sup \emptyset = \inf \mathscr{T} = 0.$$

Example 5 Let $\mathscr{T} = \left\{\frac{n}{3} : n \in \mathscr{N}_0\right\}$ and $t = \frac{n}{3}$, $n \in \mathscr{N}_0$, be arbitrarily chosen.

1. $n \in \mathscr{N}$. Then

$$\sigma\left(\frac{n}{3}\right) = \inf\left\{\frac{l}{3}, 0 : \frac{l}{3}, 0 > \frac{n}{3}, l \in \mathscr{N}_0\right\} = \frac{n+1}{3} > \frac{n}{3},$$

$$\rho\left(\frac{n}{3}\right) = \sup\left\{\frac{l}{3}, 0 : \frac{l}{3}, 0 < \frac{n}{3}, l \in \mathscr{N}_0\right\} = \frac{n-1}{3} < \frac{n}{3}.$$

Therefore all points $t = \frac{n}{3}$, $n \in \mathscr{N}$, are right-scattered and left-scattered, i.e., all points $t = \frac{n}{3}$, $n \in \mathscr{N}$, are isolated.

2. $n = 0$. Then

$$\sigma(0) = \inf\left\{\frac{l}{3}, 0 : \frac{l}{3}, 0 > 0, l \in \mathscr{N}_0\right\} = \frac{1}{3} > 0,$$
$$\rho(0) = \sup\left\{\frac{l}{3} : \frac{l}{3}, 0 < 0, l \in \mathscr{N}_0\right\} = \sup \emptyset = \inf \mathscr{T} = 0,$$

i.e., $t = 0$ is right-scattered.

Exercise 1 Classify each point $t \in \mathscr{T} = \{\sqrt[3]{2n-1} : n \in \mathscr{N}_0\}$ as left-dense, left-scattered, right-dense, or right-scattered.

Answer. The points $\sqrt[3]{2n-1}$, $n \in \mathscr{N}$, are isolated, the point -1 is right-scattered.

Definition 6 The numbers

$$H_0 = 0, \quad H_n = \sum_{k=1}^{n} \frac{1}{k}, \quad n \in \mathscr{N},$$

will be called harmonic numbers.

Exercise 2 Let

$$\mathscr{H} = \{H_n : n \in \mathscr{N}_0\}.$$

Prove that $\mathscr{H}$ is a time scale. Find $\sigma(t)$ and $\rho(t)$.

Answer. $\sigma(H_n) = H_{n+1}, n \in \mathscr{N}_0,\ \rho(H_n) = H_{n-1},\ n \in \mathscr{N},\ \rho(H_0) = H_0$.

Definition 7 The *graininess function* $\mu : \mathscr{T} \longmapsto [0, \infty)$ is defined by

$$\mu(t) = \sigma(t) - t.$$

Example 6 Let $\mathscr{T} = \left\{2^{n+1} : n \in \mathscr{N}\right\}$. Let also, $t = 2^{n+1} \in \mathscr{T}$ for some $n \in \mathscr{N}$. Then

$$\sigma(t) = \inf\left\{2^{l+1} : 2^{l+1} > 2^{n+1}, l \in \mathscr{N}\right\} = 2^{n+2} = 2t.$$

Hence,

$$\mu(t) = \sigma(t) - t = 2t - t = t \quad \text{or} \quad \mu\left(2^{n+1}\right) = 2^{n+1}, \quad n \in \mathscr{N}.$$

Example 7 Let $\mathscr{T} = \left\{\sqrt{n+1} : n \in \mathscr{N}\right\}$. Let also, $t = \sqrt{n+1}$ for some $n \in \mathscr{N}$. Then $n = t^2 - 1$ and

$$\sigma(t) = \left\{\sqrt{l+1} : \sqrt{l+1} > \sqrt{n+1}, l \in \mathscr{N}\right\} = \sqrt{n+2} = \sqrt{t^2+1}.$$

Hence,

$$\mu(t) = \sigma(t) - t = \sqrt{t^2+1} - t \quad \text{or} \quad \mu(\sqrt{n+1}) = \sqrt{n+2} - \sqrt{n+1}, \quad n \in \mathscr{N}.$$

Example 8 Let $\mathscr{T} = \left\{\frac{n}{2} : n \in \mathscr{N}_0\right\}$. Let also, $t = \frac{n}{2}$ for some $n \in \mathscr{N}_0$. Then $n = 2t$ and

$$\sigma(t) = \inf\left\{\frac{l}{2} : \frac{l}{2} > \frac{n}{2}, l \in \mathscr{N}_0\right\} = \frac{n+1}{2} = t + \frac{1}{2}.$$

Hence,

$$\mu(t) = \sigma(t) - t = t + \frac{1}{2} - t = \frac{1}{2} \quad \text{or} \quad \mu\left(\frac{n}{2}\right) = \frac{1}{2}.$$

Example 9 Suppose that $\mathcal{T}$ consists of finitely many different points: $t_1, t_2, \ldots, t_k$. Without loss of generality we can assume that

$$t_1 < t_2 < \ldots < t_k.$$

For $i = 1, 2, \ldots, k-1$ we have

$$\sigma(t_i) = \inf\{t_l \in \mathcal{T} : t_l > t_i, l = 1, 2, \ldots, k\} = t_{i+1}.$$

Hence,

$$\mu(t_i) = t_{i+1} - t_i, \quad i = 1, 2, \ldots, k-1.$$

Also,

$$\sigma(t_k) = \inf\{t_l \in \mathcal{T} : t_l > t_k, l = 1, 2, \ldots, k\} = \inf \emptyset = \sup \mathcal{T} = t_k.$$

Therefore

$$\mu(t_k) = \sigma(t_k) - t_k = t_k - t_k = 0.$$

From here,

$$\sum_{i=1}^{k} \mu(t_i) = \sum_{i=1}^{k-1} \mu(t_i) + \mu(t_k) = \sum_{i=1}^{k-1} (t_{i+1} - t_i) = t_k - t_1.$$

Exercise 3 Let $\mathcal{T} = \left\{\sqrt[3]{n+2} : n \in \mathcal{N}_0\right\}$. Find $\mu(t)$, $t \in \mathcal{T}$.

Answer. $\mu\left(\sqrt[3]{n+2}\right) = \sqrt[3]{n+3} - \sqrt[3]{n+2}$.

Definition 8 If $f : \mathcal{T} \longmapsto \mathcal{R}$ is a function, then we define the function $f^\sigma : \mathcal{T} \longmapsto \mathcal{R}$ by

$$f^\sigma(t) = f(\sigma(t)) \quad \text{for any} \quad t \in \mathcal{T}. \quad \text{i.e.,} \quad f^\sigma = f \circ \sigma.$$

Below, for convenience, we will use the following notation $\sigma^k(t) = (\sigma(t))^k$, $f^k(t) = (f(t))^k$, $k \in \mathcal{R}$.

Example 10 Let $\mathcal{T} = \left\{t = 2^{n+2} : n \in \mathcal{N}\right\}$, $f(t) = t^2 + t - 1$. Then

$$\sigma(t) = \inf\left\{2^{l+2} : 2^{l+2} > 2^{n+2}, l \in \mathcal{N}\right\} = 2^{n+3} = 2t.$$

Hence,

$$f^{\sigma}(t) = f(\sigma(t)) = \sigma^2(t) + \sigma(t) - 1 = (2t)^2 + 2t - 1 = 4t^2 + 2t - 1, \quad t \in \mathscr{T}.$$

Example 11 Let $\mathscr{T} = \{t = \sqrt{n+3} : n \in \mathscr{N}\}$, $f(t) = t + 3$, $t \in \mathscr{T}$. Then $n = t^2 - 3$ and

$$\sigma(t) = \inf\{\sqrt{l+3} : \sqrt{l+3} > \sqrt{n+3}, l \in \mathscr{N}\} = \sqrt{n+4} = \sqrt{t^2+1}.$$

Hence,

$$f(\sigma(t)) = \sigma(t) + 3 = \sqrt{t^2+1} + 3.$$

Example 12 Let $\mathscr{T} = \left\{\frac{1}{n} : n \in \mathscr{N}\right\} \cup \{0\}$, $f(t) = t^3 - t, t \in \mathscr{T}$.

1. $t = \frac{1}{n}, n \geq 2$. Then $n = \frac{1}{t}$ and

$$\sigma(t) = \inf\left\{\frac{1}{l}, 0 : \frac{1}{l}, 0 > \frac{1}{n}, l \in \mathscr{N}\right\} = \frac{1}{n-1} = \frac{t}{1-t}.$$

Hence,

$$\begin{aligned} f(\sigma(t)) = \sigma^3(t) - \sigma(t) &= \left(\frac{t}{1-t}\right)^3 - \frac{t}{1-t} \\ &= \frac{t^3}{(1-t)^3} - \frac{t}{1-t} = \frac{t^3 - t(1-t)^2}{(1-t)^3} \\ &= \frac{t^3 - t(1-2t+t^2)}{(1-t)^3} = \frac{t^3 - t + 2t^2 - t^3}{(1-t)^3} = \frac{t(2t-1)}{(1-t)^3}. \end{aligned}$$

2. $t = 1$. Then

$$\sigma(1) = \inf\left\{\frac{1}{l}, 0 : \frac{1}{l}, 0 > 1, l \in \mathscr{N}\right\} = \inf \emptyset = \sup \mathscr{T} = 1,$$
$$f(\sigma(1)) = \sigma^3(1) - \sigma(1) = 1 - 1 = 0.$$

3. $t = 0$. Then

$$\sigma(0) = \inf\left\{\frac{1}{l}, 0 : \frac{1}{l}, 0 > 0\right\} = 0,$$
$$f(\sigma(0)) = \sigma^3(0) - \sigma(0) = 0.$$

Exercise 4 Let $\mathscr{T} = \left\{t = \sqrt[3]{n+2} : n \in \mathscr{N}\right\}$, $f(t) = 1 - t^3, t \in \mathscr{T}$. Find $f(\sigma(t))$, $t \in \mathscr{T}$.

Answer. $-t^3$.

Definition 9 We define the set

$$\mathscr{T}^{\kappa} = \begin{cases} \mathscr{T}\backslash(\rho(\sup \mathscr{T}), \sup \mathscr{T}] & \text{if} \quad \sup \mathscr{T} < \infty \\ \mathscr{T} \quad \text{otherwise.} \end{cases}$$

Example 13 Let $\mathscr{T} = \left\{ \frac{1}{n} : n \in \mathscr{N} \right\} \cup \{0\}$. Then $\sup \mathscr{T} = 1$ and

$$\rho(1) = \sup\left\{ \frac{1}{l}, 0 : \frac{1}{l}, 0 < 1, l \in \mathscr{N} \right\} = \frac{1}{2}.$$

Therefore

$$\mathscr{T}^{\kappa} = \mathscr{T}\backslash\left(\frac{1}{2}, 1\right] = \left\{ \frac{1}{n} : n \in \mathscr{N}, n \geq 2 \right\} \cup \{0\}.$$

Example 14 Let $\mathscr{T} = \{2n : n \in \mathscr{N}\}$. Then $\sup \mathscr{T} = \infty$ and $\mathscr{T}^{\kappa} = \mathscr{T}$.

Example 15 Let $\mathscr{T} = \left\{ \frac{1}{n^2+3} : \quad n \in \mathscr{N} \right\} \cup \{0\}$. Then $\sup \mathscr{T} = \frac{1}{4} < \infty$,

$$\rho\left(\frac{1}{4}\right) = \sup\left\{ \frac{1}{l^2+3}, 0 : \frac{1}{l^2+3}, 0 < \frac{1}{4}, l \in \mathscr{N} \right\} = \frac{1}{7}.$$

Hence,

$$\mathscr{T}^{\kappa} = \mathscr{T}\backslash\left(\frac{1}{7}, \frac{1}{4}\right] = \left\{ \frac{1}{n^2+3} : n \geq 2 \right\} \cup \{0\}.$$

Definition 10 We assume that $a \leq b$. We define the interval $[a, b]$ in $\mathscr{T}$ by

$$[a, b] = \{t \in \mathscr{T} : a \leq t \leq b\}.$$

Open intervals, half-open intervals and so on, are defined accordingly.

Example 16 Let $[a, b]$ be an interval in $\mathscr{T}$ and b be a left-dense point. Then $\sup[a, b] = b$ and since b is a left-dense point, we have that $\rho(b) = b$. Hence,

$$[a, b]^{\kappa} = [a, b]\backslash(b, b] = [a, b]\backslash\emptyset = [a, b].$$

Example 17 Let $[a, b]$ be an interval in $\mathscr{T}$ and b be a left-scattered point. Then $\sup[a, b] = b$ and since b is a left-scattered point, we have that $\rho(b) < b$. We assume that there is $c \in (\rho(b), b]$, $c \in \mathscr{T}, c \neq b$. Then $\rho(b) < c \leq b$, which is a contradiction. Therefore

$$[a, b]^{\kappa} = [a, b]\backslash(\rho(b), b] = [a, b).$$

Exercise 5 Let $\mathscr{T} = \left\{ \frac{1}{2n+1} : n \in \mathscr{N} \right\} \cup \{0\}$. Find $\mathscr{T}^{\kappa}$.

Answer. $\left\{ \frac{1}{2n+1} : n \in \mathscr{N}, n \geq 2 \right\} \cup \{0\}$.

1.2 Differentiation

Definition 11 Assume that $f : \mathscr{T} \longmapsto \mathscr{R}$ is a function and let $t \in \mathscr{T}^{\kappa}$. We define $f^{\Delta}(t)$ to be the number, provided it exists, as follows: for any $\varepsilon > 0$ there is a neighbourhood U of t, $U = (t - \delta, t + \delta) \cap \mathscr{T}$ for some $\delta > 0$, such that

$$|f(\sigma(t)) - f(s) - f^{\Delta}(t)(\sigma(t) - s)| \leq \varepsilon|\sigma(t) - s| \quad \text{for} \quad \text{all} \quad s \in U, \quad s \neq \sigma(t).$$

We say $f^{\Delta}(t)$ the *delta* or *Hilger* derivative of f at t.

We say that f is *delta or Hilger differentiable*, shortly *differentiable*, in T^{κ} if $f^{\Delta}(t)$ exists for all $t \in \mathscr{T}^{\kappa}$. The function $f^{\Delta} : \mathscr{T} \longmapsto \mathscr{R}$ is said to be *delta derivative or Hilger* derivative, shortly derivative, of f in T^{κ}.

Remark 1 If $\mathscr{T} = \mathscr{R}$, then the delta derivative coincides with the classical derivative.

Theorem 1 *The delta derivative is well defined.*

Proof Let $t \in \mathscr{T}^{\kappa}$ and $f_i^{\Delta}(t)$, $i = 1, 2$, be such that

$$|f(\sigma(t)) - f(s) - f_1^{\Delta}(t)(\sigma(t) - s)| \leq \frac{\varepsilon}{2}|\sigma(t) - s|,$$
$$|f(\sigma(t)) - f(s) - f_2^{\Delta}(t)(\sigma(t) - s)| \leq \frac{\varepsilon}{2}|\sigma(t) - s|,$$

for any $\varepsilon > 0$ and any s belonging to a neighbourhood U of t, $U = (t - \delta, t + \delta) \cap \mathscr{T}$ for some $\delta > 0$, $s \neq \sigma(t)$. Hence,

$$\begin{aligned} |f_1^{\Delta}(t) - f_2^{\Delta}(t)| &= \left| f_1^{\Delta}(t) - \frac{f(\sigma(t)) - f(s)}{\sigma(t) - s} + \frac{f(\sigma(t)) - f(s)}{\sigma(t) - s} - f_2^{\Delta}(t) \right| \\ &\leq \left| f_1^{\Delta}(t) - \frac{f(\sigma(t)) - f(s)}{\sigma(t) - s} \right| + \left| \frac{f(\sigma(t)) - f(s)}{\sigma(t) - s} - f_2^{\Delta}(t) \right| \end{aligned}$$

$$\begin{aligned} &= \frac{|f(\sigma(t)) - f(s) - f_1^{\Delta}(t)(\sigma(t) - s)|}{|\sigma(t) - s|} + \frac{|f(\sigma(t)) - f(s) - f_2^{\Delta}(t)(\sigma(t) - s)|}{|\sigma(t) - s|} \\ &\leq \frac{\varepsilon}{2} + \frac{\varepsilon}{2} \\ &= \varepsilon. \end{aligned}$$

Since $\varepsilon > 0$ was arbitrarily chosen, we conclude that

$$f_1^{\Delta}(t) = f_2^{\Delta}(t),$$

which completes the proof.

Remark 2 Let us assume that $\sup \mathscr{T} < \infty$ and $f^{\Delta}(t)$ is defined at a point $t \in \mathscr{T} \backslash \mathscr{T}^{\kappa}$ with the same definition as given in Definition 11. Then the unique point $t \in \mathscr{T} \backslash \mathscr{T}^{\kappa}$ is $\sup \mathscr{T}$.

Hence, for any $\varepsilon > 0$ there is a neighbourhood $U = (t - \delta, t + \delta) \cap (\mathscr{T} \backslash \mathscr{T}^{\kappa})$, for some $\delta > 0$, such that

$$f(\sigma(t)) = f(s) = f(\sigma(\sup \mathscr{T})) = f(\sup \mathscr{T}), \quad s \in U.$$

Therefore for any $\alpha \in \mathscr{R}$ and $s \in U$ we have

$$\begin{aligned} |f(\sigma(t)) - f(s) - \alpha(\sigma(t) - s)| &= |f(\sup \mathscr{T}) - f(\sup \mathscr{T}) - \alpha(\sup \mathscr{T} - \sup \mathscr{T})| \\ &\le \varepsilon |\sigma(t) - s|, \end{aligned}$$

i.e., any $\alpha \in \mathscr{R}$ is the delta derivative of f at the point $t \in \mathscr{T} \backslash \mathscr{T}^{\kappa}$.

Example 18 Let $f(t) = \alpha \in \mathscr{R}$. We will prove that $f^{\Delta}(t) = 0$ for any $t \in \mathscr{T}^{\kappa}$.

Really, for $t \in \mathscr{T}^{\kappa}$ and for any $\varepsilon > 0$ there is a $\delta > 0$ such that $s \in (t - \delta, t + \delta) \cap \mathscr{T}$, $s \neq \sigma(t)$, implies

$$\begin{aligned} |f(\sigma(t)) - f(s) - 0(\sigma(t) - s)| &= |\alpha - \alpha| \\ &\le \varepsilon |\sigma(t) - s|. \end{aligned}$$

Example 19 Let $f(t) = t$, $t \in \mathscr{T}$. We will prove that $f^{\Delta}(t) = 1$ for any $t \in \mathscr{T}^{\kappa}$.

Really, for $t \in \mathscr{T}^{\kappa}$ and for any $\varepsilon > 0$ there is a $\delta > 0$ such that $s \in (t - \delta, t + \delta) \cap \mathscr{T}$, $s \neq \sigma(t)$, implies

$$\begin{aligned} |f(\sigma(t)) - f(s) - 1(\sigma(t) - \sigma(s))| &= |\sigma(t) - s - (\sigma(t) - s)| \\ &\le \varepsilon |\sigma(t) - s|. \end{aligned}$$

Example 20 Let $f(t) = t^2$, $t \in \mathscr{T}$. We will prove that $f^{\Delta}(t) = \sigma(t) + t$, $t \in \mathscr{T}^{\kappa}$.

Really, for $t \in \mathscr{T}^{\kappa}$ and for any $\varepsilon > 0$, and for any $s \in (t - \varepsilon, t + \varepsilon) \cap \mathscr{T}$, $s \neq \sigma(t)$, we have $|t - s| < \varepsilon$ and

$$\begin{aligned} |f(\sigma(t)) - f(s) - (\sigma(t) + t)(\sigma(t) - s)| &= |\sigma^2(t) - s^2 - (\sigma(t) + t)(\sigma(t) - s)| \\ &= |(\sigma(t) - s)(\sigma(t) + s) - (\sigma(t) + t)(\sigma(t) - s)| \\ &= |\sigma(t) - s||t - s| \\ &\le \varepsilon |\sigma(t) - s|. \end{aligned}$$

Exercise 6 Let $f(t) = \sqrt{t}, t \in \mathscr{T}, t > 0$. Prove that $f^{\Delta}(t) = \dfrac{1}{\sqrt{t} + \sqrt{\sigma(t)}}$ for $t \in \mathscr{T}^{\kappa}, t > 0$.

Exercise 7 Let $f(t) = t^3, t \in \mathscr{T}$. Prove that $f^{\Delta}(t) = \sigma^2(t) + t\sigma(t) + t^2$ for $t \in \mathscr{T}^{\kappa}$.

Theorem 2 ([2]) *Assume $f : \mathscr{T} \longmapsto \mathscr{R}$ is a function and let $t \in \mathscr{T}^{\kappa}$. Then we have the following.*

1. *If f is differentiable at t, then f is continuous at t.*
2. *If f is continuous at t and t is right-scattered, then f is differentiable at t with*

$$f^{\Delta}(t) = \frac{f(\sigma(t)) - f(t)}{\mu(t)}.$$

3. *If t is right-dense, then f is differentiable iff the limit*

$$\lim_{s \longrightarrow t} \frac{f(t) - f(s)}{t - s}$$

exists as a finite number. In this case

$$f^{\Delta}(t) = \lim_{s \longrightarrow t} \frac{f(t) - f(s)}{t - s}.$$

4. *If f is differentiable at t, then*

$$f(\sigma(t)) = f(t) + \mu(t) f^{\Delta}(t).$$

Example 21 Let $\mathscr{T} = \left\{ \dfrac{1}{2n+1} : n \in \mathscr{N}_0 \right\} \cup \{0\}, f(t) = \sigma(t), t \in \mathscr{T}$. We will find $f^{\Delta}(t), t \in \mathscr{T}$. For $t \in \mathscr{T}, t = \dfrac{1}{2n+1}, n = \dfrac{1-t}{2t}, n \geq 1$, we have

$$\sigma(t) = \inf \left\{ \frac{1}{2l+1}, 0 : \frac{1}{2l+1}, 0 > \frac{1}{2n+1}, l \in \mathscr{N}_0 \right\} = \frac{1}{2n-1}$$

$$= \frac{1}{2\frac{1-t}{2t} - 1} = \frac{t}{1-2t} > t,$$

i.e., any point $t = \dfrac{1}{2n+1}, n \geq 1$, is right-scattered. At these points

$$f^{\Delta}(t) = \frac{f(\sigma(t)) - f(t)}{\sigma(t) - t} = \frac{\sigma(\sigma(t)) - \sigma(t)}{\sigma(t) - t} = 2\frac{\sigma^2(t)}{(1 - 2\sigma(t))(\sigma(t) - t)}$$

$$= 2\frac{\left(\frac{t}{1-2t}\right)^2}{\left(1 - \frac{2t}{1-2t}\right)\left(\frac{t}{1-2t} - t\right)} = 2\frac{\frac{t^2}{(1-2t)^2}}{\frac{1-4t}{1-2t}\frac{2t^2}{1-2t}} = 2\frac{t^2}{2t^2(1-4t)} = \frac{1}{1-4t}.$$

Let $n = 0$, i.e., $t = 1$. Then

$$\sigma(1) = \inf\left\{\frac{1}{2l+1}, 0 : \frac{1}{2l+1}, 0 > 1, l \in \mathscr{N}_0\right\} = \inf \emptyset = \sup \mathscr{T} = 1,$$

i.e., $t = 1$ is a right-dense point. Also,

$$\lim_{s\to 1} \frac{f(1) - f(s)}{1-s} = \lim_{s\to 1} \frac{\sigma(1) - \sigma(s)}{1-s} = \lim_{s\to 1} \frac{1 - \frac{s}{1-2s}}{1-s} = \lim_{s\to 1} \frac{1-3s}{(1-s)(1-2s)} = +\infty.$$

Therefore $\sigma'(1)$ doesn't exist.
Let now $t = 0$. Then

$$\sigma(0) = \inf\left\{\frac{1}{2l+1}, 0 : \frac{1}{2l+1}, 0 > 0, l \in \mathscr{N}_0\right\} = 0.$$

Consequently $t = 0$ is right-dense. Also,

$$\lim_{h\longrightarrow 0} \frac{\sigma(h) - \sigma(0)}{h} = \lim_{h\longrightarrow 0} \frac{\frac{h}{1-2h} - 0}{h} = \lim_{h\longrightarrow 0} \frac{1}{1-2h} = 1.$$

Therefore $\sigma'(0) = 1$.

Example 22 Let $\mathscr{T} = \{n^2 : n \in \mathscr{N}_0\}$, $f(t) = t^2$, $g(t) = \sigma(t)$, $t \in \mathscr{T}$. We will find $f^\Delta(t)$ and $g^\Delta(t)$ for $t \in \mathscr{T}^\kappa$. For $t \in \mathscr{T}^\kappa$, $t = n^2$, $n = \sqrt{t}$, $n \in \mathscr{N}_0$, we have

$$\sigma(t) = \inf\{l^2 : l^2 > n^2, l \in \mathscr{N}_0\} = (n+1)^2 = (\sqrt{t}+1)^2 > t.$$

Therefore any points of $\mathscr{T}$ are right-scattered. We note that $f(t)$ and $g(t)$ are continuous functions in $\mathscr{T}$. Hence,

$$\begin{aligned}
f^\Delta(t) &= \frac{f(\sigma(t)) - f(t)}{\sigma(t) - t} = \frac{\sigma^2(t) - t^2}{\sigma(t) - t} = \sigma(t) + t \\
&= \left(\sqrt{t}+1\right)^2 + t = t + 2\sqrt{t} + 1 + t = 1 + 2\sqrt{t} + 2t, \\
g^\Delta(t) &= \frac{g(\sigma(t)) - g(t)}{\sigma(t) - t} = \frac{\sigma(\sigma(t)) - \sigma(t)}{\sigma(t) - t} \\
&= \frac{(\sqrt{\sigma(t)}+1)^2 - \sigma(t)}{\sigma(t) - t} = \frac{\sigma(t) + 2\sqrt{\sigma(t)} + 1 - \sigma(t)}{\sigma(t) - t} \\
&= \frac{1 + 2\sqrt{\sigma(t)}}{\sigma(t) - t} = \frac{1 + 2(\sqrt{t}+1)}{(\sqrt{t}+1)^2 - t} = \frac{3 + 2\sqrt{t}}{1 + 2\sqrt{t}}.
\end{aligned}$$

Example 23 Let $\mathscr{T} = \{\sqrt[4]{2n+1} : n \in \mathscr{N}_0\}$, $f(t) = t^4$, $t \in \mathscr{T}$. We will find $f^\Delta(t)$, $t \in \mathscr{T}^\kappa$. For $t \in \mathscr{T}$, $t = \sqrt[4]{2n+1}$, $n = \dfrac{t^4 - 1}{2}$, $n \in \mathscr{N}_0$, we have

$$\sigma(t) = \inf\{\sqrt[4]{2l+1} : \sqrt[4]{2l+1} > \sqrt[4]{2n+1}, l \in \mathcal{N}_0\} = \sqrt[4]{2n+3}$$
$$= \sqrt[4]{t^4+2} > t.$$

Therefore every point of $\mathcal{T}$ is right-scattered. We note that the function $f(t)$ is continuous in $\mathcal{T}$. Hence,

$$\begin{aligned} f^{\Delta}(t) &= \frac{f(\sigma(t)) - f(t)}{\sigma(t) - t} = \frac{\sigma^4(t) - t^4}{\sigma(t) - t} \\ &= \sigma^3(t) + t\sigma^2(t) + t^2\sigma(t) + t^3 \\ &= \sqrt[4]{(t^4+2)^3} + t^2\sqrt[4]{t^4+2} + t\sqrt{t^4+2} + t^3, t \in \mathcal{T}^{\kappa}. \end{aligned}$$

Exercise 8 Let $\mathcal{T} = \{\sqrt[5]{n+1} : n \in \mathcal{N}_0\}$, $f(t) = t + t^3$, $t \in \mathcal{T}$. Find $f^{\Delta}(t)$, $t \in \mathcal{T}^{\kappa}$.

Answer. $1 + \sqrt[5]{(t^5+1)^2} + t\sqrt[5]{t^5+1} + t^2$.

Theorem 3 ([2]) *Assume $f, g : \mathcal{T} \longmapsto \mathcal{R}$ are differentiable at $t \in \mathcal{T}^{\kappa}$. Then*

1. *the sum $f + g : \mathcal{T} \longmapsto \mathcal{R}$ is differentiable at t with*

$$(f+g)^{\Delta}(t) = f^{\Delta}(t) + g^{\Delta}(t).$$

2. *for any constant α, $\alpha f : \mathcal{T} \longmapsto \mathcal{R}$ is differentiable at t with*

$$(\alpha f)^{\Delta}(t) = \alpha f^{\Delta}(t).$$

3. *if $f(t)f(\sigma(t)) \neq 0$, we have that $\frac{1}{f} : \mathcal{T} \longmapsto \mathcal{R}$ is differentiable at t and*

$$\left(\frac{1}{f}\right)^{\Delta}(t) = -\frac{f^{\Delta}(t)}{f(t)f(\sigma(t))}.$$

4. *if $g(t)g(\sigma(t)) \neq 0$, we have that $\frac{f}{g} : \mathcal{T} \longmapsto \mathcal{R}$ is differentiable at t with*

$$\left(\frac{f}{g}\right)^{\Delta}(t) = \frac{f^{\Delta}(t)g(t) - f(t)g^{\Delta}(t)}{g(t)g(\sigma(t))}.$$

5. *the product $fg : \mathcal{T} \longmapsto \mathcal{R}$ is differentiable at t with*

$$(fg)^{\Delta}(t) = f^{\Delta}(t)g(t) + f(\sigma(t))g^{\Delta}(t) = f(t)g^{\Delta}(t) + f^{\Delta}(t)g(\sigma(t)).$$

Example 24 Let $f, g, h : \mathcal{T} \longmapsto \mathcal{R}$ be differentiable at $t \in \mathcal{T}^{\kappa}$. Then

$$\begin{aligned}(fgh)^{\Delta}(t) &= ((fg)h)^{\Delta}(t) = (fg)^{\Delta}(t)h(t) + (fg)(\sigma(t))h^{\Delta}(t)\\ &= (f^{\Delta}(t)g(t) + f(\sigma(t))g^{\Delta}(t))h(t) + f^{\sigma}(t)g^{\sigma}(t)h^{\Delta}(t)\\ &= f^{\Delta}(t)g(t)h(t) + f^{\sigma}(t)g^{\Delta}(t)h(t) + f^{\sigma}(t)g^{\sigma}(t)h^{\Delta}(t).\end{aligned}$$

Example 25 Let $f : \mathscr{T} \longmapsto \mathscr{R}$ be differentiable at $t \in \mathscr{T}^{\kappa}$. Then

$$\left(f^2\right)^{\Delta}(t) = (ff)^{\Delta}(t) = f^{\Delta}(t)f(t) + f(\sigma(t))f^{\Delta}(t) = f^{\Delta}(t)(f^{\sigma}(t) + f(t)).$$

Also,

$$\begin{aligned}(f^3)^{\Delta}(t) &= (ff^2)^{\Delta}(t) = f^{\Delta}(t)f^2(t) + f(\sigma(t))(f^2)^{\Delta}(t)\\ &= f^{\Delta}(t)f^2(t) + f^{\sigma}(t)f^{\Delta}(t)(f^{\sigma}(t) + f(t))\\ &= f^{\Delta}(t)(f^2(t) + f(t)f^{\sigma}(t) + (f^{\sigma})^2(t)).\end{aligned}$$

We assume that

$$(f^n)^{\Delta}(t) = f^{\Delta}(t)\sum_{k=0}^{n-1} f^k(t)(f^{\sigma})^{n-1-k}(t)$$

for some $n \in \mathscr{N}$.

We will prove that

$$(f^{n+1})^{\Delta}(t) = f^{\Delta}(t)\sum_{k=0}^{n} f^k(t)(f^{\sigma})^{n-k}(t).$$

Really,

$$\begin{aligned}(f^{n+1})^{\Delta}(t) &= (ff^n)^{\Delta}(t) = f^{\Delta}(t)f^n(t) + f^{\sigma}(t)(f^n)^{\Delta}(t)\\ &= f^{\Delta}(t)f^n(t) + f^{\Delta}(t)(f^{n-1}(t) + f^{n-2}(t)f^{\sigma}(t)\\ &\quad + \cdots + f(t)(f^{\sigma})^{n-2}(t) + (f^{\sigma})^{n-1}(t))f^{\sigma}(t)\\ &= f^{\Delta}(t)\Big(f^n(t) + f^{n-1}(t)f^{\sigma}(t) + f^{n-2}(t)(f^{\sigma})^2(t) + \cdots + (f^{\sigma})^n(t)\Big)\\ &= f^{\Delta}(t)\sum_{k=0}^{n} f^k(t)(f^{\sigma})^{n-k}(t).\end{aligned}$$

Example 26 Now we consider $f(t) = (t-a)^m$ for $a \in \mathscr{R}$ and $m \in \mathscr{N}$. We set $h(t) = (t-a)$. Then $h^{\Delta}(t) = 1$. Hence and the previous exercise, we get

$$\begin{aligned}f^{\Delta}(t) &= h^{\Delta}(t)\sum_{k=0}^{m-1} h^k(t)(h^{\sigma})^{m-1-k}(t)\\ &= \sum_{k=0}^{m-1}(t-a)^k(\sigma(t)-a)^{m-1-k}.\end{aligned}$$

Let now $g(t) = \dfrac{1}{f(t)}$. Then

$$g^{\Delta}(t) = -\frac{f^{\Delta}(t)}{f(\sigma(t))f(t)},$$

whereupon

$$\begin{aligned} g^{\Delta}(t) &= -\frac{1}{(\sigma(t)-a)^m(t-a)^m}\sum_{k=0}^{m-1}(t-a)^k(\sigma(t)-a)^{m-1-k} \\ &= -\sum_{k=0}^{m-1}\frac{1}{(t-a)^{m-k}}\frac{1}{(\sigma(t)-a)^{k+1}}. \end{aligned}$$

Definition 12 Let $f : \mathscr{T} \longmapsto \mathscr{R}$ and $t \in (T^{\kappa})^{\kappa} = \mathscr{T}^{\kappa^2}$. We define the second derivative of f at t, provided it exists, by

$$f^{\Delta^2} = \left(f^{\Delta}\right)^{\Delta} : \mathscr{T}^{\kappa^2} \longmapsto \mathscr{R}.$$

Similarly we define higher order derivatives $f^{\Delta^n} : \mathscr{T}^{\kappa^n} \longmapsto \mathscr{T}$.

Theorem 4 [2] (Leibniz Formula) *Let $S_k^{(n)}$ be the set consisting of all possible strings of length n, containing exactly k times σ and $n-k$ times Δ. If*

$$f^{\Lambda} \quad \text{exists for all} \quad \Lambda \in S_k^{(n)},$$

then

$$(fg)^{\Delta^n} = \sum_{k=0}^{n}\left(\sum_{\Lambda \in S_k^{(n)}} f^{\Lambda}\right) g^{\Delta^k}.$$

Example 27 Let μ is differentiable at $t \in \mathscr{T}^{\kappa}$ and t is right-scattered. Then

$$\begin{aligned} f^{\Delta\sigma}(t) &= \left(f^{\Delta}\right)^{\sigma}(t) = \left(\frac{f(\sigma(t)) - f(t)}{\sigma(t) - t}\right)^{\sigma} = \frac{f(\sigma(\sigma(t))) - f(\sigma(t))}{\sigma(\sigma(t)) - \sigma(t)} \\ &= \frac{f(\sigma(\sigma(t))) - f(\sigma(t))}{\sigma(t) - t}\frac{1}{\frac{\sigma(\sigma(t)) - \sigma(t)}{\sigma(t) - t}} \\ &= \left(f^{\sigma}\right)^{\Delta}(t)\frac{1}{\sigma^{\Delta}(t)} = f^{\sigma\Delta}(t)\frac{1}{1 + \mu^{\Delta}(t)}, \end{aligned}$$

i.e.,

$$f^{\sigma\Delta}(t) = (1 + \mu^{\Delta}(t))f^{\Delta\sigma}(t).$$

Also,

$$f^{\sigma\sigma\Delta}(t) = (f^{\sigma})^{\sigma\Delta}(t) = (1+\mu^{\Delta}(t))\left(f^{\sigma\Delta}\right)^{\sigma}(t) = (1+\mu^{\Delta}(t))f^{\sigma\Delta\sigma}(t),$$
$$f^{\sigma\Delta\sigma}(t) = (f^{\sigma})^{\Delta\sigma}(t) = \left((f^{\sigma})^{\Delta}\right)^{\sigma}(t)$$
$$= \left((1+\mu^{\Delta}(t))(f^{\Delta})^{\sigma}(t)\right)^{\sigma} = \left(1+\mu^{\Delta\sigma}(t)\right) f^{\Delta\sigma\sigma}(t).$$

Theorem 5 ([2]) (Chain Rule) *Assume $g : \mathscr{R} \longmapsto \mathscr{R}$ is continuous, $g : \mathscr{T} \longmapsto \mathscr{R}$ is delta differentiable on $\mathscr{T}^{\kappa}$, and $f : \mathscr{R} \longmapsto \mathscr{R}$ is continuously differentiable. Then there exists $c \in [t, \sigma(t)]$ with*

$$(f \circ g)^{\Delta}(t) = f'(g(c))g^{\Delta}(t).$$

Example 28 Let $\mathscr{T} = \mathscr{Z}$, $f(t) = t^3 + 1$, $g(t) = t^2$. We have that $g : \mathscr{R} \longmapsto \mathscr{R}$ is continuous, $g : \mathscr{T} \longmapsto \mathscr{R}$ is delta differentiable on $\mathscr{T}^{\kappa}$, $f : \mathscr{R} \longmapsto \mathscr{R}$ is continuously differentiable, $\sigma(t) = t + 1$. Then

$$g^{\Delta}(t) = \sigma(t) + t,$$

$$(f \circ g)^{\Delta}(1) = f'(g(c))g^{\Delta}(1) = 3g^2(c)(\sigma(1) + 1) = 9c^4. \tag{1.1}$$

Here $c \in [1, \sigma[1]] = [1, 2]$.
Also,

$$f \circ g(t) = f(g(t)) = g^3(t) + 1 = t^6 + 1,$$

$$(f \circ g)^{\Delta}(t) = \sigma^5(t) + t\sigma^4(t) + t^2\sigma^3(t) + t^3\sigma^2(t) + t^4\sigma(t) + t^5,$$

$$(f \circ g)^{\Delta}(1) = \sigma^5(1) + \sigma^4(1) + \sigma^3(1) + \sigma^2(1) + \sigma(1) + 1 = 63.$$

Hence and (1.1), we get

$$63 = 9c^4 \quad or \quad c^4 = 7 \quad or \quad c = \sqrt[4]{7} \in [1, 2].$$

Example 29 Let $\mathscr{T} = \{2^n : n \in \mathscr{N}_0\}$, $f(t) = t + 2$, $g(t) = t^2 - 1$. We note that $g : \mathscr{T} \longmapsto \mathscr{R}$ is delta differentiable, $g : \mathscr{R} \longmapsto \mathscr{R}$ is continuous and $f : \mathscr{R} \longmapsto \mathscr{R}$ is continuously differentiable.

For $t \in \mathscr{T}$, $t = 2^n$, $n \in \mathscr{N}_0$, $n = \log_2 t$, we have

$$\sigma(t) = \inf\left\{2^l : 2^l > 2^n, l \in \mathscr{N}_0\right\} = 2^{n+1} = 2t > t.$$

Therefore all points of $\mathscr{T}$ are right-scattered. Since $\sup \mathscr{T} = \infty$ we have that $\mathscr{T}^{\kappa} = \mathscr{T}$. Also, for $t \in \mathscr{T}$, we have

$$(f \circ g)(t) = f(g(t)) = g(t) + 2 = t^2 - 1 + 2 = t^2 + 1,$$
$$(f \circ g)^{\Delta}(t) = \sigma(t) + t = 2t + t = 3t.$$

Hence,

$$(f \circ g)^{\Delta}(2) = 6. \tag{1.2}$$

Now, using Theorem 5, we get that there is $c \in [2, \sigma(2)] = [2, 4]$ such that

$$(f \circ g)^{\Delta}(2) = f'(g(c))g^{\Delta}(2) = g^{\Delta}(2) = \sigma(2) + 2 = 4 + 2 = 6. \tag{1.3}$$

From (1.2) and (1.3) we find that for every $c \in [2, 4]$

$$(f \circ g)^{\Delta}(2) = f'(g(c))g^{\Delta}(2).$$

Example 30 Let $\mathscr{T} = \left\{3^{n^2} : n \in \mathscr{N}_0\right\}$, $f(t) = t^2 + 1$, $g(t) = t^3$. We note that $g : \mathscr{R} \longmapsto \mathscr{R}$ is continuous, $g : \mathscr{T} \longmapsto \mathscr{R}$ is delta differentiable and $f : \mathscr{R} \longmapsto \mathscr{R}$ is continuously differentiable.

For $t \in \mathscr{T}$, $t = 3^{n^2}$, $n \in \mathscr{N}_0$, $n = \left(\log_3 t\right)^{\frac{1}{2}}$, we have

$$\begin{aligned}\sigma(t) &= \inf\left\{3^{l^2} : 3^{l^2} > 3^{n^2}, l \in \mathscr{N}_0\right\} = 3^{(n+1)^2}\\ &= 3 \cdot 3^{n^2} \cdot 3^{2n} = 3t3^{2(\log_3 t)^{\frac{1}{2}}} > t.\end{aligned}$$

Consequently all points of $\mathscr{T}$ are right-scattered. Also, $\sup \mathscr{T} = \infty$. Then $\mathscr{T}^{\kappa} = \mathscr{T}$.

Hence, for $t \in \mathscr{T}$ we have

$$(f \circ g)(t) = f(g(t)) = g^2(t) + 1 = t^6 + 1,$$
$$(f \circ g)^{\Delta}(t) = (t^6 + 1)^{\Delta} = \sigma^5(t) + t\sigma^4(t) + t^2\sigma^3(t) + t^3\sigma^2(t) + t^4\sigma(t) + t^5,$$

$$\begin{aligned}(f \circ g)^{\Delta}(1) &= \sigma^5(1) + \sigma^4(1) + \sigma^3(1) + \sigma^2(1) + \sigma(1) + 1\\ &= 3^5 + 3^4 + 3^3 + 3^2 + 3 + 1 = 364.\end{aligned} \tag{1.4}$$

From Theorem 5, it follows that there is $c \in [1, \sigma(1)] = [1, 3]$ such that

$$(f \circ g)^{\Delta}(1) = f'(g(c))g^{\Delta}(1) = 2g(c)g^{\Delta}(1) = 2c^3g^{\Delta}(1). \tag{1.5}$$

Because all points of $\mathscr{T}$ are right-scattered, we have

$$g^{\Delta}(1) = \sigma^2(1) + \sigma(1) + 1 = 9 + 3 + 1 = 13.$$

Hence and (1.5), we find

$$(f \circ g)^{\Delta}(1) = 26c^3.$$

From the last equation and from (1.4) we obtain

$$364 = 26c^3 \quad or \quad c^3 = \frac{364}{26} = 14 \quad or \quad c = \sqrt[3]{14}.$$

Exercise 9 Let $\mathscr{T} = \mathscr{Z}$, $f(t) = t^2 + 2t + 1$, $g(t) = t^2 - 3t$. Find a constant $c \in [1, \sigma(1)]$ such that

$$(f \circ g)^{\Delta}(1) = f'(g(c))g^{\Delta}(1).$$

Answer. $\forall c \in [1, 2]$.

Theorem 6 ([2]) (Chain Rule) *Assume $v : \mathscr{T} \longmapsto \mathscr{R}$ is strictly increasing and $\tilde{T} = v(T)$ is a time scale. Let $w : \tilde{T} \longmapsto \mathscr{R}$. If $v^{\Delta}(t)$ and $w^{\tilde{\Delta}}(v(t))$ exist for $t \in \mathscr{T}^{\kappa}$, then*

$$(w \circ v)^{\Delta} = (w^{\tilde{\Delta}} \circ v)v^{\Delta}.$$

Example 31 Let $\mathscr{T} = \{2^{2n} : n \in \mathscr{N}_0\}$, $v(t) = t^2$, $w(t) = t^2 + 1$. Then $v : \mathscr{T} \longmapsto \mathscr{R}$ is strictly increasing, $\tilde{\mathscr{T}} = v(T) = \{2^{4n} : n \in \mathscr{N}_0\}$ is a time scale. For $t \in \mathscr{T}, t = 2^{2n}, n \in \mathscr{N}_0$, we have

$$\sigma(t) = \inf\{2^{2l} : 2^{2l} > 2^{2n}, l \in \mathscr{N}_0\} = 2^{2n+2} = 4t,$$
$$v^{\Delta}(t) = \sigma(t) + t = 5t.$$

For $t \in \tilde{\mathscr{T}}, t = 2^{4n}, n \in \mathscr{N}_0$, we have

$$\tilde{\sigma}(t) = \inf\{2^{4l} : 2^{4l} > 2^{4n}, l \in \mathscr{N}_0\} = 2^{4n+4} = 16t.$$

Also, for $t \in \mathscr{T}$, we have

$$(w \circ v)(t) = w(v(t)) = v^2(t) + 1 = t^4 + 1,$$
$$(w \circ v)^{\Delta}(t) = \sigma^3(t) + t\sigma^2(t) + t^2\sigma(t) + t^3$$

$$= 64t^3 + 16t^3 + 4t^3 + t^3 = 85t^3,$$
$$w^{\tilde{\Delta}} \circ v(t) = \tilde{\sigma}(v(t)) + v(t) = 16v(t) + v(t) = 17v(t) = 17t^2,$$
$$\left(w^{\tilde{\Delta}} \circ v(t)\right) v^{\Delta}(t) = 17t^2(5t) = 85t^3.$$

Consequently

$$(w \circ v)^{\Delta}(t) = (w^{\tilde{\Delta}} \circ v(t))v^{\Delta}(t), \quad t \in \mathscr{T}^{\kappa}.$$

Example 32 Let $\mathscr{T} = \{n + 1 : n \in \mathscr{N}_0\}$, $v(t) = t^2$, $w(t) = t$. Then $v : \mathscr{T} \longmapsto \mathscr{R}$ is strictly increasing, $\tilde{\mathscr{T}} = \{(n + 1)^2 : n \in \mathscr{N}_0\}$ is a time scale.

For $t \in \mathscr{T}, t = n+1, n \in \mathscr{N}_0$, we have

$$\sigma(t) = \inf\{l+1 : l+1 > n+1, l \in \mathscr{N}_0\} = n+2 = t+1,$$
$$v^{\Delta}(t) = \sigma(t) + t = t+1+t = 2t+1.$$

For $t \in \tilde{\mathscr{T}}, t = (n+1)^2, n \in \mathscr{N}_0$, we have

$$\begin{aligned}\tilde{\sigma}(t) &= \{(l+1)^2 : (l+1)^2 > (n+1)^2, l \in \mathscr{N}_0\} = (n+2)^2 \\ &= (n+1)^2 + 2(n+1) + 1 = t + 2\sqrt{t} + 1.\end{aligned}$$

Hence, for $t \in \mathscr{T}$, we get

$$(w^{\tilde{\Delta}} \circ v)(t) = 1, \quad (w^{\tilde{\Delta}} \circ v)(t) v^{\Delta}(t) = 1(2t+1) = 2t+1,$$
$$w \circ v(t) = v(t) = t^2, \quad (w \circ v)^{\Delta}(t) = \sigma(t) + t = 2t+1.$$

Consequently

$$(w \circ v)^{\Delta}(t) = (w^{\tilde{\Delta}} \circ v(t)) v^{\Delta}(t), \quad t \in \mathscr{T}^{\kappa}.$$

Example 33 Let $\mathscr{T} = \{2^n : n \in \mathscr{N}_0\}$, $v(t) = t$, $w(t) = t^2$. Then $v : \mathscr{T} \longmapsto \mathscr{R}$ is strictly increasing, $v(\mathscr{T}) = \mathscr{T}$.

For $t \in \mathscr{T}, t = 2^n, n \in \mathscr{N}_0$, we have

$$\begin{aligned}\sigma(t) &= \inf\{2^l : 2^l > 2^n, l \in \mathscr{N}_0\} = 2^{n+1} = 2t, \quad v^{\Delta}(t) = 1, \\ (w \circ v)(t) &= w(v(t)) = v^2(t) = t^2, \\ (w \circ v)^{\Delta}(t) &= \sigma(t) + t = 2t + t = 3t, \\ (w^{\Delta} \circ v)(t) &= \sigma(v(t)) + v(t) = 2v(t) + v(t) = 3v(t) = 3t, \\ (w^{\Delta} \circ v)(t) v^{\Delta}(t) &= 3t.\end{aligned}$$

Consequently

$$(w \circ v)^{\Delta}(t) = (w^{\tilde{\Delta}} \circ v(t)) v^{\Delta}(t), \quad t \in \mathscr{T}^{\kappa}.$$

Exercise 10 Let $\mathscr{T} = \{2^{3n} : n \in \mathscr{N}_0\}$, $v(t) = t^2$, $w(t) = t$. Prove

$$(w \circ v)^{\Delta}(t) = (w^{\tilde{\Delta}} \circ v(t)) v^{\Delta}(t), \quad t \in \mathscr{T}^{\kappa}.$$

Theorem 7 *(Derivative of the Inverse) Assume $v : \mathscr{T} \longmapsto \mathscr{R}$ is strictly increasing and $\tilde{\mathscr{T}} := v(\mathscr{T})$ is a time scale. Then*

$$(v^{-1})^{\tilde{\Delta}} \circ v(t) = \frac{1}{v^{\Delta}(t)}$$

for any $t \in \mathscr{T}^{\kappa}$ *such that* $v^{\Delta}(t) \neq 0$.

Example 34 Let $\mathscr{T} = \mathscr{N}$, $v(t) = t^2 + 1$. Then $\sigma(t) = t + 1$, $v : \mathscr{T} \longmapsto \mathscr{R}$ is strictly increasing and

$$v^{\Delta}(t) = \sigma(t) + t = 2t + 1.$$

Hence,

$$\left(v^{-1}\right)^{\tilde{\Delta}} \circ v(t) = \frac{1}{v^{\Delta}(t)} = \frac{1}{2t+1}.$$

Example 35 Let $\mathscr{T} = \{n + 3 : n \in \mathscr{N}_0\}$, $v(t) = t^2$. Then $v : \mathscr{T} \longmapsto \mathscr{R}$ is strictly increasing, $\sigma(t) = t + 1$,

$$v^{\Delta}(t) = \sigma(t) + t = 2t + 1.$$

Hence,

$$\left(v^{-1}\right)^{\tilde{\Delta}} \circ v(t) = \frac{1}{v^{\Delta}(t)} = \frac{1}{2t+1}.$$

Example 36 Let $\mathscr{T} = \left\{2^{n^2} : n \in \mathscr{N}_0\right\}$, $v(t) = t^3$. Then $v : \mathscr{T} \longmapsto \mathscr{R}$ is strictly increasing and for $t \in \mathscr{T}$, $t = 2^{n^2}$, $n \in \mathscr{N}_0$, $n = \left(\log_2 t\right)^{\frac{1}{2}}$, we have

$$\begin{aligned}\sigma(t) &= \inf\left\{2^{l^2} : 2^{l^2} > 2^{n^2}, l \in \mathscr{N}_0\right\} = 2^{(n+1)^2}\\ &= 2^{n^2}2^{2n+1} = t2^{2(\log_2 t)^{\frac{1}{2}}+1}.\end{aligned}$$

Then

$$v^{\Delta}(t) = \sigma^2(t) + t\sigma(t) + t^2 = t^2 2^{4(\log_2 t)^{\frac{1}{2}}+2} + t^2 2^{2(\log_2 t)^{\frac{1}{2}}+1} + t^2.$$

Hence,

$$\left(v^{-1}\right)^{\tilde{\Delta}} \circ v(t) = \frac{1}{t^2 2^{4(\log_2 t)^{\frac{1}{2}}+2} + t^2 2^{2(\log_2 t)^{\frac{1}{2}}+1} + t^2}.$$

Exercise 11 Let $\mathscr{T} = \{n + 5 : n \in \mathscr{N}_0\}$, $v(t) = t^2 + t$. Find $\left(v^{-1}\right)^{\tilde{\Delta}} \circ v(t)$.

Answer. $\dfrac{1}{2t+2}$.

1.3 Mean Value Theorems

Let $\mathscr{T}$ be a time scale and $a, b \in \mathscr{T}$, $a < b$. Let $f : \mathscr{T} \longmapsto \mathscr{R}$ be a function.

Theorem 8 *Suppose that f has delta derivative at each point of $[a, b]$. If $f(a) = f(b)$, then there exist points $\xi_1, \xi_2 \in [a, b]$ such that*

$$f^{\Delta}(\xi_2) \leq 0 \leq f^{\Delta}(\xi_1).$$

Proof Since f is delta differentiable at each point of $[a, b]$, then f is continuous on $[a, b]$. Therefore there exist $\xi_1, \xi_2 \in [a, b]$ such that

$$m = \min_{[a,b]} f(t) = f(\xi_1), \quad M = \max_{[a,b]} f(t) = f(\xi_2).$$

Because $f(a) = f(b)$ we assume that $\xi_1, \xi_2 \in [a, b)$.

1. Let $\sigma(\xi_1) > \xi_1$. Then

$$f^{\Delta}(\xi_1) = \frac{f(\sigma(\xi_1)) - f(\xi_1)}{\sigma(\xi_1) - \xi_1} \geq 0.$$

2. Let $\sigma(\xi_1) = \xi_1$. Then

$$f^{\Delta}(\xi_1) = \lim_{t \to \xi_1} \frac{f(\xi_1) - f(t)}{\xi_1 - t} \geq 0.$$

3. Let $\sigma(\xi_2) > \xi_2$. Then

$$f^{\Delta}(\xi_2) = \frac{f(\sigma(\xi_2)) - f(\xi_2)}{\sigma(\xi_2) - \xi_2} \leq 0.$$

4. Let $\sigma(\xi_2) = \xi_2$. Then

$$f^{\Delta}(\xi_2) = \lim_{t \to \xi_2} \frac{f(\xi_2) - f(t)}{\xi_2 - t} \leq 0.$$

Theorem 9 *If f is delta differentiable at t_0, then*

$$f(\sigma(t)) = f(t_0) + \left(f^{\Delta}(t_0) + E(t)\right)(\sigma(t) - t_0), \tag{1.6}$$

where $E(t)$ is defined in a neighbourhood of t_0 and

$$\lim_{t \longrightarrow t_0} E(t) = E(t_0) = 0.$$

Proof Define

$$E(t) = \begin{cases} \frac{f(\sigma(t)) - f(t_0)}{\sigma(t) - t_0} - f^{\Delta}(t_0), & t \in \mathscr{T}, \quad t \neq t_0, \\ 0, & t = t_0. \end{cases} \tag{1.7}$$

Solving (1.7) for $f(\sigma(t))$ yields (1.6) if $t \neq t_0$.

Let $t = t_0$. Then

1. $\sigma(t_0) > t_0$. Then (1.6) is obvious.
2. $\sigma(t_0) = t_0$. Then (1.6) is obvious.

Theorem 10 *Let f is delta differentiable at t_0. If $f^{\Delta}(t_0) > (<)0$, then there is a $\delta > 0$ such that*

$$f(\sigma(t)) \geq (\leq) f(t_0) \quad \text{for} \quad \forall t \in (t_0, t_0 + \delta)$$

and

$$f(\sigma(t)) \leq (\geq) f(t_0) \quad \text{for} \quad \forall t \in (t_0 - \delta, t_0).$$

Proof Using (1.6) we have, for $t \neq t_0$,

$$\frac{f(\sigma(t)) - f(t_0)}{\sigma(t) - t_0} = f^{\Delta}(t_0) + E(t). \tag{1.8}$$

Let $\delta > 0$ is chosen so that $|E(t)| \leq f^{\Delta}(t_0)$ for any $t \in (t_0 - \delta, t_0 + \delta)$. Such $\delta > 0$ exists because $\lim\limits_{t \longrightarrow t_0} E(t) = 0$. Hence, for any $t \in (t_0 - \delta, t_0 + \delta)$ we have

$$f^{\Delta}(t_0) + E(t) \geq 0.$$

If $t \in (t_0, t_0 + \delta)$, then $\sigma(t) \geq t_0$ and from (1.8) we find

$$\frac{f(\sigma(t)) - f(t_0)}{\sigma(t) - t_0} \geq 0,$$

i.e., $f(\sigma(t)) \geq f(t_0)$.

If $t \in (t_0 - \delta, t_0)$, then $t \leq \sigma(t) \leq t_0$ and from (1.8) we get $f(\sigma(t)) \leq f(t_0)$.

Theorem 11 *(Mean Value Theorem) Suppose that f is continuous on $[a, b]$ and has delta derivative at each point of $[a, b)$. Then there exist $\xi_1, \xi_2 \in [a, b)$ such that*

$$f^{\Delta}(\xi_1)(b - a) \leq f(b) - f(a) \leq f^{\Delta}(\xi_2)(b - a). \tag{1.9}$$

Proof Consider the function ϕ defined on $[a, b]$ by

$$\phi(t) = f(t) - f(a) - \frac{f(b) - f(a)}{b - a}(t - a).$$

Then ϕ is continuous on $[a, b]$ and has delta derivative at each point of $[a, b)$. Also, $\phi(a) = \phi(b) = 0$. Hence, there exist $\xi_1, \xi_2 \in [a, b)$ such that

$$\phi^{\Delta}(\xi_1) \leq 0 \leq \phi^{\Delta}(\xi_2)$$

or

$$f^{\Delta}(\xi_1) - \frac{f(b) - f(a)}{b - a} \leq 0 \leq f^{\Delta}(\xi_2) - \frac{f(b) - f(a)}{b - a},$$

whereupon we get (1.9).

Corollary 1 *Let f be continuous function on $[a, b]$ that has a delta derivative at each point of $[a, b)$. If $f^{\Delta}(t) = 0$ for all $t \in [a, b)$, then f is a constant function on $[a, b]$.*

Proof For every $t \in [a, b]$, using (1.9), we have that there exist $\xi_1, \xi_2 \in [a, b)$ such that

$$0 = f^{\Delta}(\xi_1)(t - a) \leq f(t) - f(a) \leq f^{\Delta}(\xi_2)(t - a) = 0,$$

i.e., $f(t) = f(a)$.

Corollary 2 *Let f be a continuous function on $[a, b]$ that has a delta derivative at each point of $[a, b)$. Then f is increasing, decreasing, nondecreasing and nonincreasing on $[a, b]$ if $f^{\Delta}(t) > 0$, $f^{\Delta}(t) < 0$, $f^{\Delta}(t) \geq 0$, $f^{\Delta}(t) \leq 0$ for any $t \in [a, b)$, respectively.*

Proof 1. Let $f^{\Delta}(t) > 0$ for any $t \in [a, b]$. Then for any $t_1, t_2 \in [a, b]$, $t_1 < t_2$. there exists $\xi_1 \in (t_1, t_2)$ such that

$$f(t_1) - f(t_2) \leq f^{\Delta}(\xi_1)(t_1 - t_2) < 0,$$

i.e., $f(t_1) < f(t_2)$.

2. Let $f^{\Delta}(t) < 0$ for any $t \in [a, b]$. Then for any $t_1, t_2 \in [a, b], t_1 < t_2$. there exists $\xi_1 \in (t_1, t_2)$ such that

$$f(t_1) - f(t_2) \geq f^{\Delta}(\xi_1)(t_1 - t_2) > 0,$$

i.e., $f(t_1) > f(t_2)$.

The cases $f^{\Delta}(t) \geq 0$ and $f^{\Delta}(t) \leq 0$ we leave to the reader for exercise.

1.4 Integration

Definition 13 A function $f : \mathscr{T} \longmapsto \mathscr{R}$ is called regulated provided its right-sided limits exist(finite) at all right-dense points in $\mathscr{T}$ and its left-sided limits exist (finite) at all left-dense points in $\mathscr{T}$.

Example 37 Let $\mathscr{T} = \mathscr{N}$ and

$$f(t) = \frac{t^2}{t - 1}, \quad g(t) = \frac{t}{t + 1}, \quad t \in \mathscr{T}.$$

We note that all points of $\mathscr{T}$ are right-scattered. The points $t \in \mathscr{T}, t \neq 1$, are left-scattered. The point $t = 1$ is left-dense. Also, $\lim_{t \longrightarrow 1-} f(t)$ is not finite and $\lim_{t \longrightarrow 1-} g(t)$ exists and it is finite. Therefore the function f is not regulated and the function g is regulated.

Example 38 Let $\mathscr{T} = \mathscr{R}$ and

$$f(t) = \begin{cases} 0 & \text{for} \quad t = 0 \\ \frac{1}{t} & \text{for} \quad t \in \mathscr{R} \backslash \{0\}. \end{cases}$$

We have that all points of $\mathscr{T}$ are dense and $\lim_{t \longrightarrow 0-} f(t)$ and $\lim_{t \longrightarrow 0+} f(t)$ are not finite. Therefore the function f is not regulated.

Exercise 12 Let $\mathscr{T} = \mathscr{R}$ and

$$f(t) = \begin{cases} 11 & \text{for} \quad t = 1 \\ \frac{1}{t-1} & \text{for} \quad t \in \mathscr{R} \backslash \{1\}. \end{cases}$$

Determine if f is regulated.

Answer. No.

Definition 14 A continuous function $f : \mathscr{T} \longmapsto \mathscr{R}$ is called pre-differentiable with region of differentiation D, provided

1. $D \subset \mathscr{T}^\kappa$,
2. $\mathscr{T}^\kappa \backslash D$ is countable and contains no right-scattered elements of $\mathscr{T}$,
3. f is differentiable at each $t \in D$.

Example 39 Let $\quad \mathscr{T} = P_{a,b} = \bigcup_{k=0}^{\infty} [k(a+b), k(a+b)+a] \quad$ for $\quad a > b > 0$, $f : \mathscr{T} \longmapsto \mathscr{R}$ be defined by

$$f(t) = \begin{cases} 0 \quad \text{if} \quad t \in \bigcup_{k=0}^{\infty} [k(a+b), k(a+b)+b] \\ \\ t - (a+b)k - b \quad \text{if} \quad t \in [(a+b)k+b, (a+b)k+a]. \end{cases}$$

Then f is pre-differentiable with $D \backslash \bigcup_{k=0}^{\infty} \{(a+b)k+b\}$.

Example 40 Let $\mathscr{T} = \mathscr{R}$ and

$$f(t) = \begin{cases} 0 & \text{if} \quad t = 3 \\ \frac{1}{t-3} & \text{if} \quad t \in \mathscr{R} \backslash \{3\}. \end{cases}$$

Since $f : \mathscr{T} \longmapsto \mathscr{R}$ is not continuous, then f is not pre-differentiable.

Example 41 Let $\mathscr{T} = \mathscr{N}_0 \bigcup \left\{1 - \frac{1}{n} : n \in \mathscr{N}\right\}$ and

$$f(t) = \begin{cases} 0 \;\; if \;\; t \in \mathscr{N} \\ t \;\; otherwise. \end{cases}$$

Then f is pre-differentiable with $D = \mathscr{T} \backslash \{1\}$.

Exercise 13 Let $\mathscr{T} = \mathscr{R}$ and

$$f(t) = \begin{cases} 0 \text{ if } t = -3 \\ \frac{1}{t+3} \text{ if } t \in \mathscr{R} \backslash \{-3\}. \end{cases}$$

Check if $f : \mathscr{T} \longmapsto \mathscr{R}$ is pre-differentiable and if it is, find the region of differentiation.

Answer. No.

Definition 15 A function $f : \mathscr{T} \longmapsto \mathscr{R}$ is called rd-continuous provide it is continuous at right-dense points in $\mathscr{T}$ and its left-sided limits exist(finite) at left-dense points in $\mathscr{T}$. The set of rd-continuous functions $f : \mathscr{T} \longmapsto \mathscr{R}$ will be denoted by $\mathscr{C}_{rd}(\mathscr{T})$.

The set of functions $f : \mathscr{T} \longmapsto \mathscr{R}$ that are differentiable and whose derivative is rd-continuous is denoted by $\mathscr{C}^1_{rd}(\mathscr{T})$.

Some results concerning rd-continuous and regulated functions are contained in the following theorem. Since its statements follow directly from the definitions, we leave its proof to the reader.

Theorem 12 *Assume* $f : \mathscr{T} \longmapsto \mathscr{R}$.

1. *If f is continuous, then f is rd-continuous.*
2. *If f is rd-continuous, then f is regulated.*
3. *The jump operator σ is rd-continuous.*
4. *If f is regulated or rd-continuous, then so is f^σ.*
5. *Assume f is continuous. If $g : \mathscr{T} \longmapsto \mathscr{R}$ is regulated or rd-continuous, then $f \circ g$ has that property.*

Theorem 13 *Every regulated function on a compact interval is bounded.*

Proof Assume that $f : [a, b] \longmapsto \mathscr{R}$, $[a, b] \subset \mathscr{T}$, is unbounded. Then for each $n \in \mathscr{N}$ there exists $t_n \in \mathscr{T}$ such that $|f(t_n)| > n$. Because $\{t_n\}_{n \in \mathscr{N}} \subset [a, b]$, there exists a subsequence $\{t_{n_k}\}_{k \in \mathscr{N}} \subset \{t_n\}_{n \in \mathscr{N}}$ such that

$$t_{n_k} > t_0 \;\; and \;\; \lim_{k \to \infty} t_{n_k} = t_0 \;\; or \;\; t_{n_k} < t_0 \;\; and \;\; \lim_{k \to \infty} t_{n_k} = t_0.$$

Since $\mathscr{T}$ is closed, we have that $t_0 \in \mathscr{T}$. Also, t_0 is a left-dense point or a right-dense point. Using that f is regulated, we get

$$|\lim_{k\longrightarrow\infty} f(t_{n_k}) = f(t_0)| \neq \infty,$$

which is a contradiction.

Theorem 14 (Induction Principle) *Let $t_0 \in \mathcal{T}$ and assume that*

$$\{S(t) : t \in [t_0, \infty)\}$$

is a family of statements satisfying

(i) $S(t_0)$ is true,
(ii) If $t \in [t_0, \infty)$ is right-scattered and $S(t)$ is true, then $S(\sigma(t))$ is true,
(iii) If $t \in [t_0, \infty)$ is right-dense and $S(t)$ is true, then there is a neighbourhood U of t such that $S(s)$ is true for all $s \in U \cap (t, \infty)$,
(iv) If $t \in (t_0, \infty)$ is left-dense and $S(s)$ is true for $s \in [t_0, t)$, then $S(t)$ is true.

Then $S(t)$ is true for all $t \in [t_0, \infty)$.

Proof Let

$$S^* = \{t \in [t_0, \infty) : S(t) \quad \text{is not true}\}.$$

We assume that $S^* \neq \emptyset$. Let $\inf S^* = t^*$. Because $\mathcal{T}$ is closed, we have that $t^* \in \mathcal{T}$.

1. If $t^* = t_0$, then $S(t^*)$ is true.
2. If $t^* \neq t_0$ and $t^* = \rho(t^*)$, using (iv), we get that $S(t^*)$ is true.
3. If $t^* \neq t_0$ and $\rho(t^*) < t^*$, then $\rho(t^*)$ is right-scattered. Since $S(\rho(t^*))$ is true, we get that $S(t^*)$ is true.

Consequently $t^* \notin S^*$.

If we suppose that t^* is right-scattered, then using that $S(t^*)$ is true and (ii), we conclude that $S(\sigma(t^*))$ is true, which is a contradiction. From the definition of t^* it follows that $t^* \neq \max \mathcal{T}$. Since t^* is not right-scattered and $t^* \neq \max \mathcal{T}$, we obtain that t^* is right-dense. Because $S(t^*)$ is true, using (iii), there exists a neighbourhood U of t^* such that $S(s)$ is true for all $s \in U \bigcap (t^*, \infty)$, which is a contradiction.

Consequently $S^* = \emptyset$.

Theorem 15 (Dual Version of Induction Principle) *Let $t_0 \in \mathcal{T}$ and assume that*

$$\{S(t) : t \in (-\infty, t_0]\}$$

is a family of statements satisfying

(i) $S(t_0)$ is true,
(ii) If $t \in (-\infty, t_0]$ is left-scattered and $S(t)$ is true, then $S(\rho(t))$ is true,
(iii) If $t \in (-\infty, t_0]$ is left-dense and $S(t)$ is true, then there is neighbourhood U of t such that $S(s)$ is true for all $s \in U \cap (-\infty, t)$,
(iv) If $t \in (-\infty, t_0)$ is right-dense and $S(s)$ is true for $s \in (t, t_0)$, then $S(t)$ is true.

Then $S(t)$ is true for all $t \in (-\infty, t_0]$.

Theorem 16 *Let f and g be real-valued functions defined on $\mathscr{T}$, both pre-differentiable with D. Then*

$$|f^{\Delta}(t)| \leq |g^{\Delta}(t)| \quad \text{for all} \quad t \in D$$

implies

$$|f(s) - f(r)| \leq g(s) - g(r) \quad \text{for all } r, s, \in \mathscr{T}, \quad r \leq s. \tag{1.10}$$

Proof Let $r, s \in \mathscr{T}$ with $r \leq s$. Let also,

$$[r, s) \backslash D = \{t_n : n \in \mathscr{N}\}.$$

We take $\varepsilon > 0$ arbitrarily. We consider the statements

$$S(t) : |f(t) - f(r)| \leq g(t) - g(r) + \varepsilon \left(t - r + \sum_{t_n < t} 2^{-n} \right)$$

for $t \in [r, s]$.

We will prove, using the induction principle, that $S(t)$ is true for all $t \in [r, s]$.

1. $S(r) : 0 \leq \varepsilon \sum_{t_n < r} 2^{-n}$ is true.
2. Let $t \in [r, s]$ is right-scattered and $S(t)$ holds. Then for $t \in D$ we have

$$\begin{aligned}
|f(\sigma(t)) - f(r)| &= |f(t) + \mu(t) f^{\Delta}(t) - f(r)| \\
&\leq \mu(t)|f^{\Delta}(t)| + |f(t) - f(r)| \\
&\leq \mu(t)|f^{\Delta}(t)| + g(t) - g(r) + \varepsilon \left(t - r + \textstyle\sum_{t_n < t} 2^{-n}\right) \\
&\leq \mu(t) g^{\Delta}(t) + g(t) - g(r) + \varepsilon \left(t - r + \textstyle\sum_{t_n < t} 2^{-n}\right) \\
&\leq g(\sigma(t)) - g(r) + \varepsilon \left(t - r + \textstyle\sum_{t_n < t} 2^{-n}\right) \quad (t < \sigma(t)) \\
&< g(\sigma(t)) - g(r) + \varepsilon \left(\sigma(t) - r + \textstyle\sum_{t_n < \sigma(t)} 2^{-n}\right),
\end{aligned}$$

   i.e., $S(\sigma(t))$ holds.
3. Let $t \in [r, s)$ and t is right-dense.

   1. case. $t \in D$. Then f and g are differentiable at t. Then there exists a neighbourhood U of t such that

   $$|f(t) - f(\tau) - f^{\Delta}(t)(t - \tau)| \leq \frac{\varepsilon}{2}|t - \tau|$$

   for all $\tau \in U$, and

   $$|g(t) - g(\tau) - g^{\Delta}(t)(t - \tau)| \leq \frac{\varepsilon}{2}|t - \tau|$$

for all $\tau \in U$. Thus

$$|f(t) - f(\tau)| \leq \left(|f^{\Delta}(t)| + \frac{\varepsilon}{2}\right)|t - \tau|$$

for all $\tau \in U$, and

$$g(\tau) - g(t) + g^{\Delta}(t)(t - \tau) \geq -\frac{\varepsilon}{2}|t - \tau|$$

for all $\tau \in U$ or

$$g(\tau) - g(t) - g^{\Delta}(t)(\tau - t) \geq -\frac{\varepsilon}{2}|t - \tau|$$

for all $\tau \in U$.
Hence for all $\tau \in U \cap (t, \infty)$

$$\begin{aligned} |f(\tau) - f(r)| &= |f(\tau) - f(t) + f(t) - f(r)| \\ &\leq |f(\tau) - f(t)| + |f(t) - f(r)| \\ &\leq \left(|f^{\Delta}(t)| + \tfrac{\varepsilon}{2}\right)|t - \tau| + g(t) - g(r) \\ &\quad + \varepsilon\left(t - r + \textstyle\sum_{t_n < t} 2^{-n}\right) \\ &\leq \left(g^{\Delta}(t) + \tfrac{\varepsilon}{2}\right)|t - \tau| + g(t) - g(r) \\ &\quad + \varepsilon\left(t - r + \textstyle\sum_{t_n < t} 2^{-n}\right) \\ &= g^{\Delta}(t)(\tau - t) + \tfrac{\varepsilon}{2}(\tau - t) + g(t) - g(r) \\ &\quad + \varepsilon\left(t - r + \textstyle\sum_{t_n < t} 2^{-n}\right) \\ &\leq g(\tau) - g(t) + \tfrac{\varepsilon}{2}|t - \tau| + \tfrac{\varepsilon}{2}(\tau - t) + g(t) - g(r) \\ &\quad + \varepsilon\left(t - r + \textstyle\sum_{t_n < t} 2^{-n}\right) \\ &= g(\tau) - g(r) + \varepsilon(\tau - t) + \varepsilon\left(t - r + \textstyle\sum_{t_n < t} 2^{-n}\right) \\ &= g(\tau) - g(r) + \varepsilon\left(\tau - r + \textstyle\sum_{t_n < \tau} 2^{-n}\right), \end{aligned}$$

so $S(\tau)$ follows for all $\tau \in U \cap (t, \infty)$.

2. case. $t \notin D$. Then $t = t_m$ for some $m \in \mathcal{N}$. Since f and g are predifferentiable, then they both are continuous. Therefore there exists a neighbourhood U of t such that

$$|f(\tau) - f(t)| \leq \frac{\varepsilon}{2}2^{-m} \quad \text{for all} \quad \tau \in U$$

and

$$|g(\tau) - g(t)| \leq \frac{\varepsilon}{2}2^{-m} \quad \text{for all} \quad \tau \in U.$$

Therefore

$$g(\tau) - g(t) \geq -\frac{\varepsilon}{2}2^{-m} \quad \text{for all} \quad \tau \in U.$$

Consequently

$$\begin{aligned}|f(\tau)-f(r)| &= |f(\tau)-f(t)+f(t)-f(r)|\\ &\le |f(\tau)-f(t)|+|f(t)-f(r)|\\ &\le \tfrac{\varepsilon}{2}2^{-m}+g(t)-g(r)+\varepsilon\left(t-r+\textstyle\sum_{t_n<t}2^{-n}\right)\\ &\le \tfrac{\varepsilon}{2}2^{-m}+g(\tau)+\tfrac{\varepsilon}{2}2^{-m}-g(r)+\varepsilon\left(t-r+\textstyle\sum_{t_n<t}2^{-n}\right)\\ &= \varepsilon 2^{-m}+g(\tau)-g(r)+\varepsilon\left(\tau-r+\textstyle\sum_{t_n<t}2^{-n}\right)\\ &\le \varepsilon 2^{-m}+g(\tau)-g(r)+\varepsilon\left(\tau-r+\textstyle\sum_{t_n<\tau}2^{-n}\right),\end{aligned}$$

so $S(\tau)$ follows for all $\tau \in U \cap (t, \infty)$.

3. Let t is left-dense and $S(t)$ is true for $\tau < t$. Then

$$\begin{aligned}&\lim_{\tau\longrightarrow t-}|f(\tau)-f(r)| \le \lim_{\tau\longrightarrow t-}\left\{g(\tau)-g(r)+\varepsilon\left(\tau-r+\textstyle\sum_{t_n<\tau}2^{-n}\right)\right\}\\ &\le \lim_{\tau\longrightarrow t-}\left\{g(\tau)-g(r)+\varepsilon\left(\tau-r+\textstyle\sum_{t_n<t}2^{-n}\right)\right\}\end{aligned}$$

implies $S(t)$ since f and g are continuous at t.

Hence and the induction principle it follows that $S(t)$ is true for all $t \in [r, s]$. Consequently (1.10) holds for all $r \le s$, $r, s \in \mathscr{T}$.

Theorem 17 *Suppose $f : \mathscr{T} \longmapsto \mathscr{R}$ is pre-differentiable with D. If U is a compact interval with endpoints $r, s \in \mathscr{T}$, then*

$$|f(s)-f(r)| \le \left\{\sup_{t\in U^k\cap D}|f^{\Delta}(t)|\right\}|s-r|.$$

Proof Without loss of generality we suppose that $r \le s$. We set

$$g(t) = \left\{\sup_{t\in U^\kappa\cap D}|f^{\Delta}(t)\right\}|(t-r), \quad t \in \mathscr{T}.$$

Then

$$g^{\Delta}(t) = \left\{\sup_{t\in U^k\cap D}|f^{\Delta}(t)|\right\} \ge |f^{\Delta}(t)|$$

for all $t \in D \cap [r, s]^{\kappa}$.

Hence and Theorem 16 it follows that

$$|f(t)-f(r)| \le g(t)-g(r) \quad \text{for all} \quad t \in [r, s],$$

whereupon

$$|f(s)-f(r)| \le g(s)-g(r) = g(s) = \left\{\sup_{t\in U^\kappa\cap D}|f^{\Delta}(t)|\right\}(s-r).$$

Theorem 18 *Let f is pre-differentiable with D. If $f^{\Delta}(t) = 0$ for all $t \in D$, then $f(t)$ is a constant function.*

Proof From Theorem 17 it follows that for all $r, s \in \mathscr{T}$

$$|f(s) - f(r)| \leq \left\{ \sup_{t \in U^{\kappa} \cap D} |f^{\Delta}(t)| \right\} |s - r| = 0,$$

i.e., $f(s) = f(r)$. Therefore f is a constant function.

Theorem 19 *Let f and g are pre-differentiable with D and $f^{\Delta}(t) = g^{\Delta}(t)$ for all $t \in D$. Then*

$$g(t) = f(t) + C \quad \text{for all} \quad t \in \mathscr{T},$$

where C is a constant.

Proof Let $h(t) = f(t) - g(t), t \in \mathscr{T}$. Then

$$h^{\Delta}(t) = f^{\Delta}(t) - g^{\Delta}(t) = 0 \quad \text{for all} \quad t \in D.$$

Hence and Theorem 18 it follows that $h(t)$ is a constant function.

Theorem 20 *Suppose $f_n : \mathscr{T} \longmapsto \mathscr{R}$ is pre-differentiable with D for each $n \in \mathscr{N}$. Assume that for each $t \in \mathscr{T}^{\kappa}$ there exists a compact interval $U(t)$ such that the sequence $\{f_n^{\Delta}\}_{n \in \mathscr{N}}$ converges uniformly on $U(t) \cap D$.*

(i) If $\{f_n\}_{n \in \mathscr{N}}$ converges at some $t_0 \in U(t)$ for some $t \in \mathscr{T}^{\kappa}$, then it converges uniformly on $U(t)$.
(ii) If $\{f_n\}_{n \in \mathscr{N}}$ converges at some $t_0 \in \mathscr{T}$, then it converges uniformly on $U(t)$ for all $t \in \mathscr{T}^{\kappa}$.
(iii) The limit mapping $f = \lim_{n \longrightarrow \infty} f_n$ is pre-differentiable with D and we have

$$f^{\Delta}(t) = \lim_{n \longrightarrow \infty} f_n^{\Delta}(t) \quad \text{for all} \quad t \in D.$$

Proof (i) Since $\{f_n^{\Delta}\}_{n \in \mathscr{N}}$ converges uniformly on $U(t) \cap D$, then there exists $N \in \mathscr{N}$ such that

$$\sup_{s \in U(t) \cap D} |(f_m - f_n)^{\Delta}(s)|$$

is finite for all $m, n \geq N$.
Let $m, n \geq N$ and $r \in U(t)$. Then

$$\begin{aligned}|f_n(r) - f_m(r)| &= |f_n(r) - f_m(r) - (f_n(t_0) - f_m(t_0)) + (f_n(t_0) - f_m(t_0))| \\ &\leq |f_n(t_0) - f_m(t_0)| + \left\{\sup_{s \in U(t) \cap D} |(f_n - f_m)^{\Delta}(s)|\right\} |r - t_0|.\end{aligned}$$

Hence, $\{f_n\}_{n \in \mathscr{N}}$ converges uniformly on $U(t)$, i.e., $\{f_n\}_{n \in \mathscr{N}}$ is locally uniformly convergent sequence.

(ii) Let $\{f_n(t_0)\}_{n \in \mathscr{N}}$ converges for some $t_0 \in \mathscr{T}$. Let

$$S(t) : \{f_n(t)\}_{n \in \mathscr{N}} \quad \text{converges.}$$

1. $S(t_0) : \{f_n(t_0)\}$ converges is true.
2. Let t is right-scattered and $S(t)$ holds. Then

$$f_n(\sigma(t)) = f_n(t) + \mu(t) f_n^{\Delta}(t)$$

converges by the assumption, i.e., $S(\sigma(t))$ holds.
3. Let t is right-dense and $S(t)$ holds. Then, by (i), $\{f_n\}_{n\in\mathcal{N}}$ converges on $U(t)$ and so $S(r)$ holds for all $r \in U(t) \cap (t, \infty)$.
4. Let t is left-dense and $S(r)$ holds for all $t_0 \le r < t$. Since $U(t) \cap [t_0, t) \neq \emptyset$, using again part (i), we have that $\{f_n\}_{n\in\mathcal{N}}$ converges on $U(t)$, in particular $S(t)$ is true.

Consequently $S(t)$ is true for all $t \in [t_0, \infty)$. Using the dual version of the induction principle for the negative direction, we have that $S(t)$ is also true for all $t \in (-\infty, t_0]$ (We note that the first part of this has already been shown, the second part follows by $f_n(\rho(t)) = f_n(t) - \mu(\rho(t)) f_n^{\Delta}(\rho(t))$, the third part and the fourth part follow again by (i)).

(iii) Let $t \in D$. Without loss of generality we can assume that $\sigma(t) \in U(t)$. We take $\varepsilon > 0$ arbitrarily. Using (i), there exists $N \in \mathcal{N}$ such that

$$|(f_n - f_m)(r) - (f_n - f_m)(\sigma(t))| \le \left\{\sup_{s\in U(t)\cap D} |(f_n - f_m)^{\Delta}(s)|\right\} |r - \sigma(t)|$$

for all $r \in U(t)$ and all $m, n \ge N$. Since $\{f_n^{\Delta}\}_{n\in\mathcal{N}}$ converges uniformly on $U(t) \cap D$, there exists $N_1 \ge N$ such that

$$\sup_{s\in U(t)\cap D} |(f_n - f_m)^{\Delta}(s)| \le \frac{\varepsilon}{3} \quad \text{for all} \quad m, n \ge N_1.$$

Hence,

$$|(f_n - f_m)(r) - (f_n - f_m)(\sigma(t))| \le \frac{\varepsilon}{3}|r - \sigma(t)|$$

for all $r \in U(t)$ and for all $m, n \ge N_1$. Now, letting $m \longrightarrow \infty$,

$$|(f_n - f)(r) - (f_n - f)(\sigma(t))| \le \frac{\varepsilon}{3}|r - \sigma(t)|$$

for all $r \in U(t)$ and all $n \ge N_1$. Let

$$g(t) = \lim_{n\longrightarrow\infty} f_n^{\Delta}(t).$$

Then there exists $M \ge N_1$ such that

$$|f_M^{\Delta}(t) - g(t)| \le \frac{\varepsilon}{3}$$

and since f_M is differentiable at t, there also exists a neighbourhood W of t with

$$|f_M(\sigma(t)) - f_M(r) - f_M^{\Delta}(t)(\sigma(t) - r)| \le \frac{\varepsilon}{3}|\sigma(t) - r|$$

for all $r \in W$.
From here, for all $r \in U(t) \cap W$, we have

$$\begin{aligned}|f(\sigma(t)) - f(r) - g(t)|\sigma(t) - r|| &\le |(f_M - f)(\sigma(t) - (f_M - f)(r)| \\ &\quad + |f_M^{\Delta}(t) - g(t)||\sigma(t) - r| \\ &\quad + |f_M(\sigma(t)) - f_M(r) - f_M^{\Delta}(t)|\sigma(t) - r|| \\ &\le \tfrac{\varepsilon}{3}|\sigma(t) - r| + \tfrac{\varepsilon}{3}|\sigma(t) - r| \\ &\quad + \tfrac{\varepsilon}{3}|\sigma(t) - r| \\ &= \varepsilon|\sigma(t) - r|.\end{aligned}$$

Consequently f is differentiable at t with $f^{\Delta}(t) = g(t)$.

Theorem 21 *Let $t_0 \in \mathscr{T}$, $x_0 \in \mathscr{R}$, $f : \mathscr{T}^{\kappa} \longmapsto \mathscr{R}$ be given regulated map. Then there exists exactly one pre-differentiable function F satisfying*

$$F^{\Delta}(t) = f(t) \quad \text{for all} \quad t \in D, \quad F(t_0) = x_0.$$

Proof Let $n \in \mathscr{N}$ and

$$S(t) : \begin{cases} \text{there exists a pre} - \text{differentiable} \quad (F_{nt}, D_{nt}), \\ F_{nt} : [t_0, t] \longmapsto \mathscr{R} \quad \text{with} \quad F_{nt}(t_0) = x_0 \quad \text{and} \\ |F_{nt}^{\Delta}(s) - f(s)| \le \frac{1}{n} \quad \text{for} \quad s \in D_{nt}. \end{cases}$$

1. $t = t_0$. Then $D_{nt_0} = \emptyset$ and $F_{nt_0}(t_0) = x_0$. Then the statement $S(t_0)$ follows.
2. Let t is right-scattered and $S(t)$ is true. Define

$$D_{n\sigma(t)} = D_{nt} \cup \{t\}$$

and $F_{n\sigma(t)}$ on $[t_0, \sigma(t)]$ by

$$F_{n\sigma(t)}(s) = \begin{cases} F_{nt}(s) \quad \text{if} \quad s \in [t_0, t] \\ F_{nt}(t) + \mu(t) f(t) \quad \text{if} \quad s = \sigma(t). \end{cases}$$

Then

$$F_{n\sigma(t)}(t_0) = F_{nt}(t_0) = x_0,$$
$$|F_{n\sigma(t)}^{\Delta}(s) - f(s)| = |F_{nt}^{\Delta}(s) - f(s)| \le \frac{1}{n} \quad \text{for} \quad s \in D_{nt}$$

and

$$\begin{aligned}|F^{\Delta}_{n\sigma(t)}(t) - f(t)| &= \left|\frac{F_{n\sigma(t)}(\sigma(t)) - F_{n\sigma(t)}(t)}{\mu(t)} - f(t)\right| \\ &= \left|\frac{F_{nt}(t) + \mu(t)f(t) - F_{n\sigma(t)}(t)}{\mu(t)} - f(t)\right| \\ &= \left|\frac{F_{nt}(t) + \mu(t)f(t) - F_{nt}(t)}{\mu(t)} - f(t)\right| \\ &= \left|\frac{\mu(t)f(t)}{\mu(t)} - f(t)\right| = 0 \le \frac{1}{n}\end{aligned}$$

and therefore the statement $S(\sigma(t))$ is valid.

3. Suppose t is right-dense and $S(t)$ is true. Since t is right-dense and $f(t)$ is regulated,

$$f(t^+) = \lim_{s \longrightarrow t, s>t} f(s) \quad \text{exists.}$$

Hence, there is a neighbourhood U of t with

$$|f(s) - f(t^+)| \le \frac{1}{n} \quad \text{for all} \quad s \in U \cap (t, \infty). \tag{1.11}$$

Let $r \in U \cap (t, \infty)$. Define

$$D_{nr} = [D_{nt} \backslash \{t\}] \cup [t, r]^{\kappa}$$

and F_{nr} on $[t_0, r]$ by

$$F_{nr}(s) = \begin{cases} F_{nt}(s) & \text{if} \quad s \in [t_0, t] \\ F_{nt}(t) + f(t^+)(s - t) & \text{if} \quad s \in (t, r]. \end{cases}$$

Then F_{nr} is continuous at t and hence on $[t_0, r]$. Also, F_{nr} is differentiable on $(t, r]^{\kappa}$ with

$$F^{\Delta}_{nr}(s) = f(t^+) \quad \text{for all} \quad s \in (t, r]^{\kappa}.$$

Hence F_{nr} is pre-differentiable on $[t_0, t)$. Since t is right-dense, we have that F_{nr} is pre-differentiable with D_{nr}. From here and from (1.11), we also have

$$|F^{\Delta}_{nr}(s) - f(s)| \le \frac{1}{n} \quad \text{for all} \quad s \in D_{nr}.$$

Therefore the statement $S(r)$ is true for all $r \in U \cap (t, \infty)$.

4. Now we suppose that t is left-dense and the statement $S(r)$ is true for $r < t$. Since $f(t)$ is regulated,

$$f(t^-) = \lim_{s \longrightarrow t, s<t} f(s) \quad \text{exists.} \tag{1.12}$$

Hence there exists a neighbourhood U of t with

$$|f(s) - f(t^-)| \leq \frac{1}{n} \quad \text{for all } s \in U \cap (-\infty, t).$$

Fix some $r \in U \cap (-\infty, t)$ and define

$$D_{nt} = \begin{cases} D_{nr} \cup (r, t) & \text{if } r \text{ is right} - \text{dense} \\ D_{nr} \cup [r, t) & \text{if } r \text{ is right} - \text{scattered} \end{cases}$$

and F_{nt} on $[t_0, t]$ by

$$F_{nt}(s) = \begin{cases} F_{nr}(s) \quad \text{if} \quad s \in (t_0, r] \\ F_{nr}(t) + f(t^-)(s - r) \quad \text{if} \quad s \in (r, t]. \end{cases}$$

We note that F_{nt} is continuous at r and hence in $[t_0, t]$. Since

$$F_{nt}^{\Delta}(s) = f(t^-) \quad \text{for all} \quad s \in (r, t],$$

F_{nt} is pre-differentiable with D_{nt} and

$$|F_{nt}^{\Delta}(s) - f(s)| \leq \frac{1}{n} \quad \text{for all} \quad s \in D_{nt}.$$

Hence, the statement $S(t)$ holds.

By the induction principle $S(t)$ is true for all $t \geq t_0, t \in \mathscr{T}$. Similarly, we can show that $S(t)$ is valid for $t \geq t_0$. Hence F_n is pre-differentiable with D_n, $F_n(t_0) = x_0$ and

$$|F_n^{\Delta}(t) - f(t)| \leq \frac{1}{n} \quad \text{for all} \quad t \in D_n.$$

Now let

$$F = \lim_{n \longrightarrow \infty} F_n \quad \text{and} \quad D = \bigcap_{n \in \mathscr{N}} D_n.$$

Then $F(t_0) = x_0$, F is pre-differentiable on D and using Theorem 20

$$F^{\Delta}(t) = \lim_{n \longrightarrow \infty} F_n^{\Delta}(t) = f(t) \quad \text{for all} \quad t \in D.$$

Definition 16 Assume $f : \mathscr{T} \longmapsto \mathscr{R}$ is a regulated function. Any function F by Theorem 21 is called a pre-antiderivative of f. We define the indefinite integral of a regulated function f by

$$\int f(t)\Delta t = F(t) + c,$$

where c is an arbitrary constant and F is a pre-antiderivative of f. We define the Cauchy integral by

$$\int_\tau^s f(t)\Delta t = F(s) - F(\tau) \quad \text{for all} \quad r, s \in \mathscr{T}.$$

A function $F : \mathscr{T} \longmapsto \mathscr{R}$ is called an antiderivative of $f : \mathscr{T} \longmapsto \mathscr{R}$ provided

$$F^\Delta(t) = f(t) \quad \text{holds for all} \quad t \in \mathscr{T}^\kappa.$$

Example 42 Let $\mathscr{T} = \mathscr{Z}$. Then $\sigma(t) = t + 1, t \in \mathscr{T}$. Let also, $f(t) = 3t^2 + 5t + 2$, $g(t) = t^3 + t^2, t \in \mathscr{T}$. Since

$$\begin{aligned} g^\Delta(t) &= \sigma^2(t) + t\sigma(t) + t^2 + \sigma(t) + t \\ &= (t+1)^2 + t(t+1) + t^2 + t + 1 + t \\ &= t^2 + 2t + 1 + t^2 + t + t^2 + 2t + 1 \\ &= 3t^2 + 5t + 2, \end{aligned}$$

we conclude that

$$\int (3t^2 + 5t + 2)\Delta t = t^3 + t^2 + c.$$

Example 43 Let $\mathscr{T} = 2^{\mathscr{N}}$, $f : \mathscr{T} \longmapsto \mathscr{R}$ is defined by $f(t) = 2\sin\frac{t}{2}\cos\frac{3t}{2}$, $t \in \mathscr{T}$. Let also, $g(t) = \sin t, t \in \mathscr{T}$. In this case we have that $\sigma(t) = 2t$. Since

$$\begin{aligned} g^\Delta(t) &= \tfrac{\sin(\sigma(t)) - \sin t}{\sigma(t) - t} \\ &= \tfrac{\sin(2t) - \sin t}{t} \\ &= \tfrac{2}{t}\sin\tfrac{t}{2}\cos\tfrac{3t}{2}, \end{aligned}$$

we get

$$\int \frac{2}{t}\sin\frac{t}{2}\cos\frac{3t}{2}\Delta t = \sin t + c.$$

Example 44 Let $\mathscr{T} = \mathscr{N}_0^2$, $f : \mathscr{T} \longmapsto \mathscr{R}$ is defined by $f(t) = \frac{1}{1 + 2\sqrt{t}}$ $\log\left(\frac{(\sqrt{t}+1)^2}{t}\right)$, $t \in \mathscr{T}$. Let also, $g(t) = \log t, t \in \mathscr{T}$. Since $\sigma(t) = (\sqrt{t}+1)^2$ and

$$\begin{aligned} g^\Delta(t) &= \tfrac{\log(\sigma(t)) - \log t}{\sigma(t) - t} \\ &= \frac{\log\left((\sqrt{t}+1)^2\right) - \log t}{(\sqrt{t}+1)^2 - t} \\ &= \tfrac{1}{1+2\sqrt{t}}\log\left(\tfrac{(\sqrt{t}+1)^2}{t}\right), \end{aligned}$$

we get

$$\int \frac{1}{1 + 2\sqrt{t}}\log\frac{(1+\sqrt{t})^2}{t}\Delta t = \log t + c.$$

Exercise 14 Let $\mathscr{T} = \mathscr{N}_0^3$. Prove that

$$\int (2t + 3\sqrt[3]{t^2} + 3\sqrt[3]{t} + 2)\Delta t = t^2 + t + c.$$

Theorem 22 *Every rd-continuous function $f : \mathscr{T} \longmapsto \mathscr{R}$ has an antiderivative. In particular, if $t_0 \in \mathscr{T}$, then F defined by*

$$F(t) = \int_{t_0}^{t} f(\tau)\Delta\tau \quad \text{for} \quad t \in \mathscr{T},$$

is an antiderivative of f.

Proof Since f is rd-continuous, then it is regulated. Let F be a function guaranteed to exist by Theorem 21, together with D, satisfying

$$F^{\Delta}(t) = f(t) \quad \text{for} \quad t \in D.$$

We have that F is pre-differentiable with D.

Let $t \in \mathscr{T}^{\kappa} \backslash D$. Then t is right-dense. Since f is rd-continuous, then f is continuous at t. Let $\varepsilon > 0$ be arbitrarily chosen. Then there exists a neighbourhood U of t such that

$$|f(s) - f(t)| \leq \varepsilon \quad \text{for all} \quad s \in U.$$

We define

$$h(\tau) = F(\tau) - f(t)(\tau - t_0) \quad \text{for} \quad \tau \in \mathscr{T}.$$

Then h is pre-differentiable with D and

$$h^{\Delta}(\tau) = F^{\Delta}(\tau) - f(t) = f(\tau) - f(t) \quad \text{for all} \quad \tau \in D.$$

Hence

$$|h^{\Delta}(s)| = |f(s) - f(t)| \leq \varepsilon \quad \text{for all} \quad s \in D \cap U.$$

Therefore

$$\sup_{s \in D \cap U} |h^{\Delta}(s)| \leq \varepsilon,$$

whereupon

$$\begin{aligned} |F(t) - F(r) - f(t)(t - r)| &= |h(t) + f(t)(t - t_0) - (h(r) \\ &\quad + f(t)(r - t_0)) - f(t)(t - r)| \\ &= |h(t) - h(r)| \\ &\leq \left\{ \sup_{s \in D \cap U} |h^{\Delta}(s)| \right\} |t - r| \\ &\leq \varepsilon |t - r|, r \in D \cap U, r \neq t, \end{aligned}$$

which shows that F is differentiable at t and $F^{\Delta}(t) = f(t)$.

Theorem 23 *If $f \in \mathscr{C}_{rd}(\mathscr{T})$ and $t \in \mathscr{T}^{\kappa}$, then*

$$\int_t^{\sigma(t)} f(\tau)\Delta\tau = \mu(t) f(t).$$

Proof Since $f \in \mathscr{C}_{rd}(\mathscr{T})$, there exists an antiderivative F of f, and

$$\begin{aligned}\int_t^{\sigma(t)} f(\tau)\Delta\tau &= F(\sigma(t)) - F(t)\\ &= \mu(t)F^{\Delta}(t)\\ &= \mu(t) f(t),\end{aligned}$$

so that the conclusion follows.

Theorem 24 *If $f^{\Delta} \geq 0$, then f is nondecreasing.*

Proof Let $f^{\Delta} \geq 0$ and let $s, t \in \mathscr{T}$ with $a \leq s \leq t \leq b$. Then

$$f(t) = f(s) + \int_s^t f^{\Delta}(\tau)\Delta\tau \geq f(s)$$

so that the conclusion follows.

Theorem 25 *If $a, b, c \in \mathscr{T}$, $\alpha \in \mathscr{R}$ and $f, g \in \mathscr{C}_{rd}(\mathscr{T})$, then*

(i) $\displaystyle\int_a^b (f(t) + g(t))\Delta t = \int_a^b f(t)\Delta t + \int_a^b g(t)\Delta t,$

(ii) $\displaystyle\int_a^b (\alpha f)(t)\Delta t = \alpha \int_a^b f(t)\Delta t,$

(iii) $\displaystyle\int_a^b f(t)\Delta t = -\int_b^a f(t)\Delta t,$

(iv) $\displaystyle\int_a^b f(t)\Delta t = \int_a^c f(t)\Delta t + \int_c^b f(t)\Delta t,$

(v) $\displaystyle\int_a^b f(\sigma(t))g^{\Delta}(t)\Delta t = (fg)(b) - (fg)(a) - \int_a^b f^{\Delta}(t)g(t)\Delta t,$

(vi) $\displaystyle\int_a^b f(t)g^{\Delta}(t)\Delta t = (fg)(b) - (fg)(a) - \int_a^b f^{\Delta}(t)g(\sigma(t))\Delta t,$

(vii) $\displaystyle\int_a^a f(t)\Delta t = 0,$

(viii) if $|f(t)| \leq g(t)$ on $[a, b]$, then

$$\left|\int_a^b f(t)\Delta t\right| \leq \int_a^b g(t)\Delta t,$$

(ix) if $f(t) \geq 0$ *for all* $a \leq t < b$*, then* $\int_a^b f(t)\Delta t \geq 0$.

Proof Since $f, g \in \mathscr{C}_{rd}(\mathscr{T})$, they possess antiderivatives F and G, respectively. We have

$$F^{\Delta}(t) = f(t) \quad \text{and} \quad G^{\Delta}(t) = g(t) \quad \text{for all} \quad t \in \mathscr{T}^{\kappa}.$$

(i) For all $t \in \mathscr{T}^{\kappa}$ we have

$$(F+G)^{\Delta}(t) = F^{\Delta}(t) + G^{\Delta}(t).$$

Hence,

$$\begin{aligned}\int_a^b (f(t)+g(t))\Delta t &= (F+G)(b) - (F+G)(a)\\ &= F(b) - F(a) + G(b) - G(a)\\ &= \int_a^b f(t)\Delta t + \int_a^b g(t)\Delta t.\end{aligned}$$

(ii) Since

$$\alpha F^{\Delta}(t) = (\alpha F)^{\Delta}(t) = \alpha f(t) \quad \textit{for all} \quad t \in \mathscr{T}^{\kappa},$$

we get

$$\begin{aligned}\int_a^b \alpha f(t)\Delta t &= (\alpha F)(b) - (\alpha F)(a)\\ &= \alpha(F(b) - F(a))\\ &= \alpha \int_a^b f(t)\Delta t.\end{aligned}$$

(iii) We have

$$\begin{aligned}\int_a^b f(t)\Delta t &= F(b) - F(a)\\ &= -(F(a) - F(b))\\ &= -\int_b^a f(t)\Delta t.\end{aligned}$$

(iv) We have

$$\begin{aligned}\int_a^b f(t)\Delta t &= F(b) - F(a)\\ &= F(c) - F(a) + F(b) - F(c)\\ &= \int_a^c f(t)\Delta t + \int_c^b f(t)\Delta t.\end{aligned}$$

(v) For all $t \in \mathscr{T}^\kappa$ we have

$$(fg)^\Delta(t) = f^\Delta(t)g(t) + f(\sigma(t))g^\Delta(t)$$

or

$$f(\sigma(t))g^\Delta(t) = (fg)^\Delta(t) - f^\Delta(t)g(t).$$

Hence and (i), (ii), we get

$$\begin{aligned}\int_a^b f(\sigma(t))g^\Delta(t)\Delta t &= \int_a^b \left((fg)^\Delta(t) - f^\Delta(t)g(t)\right)\Delta t\\ &= \int_a^b (fg)^\Delta(t)\Delta t - \int_a^b f^\Delta(t)g(t)\Delta t\\ &= (fg)(b) - (fg)(a) - \int_a^b f^\Delta(t)g(t)\Delta t.\end{aligned}$$

(vi) For all $t \in \mathscr{T}^\kappa$ we have

$$(fg)^\Delta(t) = f(t)g^\Delta(t) + f^\Delta(t)g(\sigma(t))$$

or

$$f(t)g^\Delta(t) = (fg)^\Delta(t) - f^\Delta(t)g(\sigma(t)).$$

Hence and (i), (ii), we find

$$\begin{aligned}\int_a^b f(t)g^\Delta(t)\Delta t &= \int_a^b (fg)^\Delta(t)\Delta t - \int_a^b f^\Delta(t)g(\sigma(t))\Delta t\\ &= (fg)(b) - (fg)(a) - \int_a^b f^\Delta(t)g(\sigma(t))\Delta t.\end{aligned}$$

(vii)

$$\int_a^a f(t)\Delta t = F(a) - F(a) = 0.$$

(viii) We note that

$$|F^\Delta(t)| \le G^\Delta(t) \quad on \quad [a, b].$$

Hence and Theorem 16, we get

$$|F(b) - F(a)| \leq G(b) - G(a)$$

or

$$\left| \int_a^b f(t)\Delta t \right| \leq \int_a^b g(t)\Delta t.$$

(ix) This property follows directly from the property (viii).

Exercise 15 Let $a, b \in \mathscr{T}$ and $f \in \mathscr{C}_{rd}(\mathscr{T})$.

(i) If $\mathscr{T} = \mathscr{R}$, prove

$$\int_a^b f(t)\Delta t = \int_a^b f(t)dt,$$

where the integral on the right is the usual Riemann integral from calculus.

(ii) If $[a, b]$ consists only isolated points, then

$$\int_a^b f(t)\Delta t = \begin{cases} \sum_{t\in[a,b)} \mu(t)f(t) & \text{if} \quad a < b \\ 0 \ \text{if} \quad a = b \\ -\sum_{t\in[b,a)} \mu(t)f(t) & \text{if} \quad a > b. \end{cases}$$

Definition 17 (*Improper Integral*) If $a \in \mathscr{T}$, $\sup \mathscr{T} = \infty$, and f is rd-continuous on $[a, \infty)$, then we define improper integral by

$$\int_a^\infty f(t)\Delta t = \lim_{b\longrightarrow\infty} \int_a^b f(t)\Delta t$$

provided this limit exists, and we say that the improper integral converges in this case. If this limit does not exist, then we say that the improper integral diverges.

Example 45 Let $\mathscr{T} = q^{\mathscr{N}_0}$, $q > 1$. Then $\sigma(t) = qt$. Since all points are isolated, then

$$\int_1^\infty \frac{1}{t^2}\Delta t = \sum_{t\in q^{\mathscr{N}_0}} \frac{1}{t^2}\mu(t) = \sum_{t\in q^{\mathscr{N}_0}} \frac{q-1}{t} = q.$$

1.5 The Exponential Function

1.5.1 Hilger's Complex Plane

Definition 18 Let $h > 0$.

1. The Hilger complex numbers we define as follows

$$\mathscr{C}_h = \{z \in \mathscr{C} : z \neq -\frac{1}{h}\}.$$

2. The Hilger real axis we define as follows

$$\mathscr{R}_h = \{z \in \mathscr{C} : z > -\frac{1}{h}\}.$$

3. The Hilger alternative axis we define as follows

$$A_h = \{z \in \mathscr{C} : z < -\frac{1}{h}\}.$$

4. The Hilger imaginary circle we define as follows

$$\mathscr{I}_h = \left\{z \in \mathscr{C} : \left|z + \frac{1}{h}\right| = \frac{1}{h}\right\}.$$

For $h = 0$ we set

$$\mathscr{C}_0 = \mathscr{C}, \quad \mathscr{R}_0 = \mathscr{R}, \quad \mathscr{A}_0 = \emptyset, \quad \mathscr{I}_0 = i\mathscr{R}.$$

Definition 19 Let $h > 0$ and $z \in \mathscr{C}_h$. We define the Hilger real part of z by

$$\mathrm{Re}_h(z) = \frac{|zh + 1| - 1}{h}$$

and the Hilger imaginary part by

$$\mathrm{Im}_h(z) = \frac{\mathrm{Arg}(zh + 1)}{h},$$

where $\mathrm{Arg}(z)$ denotes the principal argument of z, i.e.,

$$-\pi < \mathrm{Arg}(z) \leq \pi.$$

We note that

$$-\frac{1}{h} < \mathrm{Re}_h(z) < \infty \quad and \quad -\frac{\pi}{h} < \mathrm{Im}_h(z) < \frac{\pi}{h}.$$

In particular, $\mathrm{Re}_h(z) \in \mathscr{R}_h$.

Definition 20 Let $-\frac{\pi}{h} < w \leq \frac{\pi}{h}$. We define the Hilger purely imaginary number by

$$\overset{\circ}{i}\, w = \frac{e^{iwh} - 1}{h}.$$

Theorem 26 *Let $z \in \mathscr{C}_h$. Then* $\overset{\circ}{i}\ \mathrm{Im}_h(z) \in \mathscr{I}_h$.

Proof We have

$$\overset{\circ}{i}\ \mathrm{Im}_h(z) = \frac{e^{ih\mathrm{Im}_h(z)} - 1}{h}$$

and

$$\begin{aligned}\left|\overset{\circ}{i}\ \mathrm{Im}_h(z) + \frac{1}{h}\right| &= \left|\frac{e^{ih\mathrm{Im}_h(z)} - 1}{h} + \frac{1}{h}\right| \\ &= \frac{\left|e^{ih\mathrm{Im}_h(z)}\right|}{h} \\ &= \frac{1}{h}.\end{aligned}$$

Theorem 27 *We have*

$$\lim_{h\to 0}[\mathrm{Re}_h(z) + \overset{\circ}{i}\ \mathrm{Im}_h(z)] = \mathrm{Re}(z) + i\mathrm{Im}(z).$$

Proof We have

$$\begin{aligned} z &= \mathrm{Re}(z) + i\mathrm{Im}(z), \\ zh + 1 &= (\mathrm{Re}(z) + i\mathrm{Im}(z))h + 1 \\ &= h\mathrm{Re}(z) + 1 + ih\mathrm{Im}(z), \\ \mathrm{Arg}(zh+1) &= \arcsin\frac{h\mathrm{Im}(z)}{\sqrt{(h\mathrm{Re}(z)+1)^2 + h^2\mathrm{Im}^2(z)}}, \\ \mathrm{Im}_h(z) &= \frac{\mathrm{Arg}(zh+1)}{h} \\ &= \frac{1}{h}\arcsin\frac{h\mathrm{Im}(z)}{\sqrt{(h\mathrm{Re}(z)+1)^2 + h^2\mathrm{Im}^2(z)}}, \\ |zh+1| &= \sqrt{(h\mathrm{Re}(z)+1)^2 + h^2\mathrm{Im}^2(z)}, \end{aligned}$$

$$\begin{aligned} \mathrm{Re}_h(z) &= \frac{|zh+1| - 1}{h} \\ &= \frac{\sqrt{(h\mathrm{Re}(z)+1)^2 + h^2\mathrm{Im}^2(z)} - 1}{h}. \end{aligned}$$

Hence,

$$\lim_{h\to 0} \mathrm{Re}_h(z) = \lim_{h\to 0} \frac{\sqrt{(h\mathrm{Re}(z)+1)^2 + h^2\mathrm{Im}^2(z)} - 1}{h}$$
$$= \lim_{h\to 0} \frac{(h\mathrm{Re}(z)+1)\mathrm{Re}(z) + h\mathrm{Im}^2(z)}{\sqrt{(h\mathrm{Re}(z)+1)^2 + h^2\mathrm{Im}^2(z)}}$$
$$= \mathrm{Re}(z),$$

$$\lim_{h\to 0} \mathrm{Im}_h(z) = \lim_{h\to 0} \frac{1}{h} \arcsin \frac{h\mathrm{Im}(z)}{\sqrt{(h\mathrm{Re}(z)+1)^2 + h^2\mathrm{Im}^2(z)}}$$
$$= \lim_{h\to 0} \frac{1}{\sqrt{1 - \frac{h^2\mathrm{Im}^2(z)}{(h\mathrm{Re}(z)+1)^2 + h^2\mathrm{Im}^2(z)}}} \times$$
$$\times \frac{\mathrm{Im}(z)\sqrt{(h\mathrm{Re}(z)+1)^2 + h^2\mathrm{Im}^2(z)} - h\mathrm{Im}(z)\frac{(h\mathrm{Re}(z)+1)\mathrm{Re}(z) + h\mathrm{Im}^2(z)}{\sqrt{(h\mathrm{Re}(z)+1)^2 + h^2\mathrm{Im}^2(z)}}}{(h\mathrm{Re}(z)+1)^2 + h^2\mathrm{Im}^2(z)}$$
$$= \mathrm{Im}(z),$$

which completes the proof.

Theorem 28 *Let* $-\frac{\pi}{h} < w \leq \frac{\pi}{h}$. *Then*

$$|\overset{\circ}{i}\, w|^2 = \frac{4}{h^2} \sin^2\Big(\frac{wh}{2}\Big).$$

Proof We have

$$\overset{\circ}{i}\, w = \frac{e^{iwh} - 1}{h} = \frac{\cos(wh) - 1 + i\sin(wh)}{h}.$$

Hence,

$$|\overset{\circ}{i}\, w|^2 = \frac{(\cos(wh)-1)^2}{h^2} + \frac{\sin^2(wh)}{h^2}$$
$$= \frac{\cos^2(wh) - 2\cos(wh) + 1 + \sin^2(wh)}{h^2}$$
$$= \frac{2(1-\cos(wh))}{h^2}$$
$$= \frac{4}{h^2}\sin^2\Big(\frac{wh}{2}\Big).$$

Definition 21 The "circle plus" addition $\oplus$ on $\mathscr{C}_h$ is defined by

$$z \oplus w = z + w + zwh.$$

Theorem 29 *$(\mathscr{C}_h, \oplus)$ is an Abelian group.*

Proof Let $z, w \in \mathscr{C}_h$. Then $z, w \in \mathscr{C}$ and $z, w \neq -\frac{1}{h}$. Therefore $z \oplus w \in \mathscr{C}$.

Since

$$\begin{aligned} h(z \oplus w) + 1 &= h(z + w + zwh) + 1 \\ &= 1 + hz + hw + zwh^2 \\ &= 1 + hz + hw(1 + hz) \\ &= (1 + hw)(1 + hz) \\ &\neq 0, \end{aligned}$$

we conclude that $z \oplus w \in \mathscr{C}_h$.

Also,

$$0 \oplus z = z \oplus 0 = z,$$

i.e., 0 is the additive identity for $\oplus$.

For $z \in \mathscr{C}_h$ we have

$$\begin{aligned} z \oplus \left(-\frac{z}{1 + zh}\right) &= z - \frac{z}{1 + zh} - z\frac{z}{1 + zh}h \\ &= \frac{z^2 h}{1 + zh} - \frac{z^2 h}{1 + zh} \\ &= 0, \end{aligned}$$

i.e., the additive inverse of z under the addition $\oplus$ is $-\frac{z}{1 + zh}$. We note that $-\frac{z}{1 + zh} \in \mathscr{C}$ and

$$1 - \frac{zh}{1 + zh} = \frac{1}{1 + zh} \neq 0,$$

i.e., $-\frac{z}{1 + zh} \neq -\frac{1}{h}$. Therefore $-\frac{z}{1 + zh} \in \mathscr{C}_h$.

For $z, w, v \in \mathscr{C}_h$ we have

$$\begin{aligned} (z \oplus w) \oplus v &= (z + w + zwh) \oplus v \\ &= z + w + zwh + v + (z + w + zwh)vh \\ &= z + w + zwh + v + zvh + wvh + zwvh^2 \end{aligned}$$

and

$$\begin{aligned} z \oplus (w \oplus v) &= z + (w \oplus v) + z(w \oplus v)h \\ &= z + w + v + wvh + z(w + v + wvh)h \\ &= z + w + v + wvh + zwh + zvh + zwvh^2. \end{aligned}$$

Consequently

$$z \oplus (w \oplus v) = (z \oplus w) \oplus v,$$

i.e., in $(\mathscr{C}_h, \oplus)$ the associative law holds.

For $z, w \in \mathscr{C}_h$ we have

$$\begin{aligned} z \oplus w &= z + w + zwh \\ &= w + z + wzh \\ &= w \oplus z, \end{aligned}$$

which completes the proof.

Example 46 Let $z \in \mathscr{C}_h$ and $w \in \mathscr{C}$. We will simplify the expression

$$A = z \oplus \frac{w}{1 + hz}.$$

We have

$$\begin{aligned} A &= z + \frac{w}{1 + hz} + \frac{zw}{1 + hz} h \\ &= z + \frac{(1 + hz)w}{1 + hz} \\ &= z + w. \end{aligned}$$

Theorem 30 *For $z \in \mathscr{C}_h$ we have*

$$z = \mathrm{Re}_h(z) \oplus \overset{\circ}{i}\, \mathrm{Im}_h(z).$$

Proof We have

$$\begin{aligned} \mathrm{Re}_h(z) \oplus \overset{\circ}{i}\, \mathrm{Im}_h(z) &= \frac{|zh + 1| - 1}{h} \oplus \overset{\circ}{i}\, \frac{\mathrm{Arg}(zh + 1)}{h} \\ &= \frac{|zh + 1| - 1}{h} \oplus \frac{e^{i \mathrm{Arg}(zh+1)} - 1}{h} \\ &= \frac{|zh + 1| - 1}{h} + \frac{e^{i \mathrm{Arg}(zh+1)} - 1}{h} + \frac{|zh + 1| - 1}{h} \cdot \frac{e^{i \mathrm{Arg}(zh+1)} - 1}{h} h \\ &= \frac{1}{h} \Big(|zh + 1| - 1 + e^{i \mathrm{Arg}(zh+1)} - 1 + |zh + 1| e^{i \mathrm{Arg}(zh+1)} - |zh + 1| \end{aligned}$$

$$\begin{aligned}
&-e^{i\mathrm{Arg}(zh+1)}+1\Big) \\
&=\frac{1}{h}\Big(|zh+1|e^{i\mathrm{Arg}(zh+1)}-1\Big) \\
&=\frac{1}{h}(zh+1-1) \\
&=z.
\end{aligned}$$

Definition 22 Let $n \in \mathscr{N}$ and $z \in \mathscr{C}_h$. We definite "circle dot" multiplication $\odot$ by

$$n \odot z = z \oplus z \oplus z \oplus \cdots \oplus z.$$

Theorem 31 *Let $n \in \mathscr{N}$ and $z \in \mathscr{C}_h$. Then*

$$n \odot z = \frac{(zh+1)^n - 1}{h}. \tag{1.13}$$

Proof 1. Let $n = 2$. Then

$$\begin{aligned}
2 \odot z &= z \oplus z \\
&= z + z + z^2 h \\
&= 2z + z^2 h \\
&= \frac{1}{h}(z^2h^2 + 2zh) \\
&= \frac{1}{h}(z^2h^2 + 2zh + 1 - 1) \\
&= \frac{(zh+1)^2 - 1}{h}.
\end{aligned}$$

2. Assume

$$n \odot z = \frac{(zh+1)^n - 1}{h}$$

for some $n \in \mathscr{N}$.

3. We will prove that

$$(n+1) \odot z = \frac{(zh+1)^{n+1} - 1}{h}.$$

Indeed,

$$\begin{aligned}
(n+1) \odot z &= (n \odot z) \oplus z \\
&= \frac{(zh+1)^n - 1}{h} \oplus z \\
&= \frac{(zh+1)^n - 1}{h} + z + \frac{(zh+1)^n - 1}{h} zh
\end{aligned}$$

$$= \frac{(zh+1)^n - 1 + zh + zh(zh+1)^n - zh}{h}$$
$$= \frac{(zh+1)^{n+1} - 1}{h}.$$

Hence, we conclude that (1.13) holds for all $n \in \mathscr{N}$.

Definition 23 Let $z \in \mathscr{C}_h$. The additive inverse of z under the operation $\oplus$ is defined as follows

$$\ominus z = \frac{-z}{1+zh}.$$

Theorem 32 *Let $z \in \mathscr{C}_h$. Then*

$$\ominus(\ominus z) = z.$$

Proof We have

$$\begin{aligned} \ominus(\ominus z) &= -\frac{\ominus z}{1+(\ominus z)h} \\ &= -\frac{\frac{-z}{1+zh}}{1+\frac{-z}{1+zh}h} \\ &= \frac{\frac{z}{1+zh}}{\frac{1+zh-zh}{1+zh}} \\ &= z. \end{aligned}$$

Definition 24 Let $z, w \in \mathscr{C}_h$. We define "circle minus" substraction as follows

$$z \ominus w = z \oplus (\ominus w).$$

For $z, w \in \mathscr{C}_h$ we have

$$\begin{aligned} z \ominus w &= z \oplus (\ominus w) \\ &= z + (\ominus w) + z(\ominus w)h \\ &= z - \frac{w}{1+wh} - \frac{zwh}{1+wh} \\ &= \frac{z + zwh - w - zwh}{1+wh} \\ &= \frac{z-w}{1+wh}, \end{aligned}$$

i.e.,

$$z \ominus w = \frac{z-w}{1+wh}. \tag{1.14}$$

Theorem 33 *Let $z \in \mathscr{C}_h$. Then $\overline{z} = \ominus z$ iff $z \in \mathscr{I}_h$.*

Proof We have

$$\begin{aligned} &\overline{z} = \ominus z \iff \\ &\overline{z} = -\tfrac{z}{1+zh} \iff \\ &\overline{z} + \overline{z}zh = -z \iff \\ &2\mathrm{Re}(z) + |z|^2 h = 0. \end{aligned}$$

Also,

$$\begin{aligned} &\left|z + \tfrac{1}{h}\right| = \tfrac{1}{h} \iff \\ &\left|z + \tfrac{1}{h}\right|^2 = \tfrac{1}{h^2} \iff \\ &\left(\mathrm{Re}(z) + \tfrac{1}{h}\right)^2 + \mathrm{Im}^2(z) = \tfrac{1}{h^2} \iff \\ &\mathrm{Re}^2(z) + \tfrac{2}{h}\mathrm{Re}(z) + \tfrac{1}{h^2} + \mathrm{Im}^2(z) = \tfrac{1}{h^2} \iff \\ &|z|^2 + \tfrac{2}{h}\mathrm{Re}(z) = 0 \iff \\ &2\mathrm{Re}(z) + |z|^2 h = 0, \end{aligned}$$

which completes the proof.

Theorem 34 *Let $-\dfrac{\pi}{h} < w \le \dfrac{\pi}{h}$. Then*

$$\ominus(\overset{\circ}{i}\, w) = \overline{\overset{\circ}{i}\, w}.$$

Proof We have

$$\begin{aligned} \ominus(\overset{\circ}{i}\, w) &= -\frac{\overset{\circ}{i}\, w}{1 + (\overset{\circ}{i}\, w)h} \\ &= -\frac{\frac{e^{iwh}-1}{h}}{1 + \frac{e^{iwh}-1}{h}h} \\ &= -\frac{e^{iwh} - 1}{he^{iwh}} \\ &= \frac{e^{-iwh} - 1}{h} \\ &= \overline{\overset{\circ}{i}\, w}. \end{aligned}$$

Definition 25 Let $z \in \mathscr{C}_h$. The generalized square of z is defined as follows

$$z^{\odot} = -z(\ominus z).$$

We have

$$z^{\odot} = -z\frac{-z}{1+zh} = \frac{z^2}{1+zh}.$$

Theorem 35 *For $z \in \mathscr{C}_h$ we have*

$$(\ominus z)^{\odot} = z^{\odot}.$$

Proof We have

$$\begin{aligned}(\ominus z)^{\odot} &= -(\ominus z)(\ominus(\ominus z))\\ &= \frac{z}{1+zh}z\\ &= \frac{z^2}{1+zh}\\ &= z^{\odot}.\end{aligned}$$

Theorem 36 *For $z \in \mathscr{C}_h$ we have*

$$1+zh = \frac{z^2}{z^{\odot}}.$$

Proof We have

$$\begin{aligned}\frac{z^2}{z^{\odot}} &= \frac{z^2}{\frac{z^2}{1+zh}}\\ &= 1+zh.\end{aligned}$$

Theorem 37 *For $z \in \mathscr{C}_h$ we have*

$$z+(\ominus z) = z^{\odot}h.$$

Proof We have

$$z^{\odot}h = \frac{z^2}{1+zh}h,$$

$$\begin{aligned}z+(\ominus z) &= z - \frac{z}{1+zh}\\ &= \frac{z^2h}{1+zh},\end{aligned}$$

which completes the proof.

Theorem 38 *For $z \in \mathscr{C}_h$ we have*

$$z \oplus z^{\odot} = z + z^2.$$

Proof We have

$$\begin{aligned} z \oplus z^{\odot} &= z + z^{\odot} + z z^{\odot} h \\ &= z + \frac{z^2}{1+zh} + \frac{z^3 h}{1+zh} \\ &= z + \frac{z^2(1+zh)}{1+zh} \\ &= z + z^2. \end{aligned}$$

Theorem 39 *Let* $-\frac{\pi}{h} < w \leq \frac{\pi}{h}$. *Then*

$$-(\overset{\circ}{i}\, w)^{\odot} = \frac{4}{h^2} \sin^2\left(\frac{wh}{2}\right).$$

Proof We have

$$\begin{aligned} -(\overset{\circ}{i}\, w)^{\odot} &= -(\overset{\circ}{i}\, w)(\ominus \overset{\circ}{i}\, w) \\ &= (\overset{\circ}{i}\, w)\overline{\overset{\circ}{i}\, w} \\ &= |\overset{\circ}{i}\, w|^2 \\ &= \frac{4}{h^2} \sin^2\left(\frac{wh}{2}\right). \end{aligned}$$

Exercise 16 Let $z \in \mathscr{C}_h$. Prove that

$$z^{\odot} \in \mathscr{R} \quad iff \quad z \in \mathscr{R}_h \cup \mathscr{A}_h \cup \mathscr{I}_h.$$

1.5.2 Definition and Properties of the Exponential Function

For $h > 0$, we define the strip

$$\mathscr{Z}_h = \{z \in \mathscr{C} : -\frac{\pi}{h} < \mathrm{Im}(z) \leq \frac{\pi}{h}\}.$$

For $h = 0$, we set $\mathscr{Z}_0 = \mathscr{C}$.

Definition 26 For $h > 0$, we define the cylindrical transformation $\xi_h : \mathscr{C}_h \longmapsto \mathscr{Z}_h$ by

$$\xi_h(z) = \frac{1}{h}\mathrm{Log}(1+zh),$$

where Log is the principal logarithm function. For $h = 0$, we define $\xi_0(z) = z$ for all $z \in \mathscr{C}$.

We note that

$$\xi_h^{-1}(z) = \frac{e^{zh}-1}{h}$$

for $z \in \mathscr{Z}_h$.

Definition 27 We say that a function $f : \mathscr{T} \longmapsto \mathscr{R}$ is regressive provided

$$1 + \mu(t)f(t) \neq 0 \quad for\ all \quad t \in \mathscr{T}^\kappa$$

holds. The set of all regressive and rd-continuous functions $f : \mathscr{T} \longmapsto \mathscr{R}$ will be defined by $\mathscr{R}_1(\mathscr{T})$ or $\mathscr{R}_1$.

In $\mathscr{R}_1$ we define "circle plus" addition as follows

$$(f \oplus g)(t) = f(t) + g(t) + \mu(t)f(t)g(t).$$

Exercise 17 Prove that $(\mathscr{R}_1, \oplus)$ is an Abelian group.

Definition 28 The group $(\mathscr{R}_1, \oplus)$ will be called the regressive group.

Definition 29 For $f \in \mathscr{R}_1$ we define

$$(\ominus f)(t) = -\frac{f(t)}{1+\mu(t)f(t)} \quad for\ all \quad t \in \mathscr{T}^\kappa.$$

Exercise 18 Let $f \in \mathscr{R}_1$. Prove that $(\ominus f)(t) \in \mathscr{R}_1$ for all $t \in \mathscr{T}^\kappa$.

Definition 30 We define "circle minus" substraction $\ominus$ on $\mathscr{R}_1$ as follows

$$(f \ominus g)(t) = (f \oplus (\ominus g))(t) \quad for\ all \quad t \in \mathscr{T}^\kappa.$$

For $f, g \in \mathscr{R}_1$, we have

$$\begin{aligned} f \ominus g &= f \oplus (\ominus g) \\ &= f \oplus \left(-\frac{g}{1+\mu g}\right) \\ &= f - \frac{g}{1+\mu g} - \frac{\mu f g}{1+\mu g} \\ &= \frac{f-g}{1+\mu g}. \end{aligned}$$

Theorem 40 *Let* $f, g \in \mathscr{R}_1$. *Then*

1. $f \ominus f = 0,$
2. $\ominus(\ominus f) = f,$
3. $f \ominus g \in \mathscr{R}_1,$
4. $\ominus(f \ominus g) = g \ominus f,$
5. $\ominus(f \oplus g) = (\ominus f) \oplus (\ominus g),$
6. $f \oplus \dfrac{g}{1+\mu f} = f + g.$

Proof 1.

$$\begin{aligned} f \ominus f &= f \oplus (\ominus f) \\ &= f \oplus \left(-\frac{f}{1+\mu f}\right) \\ &= f - \frac{f}{1+\mu f} - \frac{f^2\mu}{1+\mu f} \\ &= \frac{f + \mu f^2 - f - \mu f^2}{1+\mu f} \\ &= 0. \end{aligned}$$

2.

$$\begin{aligned} \ominus(\ominus f) &= \ominus\left(-\frac{f}{1+\mu f}\right) \\ &= \frac{\frac{f}{1+\mu f}}{1 - \frac{\mu f}{1+\mu f}} \\ &= f. \end{aligned}$$

3.

$$\begin{aligned} 1 + \mu(f \ominus g) &= 1 + \frac{\mu f - \mu g}{1+\mu g} \\ &= \frac{1+\mu f}{1+\mu g} \neq 0. \end{aligned}$$

We note that $\dfrac{f-g}{1+\mu g}$ is rd-continuous. Therefore $f \ominus g \in \mathscr{R}_1$.

4.

$$\begin{aligned}
\ominus(f \ominus g) &= \ominus\left(\frac{f-g}{1+\mu g}\right) \\
&= -\frac{\frac{f-g}{1+\mu g}}{1+\mu\frac{f-g}{1+\mu g}} \\
&= -\frac{f-g}{1+\mu f} \\
&= \frac{g-f}{1+\mu f} \\
&= g \ominus f.
\end{aligned}$$

5.

$$\begin{aligned}
\ominus(f \oplus g) &= \ominus(f+g+\mu fg) \\
&= -\frac{f+g+\mu fg}{1+\mu f+\mu g+\mu^2 fg} \\
&= -\frac{f+g+\mu fg}{(1+\mu f)(1+\mu g)}, \\
\ominus f &= -\frac{f}{1+\mu f}, \\
\ominus g &= -\frac{g}{1+\mu g}, \\
(\ominus f) \oplus (\ominus g) &= \ominus f + (\ominus g) + \mu(\ominus f)(\ominus g) \\
&= -\frac{f}{1+\mu f} - \frac{g}{1+\mu g} + \frac{\mu fg}{(1+\mu f)(1+\mu g)} \\
&= \frac{-f(1+\mu g) - g(1+\mu f) + \mu fg}{(1+\mu f)(1+\mu g)} \\
&= -\frac{f+g+\mu fg}{(1+\mu f)(1+\mu g)}.
\end{aligned}$$

6.

$$f \oplus \frac{g}{1+\mu f} = f + \frac{g}{1+\mu f} + \frac{\mu f g}{1+\mu f}$$
$$= f + g.$$

Definition 31 If $f \in \mathscr{R}_1$, then we define the generalized exponential function by

$$e_f(t,s) = e^{\int_s^t \xi_{\mu(\tau)}(f(\tau))\Delta\tau} \quad for \quad s, t \in \mathscr{T}.$$

In fact, using the definition for the cylindrical transformation we have

$$e_f(t,s) = e^{\int_s^t \frac{1}{\mu(\tau)} \mathrm{Log}(1+\mu(\tau) f(\tau))\Delta\tau} \quad for \quad s, t \in \mathscr{T}.$$

Theorem 41 *(Semigroup Property) If $f \in \mathscr{R}_1$, then*

$$e_f(t,r)e_f(r,s) = e_f(t,s) \quad for\ all \quad t, r, s \in \mathscr{T}.$$

Proof We have

$$e_f(t,r)e_f(r,s) = e^{\int_r^t \xi_{\mu(\tau)}(f(\tau))\Delta\tau} e^{\int_s^r \xi_{\mu(\tau)}(f(\tau))\Delta\tau}$$
$$= e^{\int_r^t \xi_{\mu(\tau)}(f(\tau))\Delta\tau + \int_s^r \xi_{\mu(\tau)}(f(\tau))\Delta\tau}$$
$$= e^{\int_s^t \xi_{\mu(\tau)}(f(\tau))\Delta\tau}$$
$$= e_f(t,s).$$

Exercise 19 Let $f \in \mathscr{R}_1$. Prove that

$$e_0(t,s) = 1 \quad and \quad e_f(t,t) = 1.$$

Theorem 42 *Let $f \in \mathscr{R}_1$ and fix $t_0 \in \mathscr{T}$. Then*

$$e_f^{\Delta}(t,t_0) = f(t) e_f(t,t_0).$$

Proof 1. Let $\sigma(t) > t$. Then

$$\begin{aligned}
e_f^{\Delta}(t,t_0) &= \frac{e_f(\sigma(t),t_0)-e_f(t,t_0)}{\mu(t)}\\
&= \frac{e^{\int_{t_0}^{\sigma(t)}\xi_{\mu(\tau)}(f(\tau))\Delta\tau}-e^{\int_{t_0}^{t}\xi_{\mu(\tau)}(f(\tau))\Delta\tau}}{\mu(t)}\\
&= \frac{e^{\int_{t_0}^{t}\xi_{\mu(\tau)}(f(\tau))\Delta\tau+\int_{t}^{\sigma(t)}\xi_{\mu(\tau)}(f(\tau))\Delta\tau}-e^{\int_{t_0}^{t}\xi_{\mu(\tau)}(f(\tau))\Delta\tau}}{\mu(t)}\\
&= \frac{e^{\int_{t}^{\sigma(t)}\xi_{\mu(\tau)}(f(\tau))\Delta\tau}-1}{\mu(t)}e^{\int_{t_0}^{t}\xi_{\mu(\tau)}(f(\tau))\Delta\tau}\\
&= \frac{e^{\mu(t)\xi_{\mu(t)}(f(t))}-1}{\mu(t)}e_f(t,t_0)\\
&= f(t)e_f(t,t_0).
\end{aligned}$$

2. Let $\sigma(t)=t$. Then

$$\begin{aligned}
&|e_f(t,t_0)-e_f(s,t_0)-f(t)e_f(t,t_0)(t-s)|\\
&= |e_f(t,t_0)-e_f(t,t_0)e_f(s,t)-f(t)e_f(t,t_0)(t-s)|\\
&= |e_f(t,t_0)||1-e_f(s,t)-f(t)(t-s)|\\
&= |e_f(t,t_0)|\left|1-\int_s^t\xi_{\mu(\tau)}(f(\tau))\Delta\tau-e_f(s,t)+\int_s^t\xi_{\mu(\tau)}(f(\tau))\Delta\tau-f(t)(t-s)\right|\\
&\le |e_f(t,t_0)|\Big(\left|1-\int_s^t\xi_{\mu(\tau)}(f(\tau))\Delta\tau-e_f(s,t)\right|\\
&+\left|\int_s^t\xi_{\mu(\tau)}(f(\tau))\Delta\tau-f(t)(t-s)\right|\Big)\\
&\le |e_f(t,t_0)|\Big(\left|1-\int_s^t\xi_{\mu(\tau)}(f(\tau))\Delta\tau-e_f(s,t)\right|\\
&+\left|\int_s^t(\xi_{\mu(\tau)}(f(\tau))-\xi_0(f(t)))\Delta\tau\right|\Big).
\end{aligned} \tag{1.15}$$

Since $\sigma(t)=t$ and $f\in\mathscr{C}_{rd}$, then

$$\lim_{r\to t}\xi_{\mu(r)}(f(r))=\xi_0(f(t)).$$

Therefore, there is a neighbourhood U_1 of f such that

$$|\xi_{\mu(\tau)}(f(\tau))-\xi_0(f(t))|<\frac{\varepsilon}{3|e_f(t,t_0)|}\quad for\ all\quad \tau\in U_1.$$

Let $s \in U_1$. Then

$$|e_f(t,t_0)|\left|\int_s^t (\xi_{\mu(\tau)}(f(\tau)) - \xi_0(f(t)))\Delta\tau\right| \le \frac{\varepsilon}{3}|t-s|. \tag{1.16}$$

Also, using that

$$\lim_{z\to 0}\frac{1-z-e^{-z}}{z} = 0,$$

we conclude that there is a neighbourhood U_2 of t so that if $s \in U_2$, we have

$$\left|\frac{1-\int_s^t \xi_{\mu(\tau)}(f(\tau))\Delta\tau - e_f(s,t)}{\int_s^t \xi_{\mu(\tau)}(f(\tau))\Delta\tau}\right| < \varepsilon^*,$$

where

$$\varepsilon^* = \min\left\{1, \frac{\varepsilon}{1+3|f(t)||e_f(t,t_0)|}\right\}.$$

Let $s \in U = U_1 \cap U_2$. Then

$$\begin{aligned}
&|e_f(t,t_0)|\left|1-\int_s^t \xi_{\mu(\tau)}(f(\tau))\Delta\tau - e_f(s,t)\right| \\
&= |e_f(t,t_0)|\frac{\left|1-\int_s^t \xi_{\mu(\tau)}(f(\tau))\Delta\tau - e_f(s,t)\right|}{\left|\int_s^t \xi_{\mu(\tau)}(f(\tau))\Delta\tau\right|}\left|\int_s^t \xi_{\mu(\tau)}(f(\tau))\Delta\tau\right| \\
&\le |e_f(t,t_0)|\varepsilon^*\left|\int_s^t \xi_{\mu(\tau)}(f(\tau))\Delta\tau\right| \\
&\le |e_f(t,t_0)|\varepsilon^*\left\{\left|\int_s^t (\xi_{\mu(\tau)}(f(\tau)) - \xi_0(f(t)))\Delta\tau\right| + |f(t)||t-s|\right\} \\
&\le |e_f(t,t_0)|\left|\int_s^t (\xi_{\mu(\tau)}(f(\tau)) - \xi_0(f(t)))\Delta\tau\right| + |e_f(t,t_0)|\varepsilon^*|f(t)||t-s| \\
&\le \tfrac{\varepsilon}{3}|t-s| + \tfrac{\varepsilon}{3}|t-s| \\
&= \tfrac{2\varepsilon}{3}|t-s|.
\end{aligned}$$

From the last inequality and from (1.15), and (1.16), we conclude that

$$\begin{aligned}
|e_f(t,t_0) - e_f(s,t_0) - f(t)e_f(t,t_0)(t-s)| &< \frac{\varepsilon}{3}|t-s| + \frac{\varepsilon}{3}|t-s| + \frac{\varepsilon}{3}|t-s| \\
&= \varepsilon|t-s|,
\end{aligned}$$

which completes the proof.

Corollary 3 *Let $f \in \mathscr{R}_1$ and fix $t_0 \in \mathscr{T}$. Then $e_f(t, t_0)$ is a solution to the Cauchy problem*

$$y^{\Delta}(t) = f(t)y(t), \quad y(t_0) = 1. \tag{1.17}$$

Corollary 4 *Let $f \in \mathscr{R}_1$ and fix $t_0 \in \mathscr{T}$. Then $e_f(t, t_0)$ is the unique solution of the problem (1.17).*

Proof Let $y(t)$ be any solution of the problem (1.17). Then

$$\begin{aligned}\left(\frac{y(t)}{e_f(t, t_0)}\right)^{\Delta} &= \frac{y^{\Delta}(t)e_f(t, t_0) - y(t)e_f^{\Delta}(t, t_0)}{e_f(\sigma(t), t_0)e_f(t, t_0)} \\ &= \frac{f(t)y(t)e_f(t, t_0) - y(t)f(t)e_f(t, t_0)}{e_f(\sigma(t), t_0)e_f(t, t_0)} \\ &= 0.\end{aligned}$$

Consequently $y(t) = ce_f(t, t_0)$, where c is a constant. Hence and

$$1 = y(t_0) = ce_f(t_0, t_0) = c,$$

we conclude that $y(t) = e_f(t, t_0)$.

Theorem 43 *Let $f \in \mathscr{R}_1$. Then*

$$e_f(\sigma(t), s) = (1 + \mu(t)f(t))e_f(t, s).$$

Proof We have

$$\begin{aligned}e_f(\sigma(t), s) - e_f(t, s) &= \mu(t)e_f^{\Delta}(t, s) \\ &= \mu(t)f(t)e_f(t, s),\end{aligned}$$

which completes the proof.

Theorem 44 *Let $f \in \mathscr{R}_1$. Then*

$$e_f(t, s) = \frac{1}{e_f(s, t)} = e_{\ominus f}(s, t).$$

Proof We have

$$
\begin{aligned}
e_f(t,s) &= e^{\int_s^t \xi_{\mu(\tau)}(f(\tau))\Delta\tau} \\
&= e^{-\int_t^s \xi_{\mu(\tau)}(f(\tau))\Delta\tau} \\
&= \frac{1}{e^{\int_t^s \xi_{\mu(\tau)}(f(\tau))\Delta\tau}} \\
&= \frac{1}{e_f(s,t)}.
\end{aligned}
$$

Now we fix $t_0 \in \mathscr{T}$ and consider the problem

$$y^{\Delta}(t) = \ominus f(t)y(t), \quad y(t_0) = 1.$$

Its solution is $e_{\ominus f}(t,s)$.
Also,

$$
\begin{aligned}
\left(\frac{1}{e_f(t,s)}\right)^{\Delta} &= -\frac{e_f^{\Delta}(t,s)}{e_f(\sigma(t),s)e_f(t,s)} \\
&= -\frac{f(t)e_f(t,s)}{(1+\mu(t)f(t))e_f(t,s)e_f(t,s)} \\
&= -\frac{f(t)}{(1+\mu(t)f(t))e_f(t,s)} \\
&= (\ominus f)(t)\frac{1}{e_f(t,s)}.
\end{aligned}
$$

Therefore

$$\frac{1}{e_f(t,s)} = e_{\ominus f}(t,s).$$

Theorem 45 *Let $f, g \in \mathscr{R}_1$. Then*

$$e_f(t,s)e_g(t,s) = e_{f\oplus g}(t,s).$$

Proof We have

$$
\begin{aligned}
e_f(t,s)e_g(t,s) &= e^{\int_s^t \xi_{\mu(\tau)}(f(\tau))\Delta\tau} e^{\int_s^t \xi_{\mu(\tau)}(g(\tau))\Delta\tau} \\
&= e^{\int_s^t (\xi_{\mu(\tau)}(f(\tau))+\xi_{\mu(\tau)}(g(\tau)))\Delta\tau} \\
&= e^{\int_s^t \frac{1}{\mu(\tau)}(\mathrm{Log}(1+\mu(\tau)f(\tau))+\mathrm{Log}(1+\mu(\tau)g(\tau)))\Delta\tau}
\end{aligned}
$$

$$= e^{\int_s^t \frac{1}{\mu(\tau)} \mathrm{Log}((1+\mu(\tau)f(\tau))(1+\mu(\tau)g(\tau)))\Delta\tau}$$

$$= e^{\int_s^t \frac{1}{\mu(\tau)} \mathrm{Log}(1+\mu(\tau)(f(\tau)+g(\tau)+\mu(\tau)f(\tau)g(\tau)))\Delta\tau}$$

$$= e^{\int_s^t \xi_{\mu(\tau)}((f\oplus g)(\tau))\Delta\tau}$$

$$= e_{f\oplus g}(t,s).$$

Theorem 46 *Let $f, g \in \mathscr{R}_1$. Then*

$$\frac{e_f(t,s)}{e_g(t,s)} = e_{f\ominus g}(t,s).$$

Proof We have

$$\frac{e_f(t,s)}{e_g(t,s)} = e_f(t,s)e_{\ominus g}(t,s)$$

$$= e_{f\oplus(\ominus g)}(t,s)$$

$$= e_{f\ominus g}(t,s).$$

Theorem 47 *Let $f \in \mathscr{R}_1$. Then*

$$e_f(t,\sigma(s))e_f(s,r) = \frac{1}{1+\mu(s)f(s)} e_f(t,r).$$

Proof We have

$$e_f(t,\sigma(s))e_f(s,r) = e^{\int_{\sigma(s)}^t \xi_{\mu(\tau)}(f(\tau))\Delta\tau} e^{\int_r^s \xi_{\mu(\tau)}(f(\tau))\Delta\tau}$$

$$= e^{\int_{\sigma(s)}^s \xi_{\mu(\tau)}(f(\tau))\Delta\tau + \int_s^t \xi_{\mu(\tau)}(f(\tau))\Delta\tau + \int_r^s \xi_{\mu(\tau)}(f(\tau))\Delta\tau}$$

$$= e^{-\xi_{\mu(s)}(f(s))\mu(s) + \int_r^t \xi_{\mu(\tau)}(f(\tau))\Delta\tau}$$

$$= e^{-\mathrm{Log}(1+f(s)\mu(s))} e^{\int_r^t \xi_{\mu(\tau)}(f(\tau))\Delta\tau}$$

$$= \frac{1}{1+\mu(s)f(s)} e_f(t,r).$$

Theorem 48 *Let $f, g \in \mathscr{R}_1$. Then*

$$e^{\Delta}_{f\ominus g}(t,t_0) = \frac{(f(t)-g(t))e_f(t,t_0)}{e_g(\sigma(t),t_0)}.$$

Proof We have

$$\begin{aligned}
e^{\Delta}_{f\ominus g}(t,t_0) &= \left(\frac{e_f(t,t_0)}{e_g(t,t_0)}\right)^{\Delta} \\
&= \frac{e^{\Delta}_f(t,t_0)e_g(t,t_0) - e_f(t,t_0)e^{\Delta}_g(t,t_0)}{e_g(t,t_0)e_g(\sigma(t),t_0)} \\
&= \frac{f(t)e_f(t,t_0)e_g(t,t_0) - g(t)e_f(t,t_0)e_g(t,t_0)}{e_g(t,t_0)e_g(\sigma(t),t_0)} \\
&= \frac{(f(t)-g(t))e_f(t,t_0)}{e_g(\sigma(t),t_0)}.
\end{aligned}$$

Theorem 49 *Let $f \in \mathscr{R}_1$ and $a, b, c \in \mathscr{T}$. Then*

$$(e_f(c,t))^{\Delta} = -f(t)(e_f(c,t))^{\sigma} = -f(t)e_f(c,\sigma(t))$$

and

$$\int_a^b f(t)e_f(c,\sigma(t))\Delta t = e_f(c,a) - e_f(c,b).$$

Proof We have

$$\begin{aligned}
f(t)e_f(c,\sigma(t)) &= f(t)e_{\ominus f}(\sigma(t),c) \\
&= f(t)(1+\mu(t)\ominus f(t))e_{\ominus f}(t,c) \\
&= f(t)\left(1-\frac{\mu(t)f(t)}{1+\mu(t)f(t)}\right)e_{\ominus f}(t,c) \\
&= \frac{f(t)}{1+\mu(t)f(t)}e_{\ominus f}(t,c) \\
&= -(\ominus f)(t)e_{\ominus f}(t,c) \\
&= -e^{\Delta}_{\ominus f}(t,c) \\
&= -e^{\Delta}_f(c,t).
\end{aligned}$$

Hence,

$$\int_a^b f(t)e_f(c,\sigma(t))\Delta t = -\int_a^b e_f^{\Delta}(c,t)\Delta t$$
$$= -e_f(c,t)\Big|_{t=a}^{t=b}$$
$$= e_f(c,a) - e_f(c,b).$$

Exercise 20 Assume $1+\mu(t)\dfrac{2}{t} \neq 0$, $1+\mu(t)\dfrac{5}{t} \neq 0$ for all $t \in \mathscr{T} \cap (0,\infty)$. Let also, $t_0 \in \mathscr{T} \cap (0,\infty)$. Evaluate the integral

$$I = \int_{t_0}^{t} \frac{e_{\frac{5}{s}}(s,t_0)}{s e^{\sigma}_{\frac{2}{s}}(s,t_0)}\Delta s.$$

Solution. We have

$$\left(\frac{5}{s}-\frac{2}{s}\right)\frac{e_{\frac{5}{s}}(s,t_0)}{e^{\sigma}_{\frac{2}{s}}(s,t_0)} = \frac{3}{s}\frac{e_{\frac{5}{s}}(s,t_0)}{e^{\sigma}_{\frac{2}{s}}(s,t_0)}$$
$$= e^{\Delta}_{\frac{5}{s}\ominus\frac{2}{s}}(s,t_0)$$
$$= e^{\Delta}_{\frac{\frac{3}{s}}{1+\mu(s)\frac{2}{s}}}(s,t_0)$$
$$= e^{\Delta}_{\frac{3}{s+2\mu(s)}}(s,t_0).$$

Hence,

$$I = \frac{1}{3}\int_{t_0}^{t} e^{\Delta}_{\frac{3}{s+2\mu(s)}}(s,t_0)\Delta s$$
$$= \frac{1}{3}e_{\frac{3}{s+2\mu(s)}}(s,t_0)\Big|_{s=t_0}^{s=t}$$
$$= \frac{1}{3}e_{\frac{3}{t+2\mu(t)}}(t,t_0) - \frac{1}{3}.$$

Exercise 21 Let $\alpha \in \mathscr{R}_1$. Let also, the exponents

$$e_{\frac{\alpha^2}{t}-\frac{(\alpha-1)^2}{\sigma(t)}}(t,t_0) \quad and \quad e_{\frac{\alpha}{t}}(t,t_0)$$

exist for all $t \in \mathscr{T} \cap (0, \infty)$. Prove that

1.
$$\frac{e_{\frac{\alpha^2}{t} - \frac{(\alpha-1)^2}{\sigma(t)}}(t, t_0)}{e_{\frac{\alpha}{t}}(t, t_0)} = e_{\frac{\alpha-1}{\sigma(t)}}(t, t_0),$$

2.
$$\frac{e_{\frac{\alpha-1}{\sigma(t)}}(t, t_0)}{e_{\frac{\alpha}{t}}(t, t_0)} = e_{-\frac{1}{\sigma(t)}}(t, t_0) = \frac{t_0}{t}$$

for all $t, t_0 \in \mathscr{T} \cap (0, \infty)$.

Solution.

1. We have

$$\begin{aligned}
\left(\frac{\alpha^2}{t} - \frac{(\alpha-1)^2}{\sigma(t)}\right) \ominus \frac{\alpha}{t} &= \left(\frac{\alpha^2}{t} - \frac{(\alpha-1)^2}{\sigma(t)}\right) \oplus \left(\ominus \frac{\alpha}{t}\right) \\
&= \frac{\alpha^2}{t} - \frac{(\alpha-1)^2}{\sigma(t)} + \left(\ominus \frac{\alpha}{t}\right) + \left(\frac{\alpha^2}{t} - \frac{(\alpha-1)^2}{\sigma(t)}\right)\left(\ominus \frac{\alpha}{t}\right)\mu(t) \\
&= \frac{\alpha^2}{t} - \frac{(\alpha-1)^2}{\sigma(t)} - \frac{\frac{\alpha}{t}}{1 + \frac{\alpha}{t}\mu(t)} \\
&\quad + \left(\frac{\alpha^2}{t} - \frac{(\alpha-1)^2}{\sigma(t)}\right)\left(-\frac{\frac{\alpha}{t}}{1 + \frac{\alpha}{t}\mu(t)}\right)\mu(t) \\
&= \frac{\alpha^2}{t} - \frac{(\alpha-1)^2}{\sigma(t)} - \frac{\alpha}{t + \alpha\mu(t)} - \left(\frac{\alpha^2}{t} - \frac{(\alpha-1)^2}{\sigma(t)}\right)\frac{\alpha}{t + \alpha\mu(t)}\mu(t) \\
&= \left(\frac{\alpha^2}{t} - \frac{(\alpha-1)^2}{\sigma(t)}\right)\left(1 - \frac{\alpha\mu(t)}{t + \alpha\mu(t)}\right) - \frac{\alpha}{t + \alpha\mu(t)} \\
&= \left(\frac{\alpha^2}{t} - \frac{(\alpha-1)^2}{\sigma(t)}\right)\frac{t}{t + \alpha\mu(t)} - \frac{\alpha}{1 + \alpha\mu(t)} \\
&= \frac{\alpha^2\sigma(t) - t(\alpha-1)^2}{t\sigma(t)} \frac{t}{t + \alpha\mu(t)} - \frac{\alpha}{t + \alpha\mu(t)} \\
&= \frac{\alpha^2(t + \mu(t)) - t\alpha^2 + 2\alpha t - t}{\sigma(t)} \frac{1}{t + \alpha\mu(t)} - \frac{\alpha}{t + \alpha\mu(t)}
\end{aligned}$$

$$= \frac{\alpha^2\mu(t) + 2\alpha t - t}{\sigma(t)(t + \alpha\mu(t))} - \frac{\alpha}{t + \alpha\mu(t)}$$

$$= \frac{\alpha^2\mu(t) + 2\alpha t - t - \alpha\mu(t) - \alpha t}{\sigma(t)(t + \alpha\mu(t))}$$

$$= \frac{\alpha\mu(t)(\alpha - 1) + (\alpha - 1)t}{\sigma(t)(t + \alpha\mu(t))}$$

$$= \frac{(\alpha - 1)(t + \alpha\mu(t))}{\sigma(t)(t + \alpha\mu(t))}$$

$$= \frac{\alpha - 1}{\sigma(t)}.$$

Hence,

$$\frac{e_{\frac{\alpha^2}{t} - \frac{(\alpha-1)^2}{\sigma(t)}}(t, t_0)}{e_{\frac{\alpha}{t}}(t, t_0)} = e_{\left(\frac{\alpha^2}{t} - \frac{(\alpha-1)^2}{\sigma(t)}\right)\ominus\frac{\alpha}{t}}(t, t_0)$$

$$= e_{\frac{\alpha-1}{t}}(t, t_0).$$

2. We have

$$\frac{\alpha - 1}{\sigma(t)} \ominus \frac{\alpha}{t} = \frac{\alpha - 1}{\sigma(t)} \oplus \left(\ominus\frac{\alpha}{t}\right)$$

$$= \frac{\alpha - 1}{\sigma(t)} + \left(\ominus\frac{\alpha}{t}\right) + \frac{\alpha - 1}{\sigma(t)}\left(\ominus\frac{\alpha}{t}\right)\mu(t)$$

$$= \frac{\alpha - 1}{\sigma(t)} - \frac{\frac{\alpha}{t}}{1 + \mu(t)\frac{\alpha}{t}} - \frac{\alpha - 1}{\sigma(t)}\frac{\frac{\alpha}{t}}{1 + \mu(t)\frac{\alpha}{t}}\mu(t)$$

$$= \frac{\alpha - 1}{\sigma(t)} - \frac{\alpha}{t + \alpha\mu(t)} - \frac{\alpha - 1}{\sigma(t)}\frac{\alpha}{t + \alpha\mu(t)}\mu(t)$$

$$= \frac{\alpha - 1}{\sigma(t)}\left(1 - \frac{\alpha\mu(t)}{t + \alpha\mu(t)}\right) - \frac{\alpha}{t + \alpha\mu(t)}$$

$$= \frac{t(\alpha - 1)}{\sigma(t)(t + \alpha\mu(t))} - \frac{\alpha}{t + \alpha\mu(t)}$$

$$= \frac{\alpha t - t - \alpha\mu(t) - \alpha t}{\sigma(t)(t + \alpha\mu(t))}$$

$$= -\frac{t + \alpha\mu(t)}{\sigma(t)(t + \alpha\mu(t))}$$

$$= -\frac{1}{\sigma(t)}.$$

Therefore

$$\begin{aligned}
\frac{e_{\frac{\alpha-1}{\sigma(t)}}(t, t_0)}{e_{\frac{\alpha}{t}}(t, t_0)} &= e_{\frac{\alpha-1}{\sigma(t)} \ominus \frac{\alpha}{t}}(t, t_0) \\
&= e_{-\frac{1}{\sigma(t)}}(t, t_0) \\
&= e^{\int_{t_0}^{t} \xi_{\mu(\tau)}\left(-\frac{1}{\sigma(\tau)}\right)\Delta\tau} \\
&= e^{\int_{t_0}^{t} \frac{1}{\mu(\tau)} \mathrm{Log}\left(1 - \frac{\mu(\tau)}{\sigma(\tau)}\right)\Delta\tau} \\
&= e^{\int_{t_0}^{t} \frac{1}{\mu(\tau)} \mathrm{Log}\frac{\tau}{\sigma(\tau)}\Delta\tau} \\
&= e^{-\mathrm{Log}\tau \Big|_{\tau=t_0}^{\tau=t}} \\
&= e^{-\mathrm{Log}\frac{t}{t_0}} \\
&= \frac{t_0}{t}.
\end{aligned}$$

1.5.3 Examples for Exponential Functions

Let $\alpha : \mathcal{T} \longmapsto \mathcal{R}$ be regulated and $1 + \alpha(t)\mu(t) \neq 0$ for all $t \in \mathcal{T}$. Let also, $t_0, t \in \mathcal{T}$, $t_0 < t$.

1. $\mathscr{T} = h\mathscr{Z}$, $h > 0$. Every point in $\mathscr{T}$ is isolated and $\mu(t) = h$ for every $t \in \mathscr{T}$. Then

$$\begin{aligned} e_\alpha(t, t_0) &= e^{\int_{t_0}^{t} \frac{1}{\mu(\tau)} \mathrm{Log}(1+\alpha(\tau)\mu(\tau))\Delta\tau} \\ &= e^{\sum_{s\in[t_0,t)} \frac{1}{\mu(s)} \mathrm{Log}(1+\alpha(s)\mu(s))\mu(s)} \\ &= e^{\sum_{s\in[t_0,t)} \mathrm{Log}(1+h\alpha(s))} \\ &= \prod_{s\in[t_0,t)} (1 + h\alpha(s)). \end{aligned}$$

If α is a constant, then

$$\begin{aligned} e_\alpha(t, t_0) &= \prod_{s\in[t_0,t)} (1 + h\alpha) \\ &= (1 + h\alpha)^{t-t_0}. \end{aligned}$$

2. $\mathscr{T} = q^{\mathscr{N}_0}$, $q > 1$. Every point of $\mathscr{T}$ is isolated and $\mu(t) = (q-1)t$ for all $t \in \mathscr{T}$. Then

$$\begin{aligned} e_\alpha(t, t_0) &= e^{\int_{t_0}^{t} \frac{1}{\mu(\tau)} \mathrm{Log}(1+\alpha(\tau)\mu(\tau))\Delta\tau} \\ &= e^{\sum_{s\in[t_0,t)} \frac{1}{\mu(s)} \mathrm{Log}(1+\alpha(s)\mu(s))\mu(s)} \\ &= e^{\sum_{s\in[t_0,t)} \mathrm{Log}(1+\alpha(s)\mu(s))} \\ &= e^{\sum_{s\in[t_0,t)} \mathrm{Log}(1+(q-1)s\alpha(s))} \\ &= \prod_{s\in[t_0,t)} (1 + (q-1)s\alpha(s)). \end{aligned}$$

Exercise 22 Let $\mathscr{T} = q^{\mathscr{N}_0} \cup \{0\}$, $0 < q < 1$. Prove that

$$e_\alpha(t, t_0) = \prod_{s\in[t_0,t)} \left(1 + \frac{1-q}{q}\alpha(s)s\right).$$

3. $\mathscr{T} = \mathscr{N}_0^k$, $k \in \mathscr{N}$. Every point of $\mathscr{T}$ is isolated and

$$\mu(t) = \left(\sqrt[k]{t} + 1\right)^k.$$

Then

$$\begin{aligned} e_\alpha(t, t_0) &= e^{\int_{t_0}^{t} \frac{1}{\mu(\tau)} \mathrm{Log}(1+\alpha(\tau)\mu(\tau))\Delta\tau} \\ &= e^{\sum_{s\in[t_0,t)} \frac{1}{\mu(s)} \mathrm{Log}(1+\alpha(s)\mu(s))\mu(s)} \\ &= e^{\sum_{s\in[t_0,t)} \mathrm{Log}(1+\alpha(s)\mu(s))} \\ &= \prod_{s\in[t_0,t)} (1+\alpha(s)\mu(s)) \\ &= \prod_{s\in[t_0,t)} \left(1 + \left(\left(\sqrt[k]{s}+1\right)^k - s\right)\alpha(s)\right). \end{aligned}$$

1.6 Hyperbolic and Trigonometric Functions

Definition 32 (*Hyperbolic Functions*) If $f \in \mathscr{C}_{rd}$ and $-\mu f^2 \in \mathscr{R}_1$, then we define the hyperbolic functions $\cosh_f$ and $\sinh_f$ by

$$\cosh_f = \frac{e_f + e_{-f}}{2} \quad \textit{and} \quad \sinh_f = \frac{e_f - e_{-f}}{2}.$$

Theorem 50 *Let $f \in \mathscr{C}_{rd}$. If $-\mu f^2 \in \mathscr{R}_1$, then we have*

1. $\cosh_f^\Delta = f \sinh_f$,
2. $\sinh_f^\Delta = f \cosh_f$,
3. $\cosh_f^2 - \sinh_f^2 = e_{-\mu f^2}$.

Proof 1. We have

$$\begin{aligned} \cosh_f^\Delta &= \left(\frac{e_f + e_{-f}}{2}\right)^\Delta \\ &= \frac{f e_f - f e_{-f}}{2} \\ &= f \sinh_f. \end{aligned}$$

2. We have

$$\sinh_f^{\Delta} = \left(\frac{e_f - e_{-f}}{2}\right)^{\Delta}$$
$$= \frac{fe_f + fe_{-f}}{2}$$
$$= f\cosh_f .$$

3. We have

$$\cosh_f^2 - \sinh_f^2 = \left(\frac{e_f + e_{-f}}{2}\right)^2 - \left(\frac{e_f - e_{-f}}{2}\right)^2$$
$$= \frac{e_f^2 + 2e_f e_{-f} + e_{-f}^2}{4} - \frac{e_f^2 - 2e_f e_{-f} + e_{-f}^2}{4}$$
$$= e_f e_{-f}$$
$$= e_{f\oplus(-f)}$$
$$= e_{-\mu f^2}.$$

Definition 33 (*Trigonometric Functions*) If $f \in \mathscr{C}_{rd}$ and $\mu f^2 \in \mathscr{R}_1$, then we define the trigonometric functions $\cos_f$ and $\sin_f$ by

$$\cos_f = \frac{e_{if} + e_{-if}}{2} \quad \text{and} \quad \sin_f = \frac{e_{if} - e_{-if}}{2i}.$$

Theorem 51 *Let $f \in \mathscr{C}_{rd}$ and $-\mu f^2 \in \mathscr{R}_1$. Then*

1. $\cos_f^{\Delta} = -f\sin_f$.
2. $\sin_f^{\Delta} = f\cos_f$.
3. $\cos_f^2 + \sin_f^2 = e_{\mu f^2}$.

Proof 1. We have

$$\cos_f^{\Delta} = \left(\frac{e_{if} + e_{-if}}{2}\right)^{\Delta}$$
$$= \frac{ife_{if} - ife_{-if}}{2}$$

$$= -f \frac{e_{if} - e_{-if}}{2i}$$

$$= -f \sin_f .$$

2. We have

$$\begin{aligned}
\sin_f^{\Delta} &= \left(\frac{e_{if} - e_{-if}}{2i} \right)^{\Delta} \\
&= \frac{ife_{if} + ife_{-if}}{2i} \\
&= f \cos_f .
\end{aligned}$$

3. We have

$$\begin{aligned}
\cos_f^2 + \sin_f^2 &= \left(\frac{e_{if} + e_{-if}}{2} \right)^2 + \left(\frac{e_{if} - e_{-if}}{2i} \right)^2 \\
&= \frac{e_{if}^2 + 2e_{if}e_{-if} + e_{-if}^2}{4} - \frac{e_{if}^2 - 2e_{if}e_{-if} + e_{-if}^2}{4} \\
&= e_{if}e_{-if} \\
&= e_{if \oplus (-if)} \\
&= e_{\mu f^2} .
\end{aligned}$$

Exercise 23 Let $f \in \mathscr{C}_{rd}$ and $\mu f^2 \in \mathscr{R}_1$. Show Euler's formula

$$e_{if} = \cos_f + i \sin_f .$$

1.7 Dynamic Equations

Theorem 52 *Let $p : \mathscr{T} \longmapsto \mathscr{R}$ be rd-continuous and $1 + \mu(t)p(t) \neq 0$ for all $t \in \mathscr{T}$. Let also, $t_0 \in \mathscr{T}$ and $\phi_0 \in \mathscr{C}$. Then the unique solution to the initial value problem*

$$\phi^{\Delta} = p(t)\phi, \quad \phi(t_0) = \phi_0, \tag{1.18}$$

is given by

$$\phi(t) = \phi_0 e_p(t, t_0) \quad on \quad \mathscr{T}. \tag{1.19}$$

Proof We have

$$\begin{aligned}\phi^{\Delta}(t) &= \left(\phi_0 e_p(t,t_0)\right)^{\Delta}\\ &= \phi_0 e_p^{\Delta}(t,t_0)\\ &= \phi_0 p(t) e_p(t,t_0)\\ &= p(t)\phi, \quad t \in \mathscr{T},\end{aligned}$$

and

$$\phi(t_0) = \phi_0 e_p(t_0,t_0) = \phi_0.$$

Consequently $\phi(t)$, defined by (1.19), satisfies (1.18).

Now we will prove that the Eq. (1.18) has unique solution. Let $\phi_1(t)$ and $\phi_2(t)$ be two solutions of (1.18). Then

$$\psi(t) = \phi_1(t) - \phi_2(t), \quad t \in \mathscr{T},$$

satisfies the problem

$$\psi^{\Delta} = p(t)\psi, \quad \psi(t_0) = 0.$$

Therefore $\psi(t) \equiv 0$ on $\mathscr{T}$.

Theorem 53 *(Variation of Constants) Let $f, p : \mathscr{T} \longmapsto \mathscr{R}$ be rd-continuous functions and $1 + \mu(t)p(t) \neq 0$ for all $t \in \mathscr{T}$, $t_0 \in \mathscr{T}$ and $\phi_0 \in \mathscr{C}$. Then the unique solution of the initial value problem*

$$\phi^{\Delta}(t) = -p(t)\phi(\sigma(t)) + f(t), \quad \phi(t_0) = \phi_0, \tag{1.20}$$

is given by

$$\phi(t) = e_{\ominus p}(t,t_0)\phi_0 + \int_{t_0}^{t} e_{\ominus p}(t,\tau) f(\tau)\Delta\tau. \tag{1.21}$$

Proof We start by showing that $\phi(t)$ given by (1.21) is a solution to (1.20).

Indeed,

$$\begin{aligned}\phi^{\Delta}(t) &= \left(e_{\ominus p}(t,t_0)\phi_0 + \int_{t_0}^{t} e_{\ominus p}(t,\tau) f(\tau)\Delta\tau\right)^{\Delta}\\ &= \left(e_{\ominus p}(t,t_0)\phi_0\right)^{\Delta} + \left(\int_{t_0}^{t} e_{\ominus p}(t,\tau) f(\tau)\Delta\tau\right)^{\Delta}\end{aligned}$$

$$= \phi_0 e^{\Delta}_{\ominus p}(t, t_0) + \int_{t_0}^{t} e^{\Delta}_{\ominus p}(t, \tau) f(\tau)\Delta\tau + e_{\ominus p}(\sigma(t), t) f(t)$$

$$= \phi_0 \ominus p(t) e_{\ominus p}(t, t_0) + \ominus p(t) \int_{t_0}^{t} e_{\ominus p}(t, \tau) f(\tau)\Delta\tau + e_{\ominus p}(\sigma(t), t) f(t)$$

$$= \phi_0 \ominus p(t) e_{\ominus p}(t, t_0) + \ominus p(t) \int_{t_0}^{t} e_{\ominus p}(t, \tau) f(\tau)\Delta\tau$$

$$+ (1 + \mu(t) \ominus p(t)) e_{\ominus p}(t, t) f(t)$$

$$= \phi_0 \ominus p(t) e_{\ominus p}(t, t_0) + \ominus p(t) \int_{t_0}^{t} e_{\ominus p}(t, \tau) f(\tau)\Delta\tau$$

$$+ \frac{1}{1 + \mu(t) p(t)} f(t)$$

$$= -\frac{p(t)}{1 + \mu(t) p(t)} \phi_0 e_{\ominus p}(t, t_0) - \frac{p(t)}{1 + \mu(t) p(t)} \int_{t_0}^{t} e_{\ominus p}(t, \tau)\Delta\tau$$

$$+ \frac{1}{1 + \mu(t) p(t)} f(t).$$

Multiplying both sides by $1 + \mu(t) p(t)$ gives

$$(1 + \mu(t) p(t)) \phi^{\Delta}(t) = -\phi_0 p(t) e_{\ominus p}(t, t_0) - p(t) \int_{t_0}^{t} e_{\ominus p}(t, \tau) f(\tau)\Delta\tau + f(t)$$

$$= -p(t) \left(\phi_0 e_{\ominus p}(t, t_0) + \int_{t_0}^{t} e_{\ominus p}(t, \tau) f(\tau)\Delta\tau \right) + f(t)$$

$$= -p(t)\phi(t) + f(t).$$

Hence,

$$\phi^{\Delta}(t) = -\mu(t) p(t) \phi^{\Delta}(t) - p(t)\phi(t) + f(t)$$

$$= -p(t) \left(\mu(t)\phi^{\Delta}(t) + \phi(t) \right) + f(t)$$

$$= -p(t)\phi(\sigma(t)) + f(t).$$

Also,

$$\phi(t_0) = e_{\ominus p}(t_0, t_0)\phi_0 + \int_{t_0}^{t_0} e_{\ominus p}(t_0, \tau) f(\tau)\Delta\tau = \phi_0.$$

Consequently $\phi(t)$ satisfies (1.20).

Now we proceed to show the uniqueness of the solution. Suppose that $\phi(t)$ is a solution of (1.20). Then

$$f(t) = \phi^{\Delta}(t) + p(t)\phi(\sigma(t)),$$

whereupon

$$\begin{aligned}
e_p(t, t_0) f(t) &= e_p(t, t_0)\left(\phi^{\Delta}(t) + p(t)\phi(\sigma(t))\right) \\
&= e_p(t, t_0)\phi^{\Delta}(t) + p(t)e_p(t, t_0)\phi(\sigma(t)) \\
&= e_p(t, t_0)\phi^{\Delta}(t) + e_p^{\Delta}(t, t_0)\phi(\sigma(t)) \\
&= \left(e_p(t, t_0)\phi(t)\right)^{\Delta},
\end{aligned}$$

whereupon

$$\begin{aligned}
\int_{t_0}^{t} e_p(\tau, t_0) f(\tau)\Delta\tau &= \int_{t_0}^{t} \left(e_p(\tau, t_0)\phi(\tau)\right)^{\Delta} \Delta\tau \\
&= e_p(\tau, t_0)\phi(\tau)\Big|_{\tau=t_0}^{\tau=t} \\
&= e_p(t, t_0)\phi(t) - \phi(t_0),
\end{aligned}$$

i.e.,

$$e_p(t, t_0)\phi(t) = \phi_0 + \int_{t_0}^{t} e_p(\tau, t_0) f(\tau)\Delta\tau.$$

Solving for $\phi(t)$ we get

$$\begin{aligned}
\phi(t) &= \phi_0 e_{\ominus p}(t, t_0) + \int_{t_0}^{t} e_{\ominus p}(t, t_0)e_p(\tau, t_0) f(\tau)\Delta\tau \\
&= \phi_0 e_{\ominus p}(t, t_0) + \int_{t_0}^{t} e_{\ominus p}(t, t_0)e_{\ominus p}(t_0, \tau) f(\tau)\Delta\tau \\
&= \phi_0 e_{\ominus p}(t, t_0) + \int_{t_0}^{t} e_{\ominus p}(t, \tau) f(\tau)\Delta\tau.
\end{aligned}$$

Consequently if $\phi_1(t)$ and $\phi_2(t)$ are solutions to the problem (1.20), we find that $\psi(t) = \phi_1(t) - \phi_2(t)$ satisfies the problem

$$\psi^{\Delta}(t) = -p(t)\psi(\sigma(t)), \quad \psi(t_0) = 0,$$

whereupon $\psi(t) \equiv 0$ on $\mathscr{T}$.

Corollary 5 *(Variation of Constants) Let $f, p : \mathscr{T} \longmapsto \mathscr{R}$ be rd-continuous functions and $1 + \mu(t)p(t) \neq 0$ for all $t \in \mathscr{T}$, $t_0 \in \mathscr{T}$ and $\phi_0 \in \mathscr{R}$. Then the unique solution of the initial value problem*

$$\phi^{\Delta}(t) = p(t)\phi(t) + f(t), \quad \phi(t_0) = \phi_0, \tag{1.22}$$

is given by

$$\phi(t) = e_p(t, t_0)\phi_0 + \int_{t_0}^{t} e_p(t, \sigma(\tau)) f(\tau)\Delta\tau.$$

Proof Let ϕ be a solution to the problem (1.22). We have

$$\phi(t) = \phi(\sigma(t)) - \mu(t)\phi^{\Delta}(t).$$

Then (1.22) takes the form

$$\begin{aligned}\phi^{\Delta}(t) &= p(t)\left(\phi(\sigma(t)) - \mu(t)\phi^{\Delta}(t)\right) + f(t)\\ &= p(t)\phi(\sigma(t)) - p(t)\mu(t)\phi^{\Delta}(t) + f(t)\end{aligned}$$

or

$$(1 + p(t)\mu(t))\phi^{\Delta}(t) = p(t)\phi(\sigma(t)) + f(t).$$

Hence,

$$\begin{aligned}\phi^{\Delta}(t) &= \frac{p(t)}{1 + \mu(t)p(t)}\phi(\sigma(t)) + \frac{f(t)}{1 + \mu(t)p(t)}\\ &= -\frac{-p(t)}{1 + \mu(t)p(t)}\phi(\sigma(t)) + \frac{f(t)}{1 + \mu(t)p(t)}\\ &= - \ominus p(t)\phi(\sigma(t)) + \frac{f(t)}{1 + p(t)\mu(t)}.\end{aligned}$$

Therefore $\phi(t)$ satisfies the problem

$$\phi^{\Delta}(t) = - \ominus p(t)\phi(\sigma(t)) + \frac{f(t)}{1 + \mu(t)p(t)}, \quad \phi(t_0) = \phi_0. \tag{1.23}$$

From here, from Theorem 53 and from $\ominus(\ominus p)(t) = p(t)$, we get that the problem (1.23) has unique solution given by

$$\begin{aligned}
\phi(t) &= \phi_0 e_p(t, t_0) + \int_{t_0}^{t} e_p(t, \tau)\frac{f(\tau)}{1+\mu(\tau)p(\tau)}\Delta\tau \\
&= \phi_0 e_p(t, t_0) + \int_{t_0}^{t} \frac{f(\tau)}{e_p(\tau, t)(1+\mu(\tau)p(\tau))}\Delta\tau \\
&= \phi_0 e_p(t, t_0) + \int_{t_0}^{t} \frac{f(\tau)}{e_p(\sigma(\tau), t)}\Delta\tau \\
&= \phi_0 e_p(t, t_0) + \int_{t_0}^{t} e_p(t, \sigma(\tau)) f(\tau)\Delta\tau,
\end{aligned}$$

which completes the proof.

Example 47 Consider the problem

$$\phi^{\Delta} = 2\phi + 3^t, \quad \phi(0) = 0, \quad \mathscr{T} = \mathscr{Z}.$$

Here

$$p(t) = 2, \quad f(t) = 3^t, \quad \phi_0 = 0,$$

$$\sigma(t) = t + 1, \quad \mu(t) = 1, \quad t \in \mathscr{T}.$$

Then, using Corollary 5, we obtain

$$\begin{aligned}
\phi(t) &= \int_0^t e_2(t, \sigma(\tau))3^{\tau}\Delta\tau \\
&= \int_0^t e_2(t, \tau+1)3^{\tau}\Delta\tau \\
&= \int_0^t 3^{t-\tau-1}3^{\tau}\Delta\tau \\
&= \int_0^t 3^{t-1}\Delta\tau \\
&= 3^{t-1}\int_0^t \Delta\tau \\
&= t3^{t-1}.
\end{aligned}$$

Example 48 Consider the problem

$$\phi^{\Delta} = 4\phi + t, \quad \phi(0) = 1, \quad \mathscr{T} = 2\mathscr{Z}.$$

Here

$$\sigma(t) = t + 2, \quad \mu(t) = 2, \quad p(t) = 4,$$

$$f(t) = t, \quad \phi_0 = 1, \quad t \in \mathscr{T}.$$

Then, using Corollary 5, we obtain

$$\begin{aligned}
\phi(t) &= e_4(t, 0) + \int_0^t e_4(t, \sigma(\tau))\tau \Delta\tau \\
&= 9^{\frac{t}{2}} + \int_0^t e_4(t, \tau + 2)\tau \Delta\tau \\
&= 3^t + \int_0^t 9^{\frac{t-\tau-2}{2}} \tau \Delta\tau \\
&= 3^t + 9^{\frac{t}{2}-1} \int_0^t 9^{-\frac{\tau}{2}} \tau \Delta\tau \\
&= 3^t + 3^{t-2} \sum_{s\in[0,t-2]} \mu(s) s 3^{-s} \\
&= 3^t + 2 \cdot 3^{t-2} \sum_{s\in[0,t-2]} s 3^{-s}.
\end{aligned}$$

Example 49 Consider the equation

$$\phi^{\Delta} = p(t)\phi + e_p(t, t_0), \quad \phi(t_0) = 0,$$

where $p : \mathscr{T} \longmapsto \mathscr{R}$ is rd-continuous and $1 + \mu(t)p(t) \neq 0$ for all $t \in \mathscr{T}$.
Using Corollary 5, we obtain

$$\begin{aligned}
\phi(t) &= \int_{t_0}^t e_p(t, \sigma(\tau)) e_p(\tau, t_0) \Delta\tau \\
&= \int_{t_0}^t \frac{1}{e_p(\sigma(\tau), t)} e_p(\tau, t_0) \Delta\tau \\
&= \int_{t_0}^t \frac{1}{(1 + p(\tau)\mu(\tau)) e_p(\tau, t)} e_p(\tau, t_0) \Delta\tau \\
&= \int_{t_0}^t \frac{e_p(t, t_0)}{1 + p(\tau)\mu(\tau)} \Delta\tau.
\end{aligned}$$

Exercise 24 Let $p : \mathscr{T} \longmapsto \mathscr{R}$ be rd-continuous and $1 + \mu(t)p(t) \neq 0$ for all $t \in \mathscr{T}$. Let also, $\phi(t)$ be a solution to the equation

$$\phi^{\Delta} - p(t)\phi = 0$$

and $\psi(t)$ be a solution to the equation

$$\psi^{\Delta} + p(t)\psi^{\sigma} = 0.$$

Prove that

$$\phi(t)\psi(t) = c, \quad t \in \mathcal{T},$$

where c is a constant.

1.8 Advanced Practical Exercises

Problem 1 Classify each points $t \in \mathcal{T} = \{\sqrt[4]{7n} : n \in \mathcal{N}_0\}$ as left-dense, left-scattered, right-dense or right-scattered.

Answer. Each points $t = \sqrt[4]{7n}$, $n \in \mathcal{N}$, are isolated, $t = 0$ is right-scattered.

Problem 2 Let $\mathcal{T} = \left\{\sqrt[4]{n+7} : n \in \mathcal{N}_0\right\}$. Find $\mu(t)$, $t \in \mathcal{T}$.

Answer. $\mu\left(\sqrt[4]{n+7}\right) = \sqrt[4]{n+8} - \sqrt[4]{n+7}$.

Problem 3 Let $\mathcal{T} = \left\{t = \sqrt[4]{2n+1} : n \in \mathcal{N}\right\}$, $f(t) = 1 + 2t^4$, $t \in \mathcal{T}$. Find $f(\sigma(t))$, $t \in \mathcal{T}$.

Answer. $5 + 2t^4$.

Problem 4 Let $\mathcal{T} = \left\{\frac{2}{4n+3} : n \in \mathcal{N}\right\} \cup \{0\}$. Find $\mathcal{T}^{\kappa}$.

Answer. $\left\{\frac{2}{4n+3} : n \in \mathcal{N}, n \geq 2\right\} \cup \{0\}$.

Problem 5 Let $f(t) = t + t^2 + t^3$, $t \in \mathcal{T}$. Prove that

$$f^{\Delta}(t) = 1 + t + t^2 + (1+t)\sigma(t) + \sigma^2(t), \quad t \in \mathcal{T}^{\kappa}.$$

Problem 6 Let $\mathcal{T} = \{n^3 : n \in \mathcal{N}_0\}$, $f(t) = t^2 + 2t$, $t \in \mathcal{T}$. Find $f^{\Delta}(t)$, $t \in \mathcal{T}^{\kappa}$.

Answer. $2 + t + \left(\sqrt[3]{t} + 1\right)^3$.

Problem 7 Let $\mathcal{T} = \{n+2 : n \in \mathcal{N}_0\}$, $f(t) = t^2 + 2$, $g(t) = t^2$. Find a constant $c \in [2, \sigma(2)]$ such that

$$(f \circ g)^{\Delta}(2) = f'(g(c))g^{\Delta}(2).$$

Answer. $c = \sqrt{\frac{13}{2}}$.

Problem 8 Let $\mathscr{T} = \{2^{4n+2} : n \in \mathscr{N}_0\}$, $v(t) = t^3$, $w(t) = t^2 + t$. Prove

$$(w \circ v)^{\Delta}(t) = (w^{\tilde{\Delta}} \circ v(t))v^{\Delta}(t), \quad t \in \mathscr{T}^{\kappa}.$$

Problem 9 Let $\mathscr{T} = \{n + 9 : n \in \mathscr{N}_0\}$, $v(t) = t^2 + 7t + 8$. Find $\left(v^{-1}\right)^{\tilde{\Delta}} \circ v(t)$.

Answer. $\frac{1}{2t + 8}$.

Problem 10 Let $\mathscr{T} = \mathscr{R}$ and

$$f(t) = \begin{cases} 1 & \text{for} \quad t = 2 \\ \frac{10}{t-2} & \text{for} \quad t \in \mathscr{R} \backslash \{2\}. \end{cases}$$

Determine if f is regulated.

Answer. No.

Problem 11 Let $\mathscr{T} = \mathscr{R}$ and

$$f(t) = \begin{cases} 0 & \text{if} \quad t = 5 \\ \frac{1}{t-5} & \text{if} \quad t \in \mathscr{R} \backslash \{5\}. \end{cases}$$

Check if $f : \mathscr{T} \longmapsto \mathscr{R}$ is pre-differentiable and if it is, find the region of differentiation.

Answer. No.

Problem 12 Let $\mathscr{T} = 3^{\mathscr{N}}$. Prove that

$$-\int \frac{1}{2t} \sin t \sin(2t) \Delta t = \cos t + c.$$

Problem 13 Let $p : \mathscr{T} \longmapsto \mathscr{R}$ be rd-continuous and $1 + \mu(t)p(t) \neq 0$ for all $t \in \mathscr{T}$. Let also, $\phi(t)$ be a nontrivial solution to the equation

$$\phi^{\Delta} - p(t)\phi = 0.$$

Prove that $\frac{1}{\phi(t)}$ is a solution to the equation

$$\psi^{\Delta} + p(t)\psi^{\sigma} = 0.$$

Chapter 2
Introductory Concepts of Integral Equations on Time Scales

A delta integral equation (or in short integral equation) is the equation in which the unknown function $\phi(x)$ appears inside a delta integral sign. The subject of integral equations is one of the most useful mathematical tools in pure and applied mathematics. Many initial and boundary value problems associated with dynamic equations on time scales can be transformed into problems of solving some approximate integral equations.

Suppose that $\mathscr{T}$ is a time scale and let σ, ρ and Δ denote the forward jump operator, the backward jump operator and the delta differentiation operator, respectively,break on $\mathscr{T}$. A standard type of integral equation in $\phi(x)$ is of the following form

$$\phi(x) = u(x) + \lambda \int_{f(x)}^{g(x)} K(x, y)\phi(y)\Delta y, \tag{2.1}$$

where $f, g : \mathscr{T} \longmapsto \mathscr{R}$ are the limits of integration, λ is a constant parameter, and $K : \mathscr{T} \times \mathscr{T} \longmapsto \mathscr{R}$ is a known function of the variables x and y, $u : \mathscr{T} \longmapsto \mathscr{R}$ is a known function. The function $K(x, y)$ in (2.1) is called the kernel or the nucleus of the integral equation. In (2.1) the unknown function $\phi(x)$ appears the integral sign. There are many other cases in which the unknown function $\phi(x)$ appears inside and outside the integral sign. The functions $u(x)$ and $K(x, y)$ are given in advance. Note that the limits of integration $f(x)$ and $g(x)$ may be both constants, variables or mixed. If the limits of integration are fixed, the Eq. (2.1) is called a generalized Fredholm integral equation given in the form

$$\phi(x) = u(x) + \lambda \int_{a}^{b} K(x, y)\phi(y)\Delta y, \tag{2.2}$$

where a and b are constants. If at least one limit is variable, the equation is called a generalized Volterra integral equation given in the form

S.G. Georgiev, *Integral Equations on Time Scales*, Atlantis Studies in Dynamical Systems 5, DOI 10.2991/978-94-6239-228-1_2

$$\phi(x) = u(x) + \lambda \int_a^x K(x, y)\phi(y)\Delta y. \tag{2.3}$$

Example 1 Let $\mathscr{T} = 2^{\mathscr{N}_0}$. Then the equation

$$\phi(x) = \sinh_f(x, 1) + \lambda \int_1^x (x - y)\phi(y)\Delta y,$$

where $f \in \mathscr{R}_1$, is an example for a generalized Volterra integral equation.

If the unknown function $\phi(x)$ appears only under the integral sign of generalized Fredholm or generalized Volterra equation, the integral equation is called a first kind generalized Fredholm or generalized Volterra integral equation, respectively.

Example 2 Let $\mathscr{T} = \mathscr{Z}$. The equation

$$\cos_f(x, 0) = \int_1^2 y^2\phi(y)\Delta y,$$

where $f \in \mathscr{R}_1$, is a first kind generalized Fredholm integral equation.

If the unknown function $\phi(x)$ appears both inside and outside the integral sign of generalized Fredholm integral equation or generalized Volterra integral equation, the integral equation is called a second kind generalized Fredholm or generalized Volterra integral equation, respectively.

Example 3 Let $\mathscr{T} = 2^{\mathscr{N}_0}$. The equation

$$\phi(x) = x^2 - 1 + 2\int_1^4 (x - y)\phi(y)\Delta y$$

is a second kind generalized Fredholm integral equation.

If in the Eq. (2.2) or (2.3) the function $u(x)$ is identically zero, the resulting equation

$$\phi(x) = \lambda \int_a^b K(x, y)\phi(y)\Delta y$$

or

$$\phi(x) = \lambda \int_a^x K(x, y)\phi(y)\Delta y$$

is called homogeneous generalized Fredholm or homogeneous generalized Volterra integral equation, respectively. Any equation that includes both (delta-)integrals and (delta-)derivatives of the unknown function $\phi(x)$ is called delta-integro-delta-differential equation (or in short integro-differential equation). The Fredholm integro-differential equation is of the form

$$\phi^{\Delta^k}(x) = u(x) + \lambda \int_a^b K(x,y)\phi(y)\Delta y$$

However, the Volterra integro-differential equation is of the form

$$\phi^{\Delta^k}(x) = u(x) + \lambda \int_a^x K(x,y)\phi(y)\Delta y.$$

The equation

$$u(x) = \int_0^x \frac{1}{(x-y)^\alpha}\phi(y)\Delta y, \quad 0 < \alpha < 1,$$

is called generalized Abel's integral equation. The equation

$$\phi(x) = u(x) + \int_0^x \frac{1}{(x-y)^\alpha}\phi(y)\Delta y, \quad 0 < \alpha < 1,$$

is called generalized weakly singular integral equation. If the unknown function $\phi(x)$ inside the integral sign is one, the integral equation or the integro-differential equation is called linear.

Example 4 Let $\mathscr{T} = \mathscr{Z}$. The equation

$$x = \int_1^3 (x-2y)\phi(y)\Delta y$$

is a linear equation.

If the equation contains nonlinear function of the unknown function $\phi(x)$, the integral equation or the integro-differential equation is called nonlinear.

Example 5 Let $\mathscr{T} = 2^{\mathscr{N}_0}$. Then

$$u(x) = x^2 - \int_1^x (x-y)^2 u^2(y)\Delta y$$

is a nonlinear equation.

The main objective of this text is to determine the unknown function $\phi(x)$ that will satisfy (2.1) using a number of solution techniques. We shall explain these methods to find solutions of the unknown function.

2.1 Reducing Double Integrals to Single Integrals

It will be seen later that we can convert initial value problems and other problems to integral equations. It is useful to outline the formula that will reduce double integrals to single integrals.

Theorem 1 *Let $f : \mathscr{T} \longmapsto \mathscr{R}$ be integrable and $a \in \mathscr{T}$. Then*

$$\int_a^x \int_a^{x_1} f(t)\Delta t \Delta x_1 = \int_a^x (x - \sigma(t))f(t)\Delta t \quad for \quad x \in \mathscr{T}. \tag{2.4}$$

Proof Using integration by parts, we have

$$\begin{aligned} \int_a^x \int_a^{x_1} f(t)\Delta t \Delta x_1 &= \int_a^x (x_1 - a)^{\Delta} \int_a^{x_1} f(t)\Delta t \Delta x_1 \\ &= (x_1 - a) \int_a^{x_1} f(t)\Delta t \Big|_{x_1=a}^{x_1=x} - \int_a^x (\sigma(x_1) - a)f(x_1)\Delta x_1 \\ &= (x - a) \int_a^x f(t)\Delta t - \int_a^x (\sigma(t) - a)f(t)\Delta t \\ &= \int_a^x (x - \sigma(t))f(t)\Delta t. \end{aligned}$$

Example 6 Let $\mathscr{T} = \mathscr{Z}$. Then $\sigma(t) = t + 1$, $t \in \mathscr{T}$, and

$$\begin{aligned} \int_0^x \int_0^{x_1} t^2 \Delta t \Delta x_1 &= \int_0^x (x - \sigma(t))t^2 \Delta t \\ &= \int_0^x (x - t - 1)t^2 \Delta t. \end{aligned}$$

Example 7 Let $\mathscr{T} = 2^{\mathscr{N}_0}$. Then $\sigma(t) = 2t$, $t \in \mathscr{T}$, and

$$\begin{aligned} \int_1^x \int_1^{x_1} e_t(t, 1)\Delta t \Delta x_1 &= \int_1^x (x - \sigma(t))e_t(t, 1)\Delta t \\ &= \int_1^x (x - 2t)e_t(t, 1)\Delta t. \end{aligned}$$

Example 8 Let $\mathcal{T} = 3\mathcal{Z}$. Then $\sigma(t) = t + 3$, $t \in \mathcal{T}$,and

$$\int_3^x \int_3^{x_1} (\sin_t(t,0) + t^3)\Delta t \Delta x_1 = \int_3^x (x - \sigma(t))(\sin_t(t,0) + t^3)\Delta t$$
$$= \int_3^x (x - t - 3)(\sin_t(t,0) + t^3)\Delta t.$$

Exercise 1 Convert the following double integrals to single integrals.

1. $\int_0^x \int_0^{x_1} (t^2 - t)\Delta t \Delta x_1,\ \mathcal{T} = \mathcal{Z}.$
2. $\int_1^x \int_1^{x_1} \frac{t^2+1}{t^4+1}\Delta t \Delta x_1,\ \mathcal{T} = 2^{\mathcal{N}_0}.$
3. $\int_2^x \int_2^{x_1} (3t - 2)\Delta t \Delta x_1,\ \mathcal{T} = 2\mathcal{N}.$

Answer

1. $\int_0^x (x - t - 1)(t^2 - t)\Delta t,$
2. $\int_1^x (x - 2t)\frac{t^2+1}{t^4+1}\Delta t,$
3. $\int_2^x (x - t - 2)(3t - 2)\Delta t.$

As a result to (2.11) we can show the following corollary.

Corollary 1 *Let $f : \mathcal{T} \longmapsto \mathcal{R}$ be integrable and $a \in \mathcal{T}$. Then*

$$\int_a^x \int_a^{x_1} (x - \sigma(t))f(t)\Delta t \Delta x_1 = \int_a^x (x - \sigma(t))^2 f(t)\Delta t \quad \textit{for} \quad x \in \mathcal{T}.$$

Proof By Theorem 1, we have

$$\int_a^x \int_a^{x_1} (x - \sigma(t))f(t)\Delta t \Delta x_1 = x\int_a^x \int_a^{x_1} f(t)\Delta t \Delta x_1 - \int_a^x \int_a^{x_1} \sigma(t)f(t)\Delta t \Delta x_1$$
$$= x\int_a^x (x - \sigma(t))f(t)\Delta t - \int_a^x (x - \sigma(t))\sigma(t)f(t)\Delta t$$
$$= \int_a^x (x - \sigma(t))^2 f(t)\Delta t.$$

Example 9 Let $\mathcal{T} = 2\mathcal{Z}$. Then $\sigma(t) = t + 2$, $t \in \mathcal{T}$, and

$$\int_{-2}^x \int_{-2}^{x_1} (x - t - 2)(t^3 + 1)\Delta t \Delta x_1 = \int_{-2}^x (x - t - 2)^2(t^3 + 1)\Delta t.$$

Example 10 Let $\mathscr{T} = 4^{\mathscr{N}_0}$. Then $\sigma(t) = 4t$, $t \in \mathscr{T}$, and

$$\int_{16}^{x}\int_{16}^{x_1}(x-4t)(t^2+t)\Delta t\Delta x_1 = \int_{16}^{x}(x-4t)^2(t^2+t)\Delta t.$$

Example 11 Let $\mathscr{T} = \mathscr{N}_0^2$. Then $\sigma(t) = (\sqrt{t}+1)^2$, $t \in \mathscr{T}$. Then

$$\int_0^x\int_0^{x_1}(x-(\sqrt{t}+1)^2)\sqrt{t}\Delta t\Delta x_1 = \int_0^x(x-(\sqrt{t}+1)^2)^2\sqrt{t}\Delta t.$$

Exercise 2 Convert the following double integrals to single integrals.

1. $\int_0^x\int_0^{x_1}(x-t-2)t^2\Delta t\Delta x_1$, $\mathscr{T} = 2^{\mathscr{N}_0}\cup\{0\}$,
2. $\int_0^x\int_0^{x_1}(x-5t)(t^2+2t+3)\Delta t\Delta x_1$, $\mathscr{T} = 5^{\mathscr{N}_0}\cup\{0\}$.
3. $\int_1^x\int_1^{x_1}(x-t-1)t\Delta t\Delta x_1$, $\mathscr{T} = \mathscr{N}_0$,

where for $q > 1$ with $q^{\mathscr{N}_0}\cup\{0\}$ we will denote in all places in this book the set

$$0, \ldots, \frac{1}{q^k}, \frac{1}{q^{k-1}}, \ldots, 1, q, q^2, \ldots.$$

Answer

1. $\int_0^x(x-t-2)^2t^2\Delta t$,
2. $\int_0^x(x-5t)^2(t^2+2t+3)\Delta t$,
3. $\int_1^x(x-t-1)^2t\Delta t$.

2.2 Converting IVP to Generalized Volterra Integral Equations

In this section we will convert an initial value problem (IVP) to an equivalent generalized Volterra integral equation and generalized Volterra integro-differential equation. We will apply this process to a first order IVP

$$z^{\Delta}(x) + a(x)z(x) = u(x) \tag{2.5}$$

subject to the initial condition

$$z(x_0) = z_0, \tag{2.6}$$

where $x_0 \in \mathscr{T}$, $z_0 \in \mathscr{R}$, $a, u \in \mathscr{C}_{rd}(\mathscr{T})$, and to a second order IVP

$$z^{\Delta^2}(x) + a(x)z^{\Delta}(x) + b(x)z(x) = u(x) \tag{2.7}$$

subject to the initial conditions

$$z(x_0) = z_0, \quad z^{\Delta}(x_0) = z_0^{\Delta}, \tag{2.8}$$

where $x_0 \in \mathscr{T}$, $z_0, z_0^{\Delta} \in \mathscr{R}$, and $a, b, u \in \mathscr{C}_{rd}(\mathscr{T})$.
Firstly, we will consider the problem (2.5), (2.6). Let

$$\phi(x) = z^{\Delta}(x). \tag{2.9}$$

Now, using (2.6), we get

$$\begin{aligned}\int_{x_0}^{x} \phi(y)\Delta y &= \int_{x_0}^{x} z^{\Delta}(y)\Delta y\\ &= z(y)\Big|_{y=x_0}^{y=x}\\ &= z(x) - z(x_0)\\ &= z(x) - z_0,\end{aligned}$$

i.e.,

$$z(x) = z_0 + \int_{x_0}^{x} \phi(y)\Delta y. \tag{2.10}$$

Substituting (2.9) and (2.10) into (2.5) yields the following generalized Volterra integral equation

$$\phi(x) + a(x)\left(z_0 + \int_{x_0}^{x} \phi(y)\Delta y\right) = u(x).$$

The last equation can be written as standard generalized Volterra integral equation in the following way

$$\phi(x) = u(x) - z_0 a(x) - \int_{x_0}^{x} a(x)\phi(y)\Delta y.$$

Example 12 Let $\mathscr{T} = \mathscr{N}$. Consider populations with a fixed interval between generations or possibly a fixed interval between measurements. With x_0 we will denote the initial population size and with $x(t)$ we will denote the population size at time t.

Suppose the population changes only through births and deaths and suppose further that the birth and death rates are constants b and d, respectively. Then

$$x^{\Delta}(t) = (b - d)x(t), \quad t \in \mathscr{T}, \tag{2.11}$$

$$x(0) = x_0 \tag{2.12}$$

determines the population size in each generation. We integrate the Eq. (2.11) from 0 to t, and using (2.12), we get the integral equation

$$x(t) = x_0 + (b - d) \int_0^t x(s)\Delta s, \quad t \in \mathscr{T}.$$

Example 13 (*Verhulst difference equation*) Let $\mathscr{T} = \mathscr{N}$. The dynamic equation

$$x^{\Delta}(t) = \frac{(r - A)x(t) - x^2(t)}{A + x(t)}, \quad t \in \mathscr{T}, \tag{2.13}$$

describes a population that die out completely in each generation and has birth rates that saturate for large population sizes. Here A and r are positive constants. If we suppose that $x(0) = x_0$, then the Eq. (2.13) can be converted to a generalized Volterra integral equation

$$x(t) = x_0 + \int_0^t \frac{(r - A)x(s) - x^2(s)}{A + x(s)} \Delta s.$$

Exercise 3 Let $\mathscr{T} = \mathscr{H}$. Consider a simple electric circuit. The total charge $Q(t)$ on the capacitor at $t \in \mathscr{H}$ is given by the equation

$$Q^{\Delta}(t) = bQ(t), \quad b = \text{const}.$$

Reduce it to an integral equation if $Q(t_0) = Q_0$ for some $t_0 \in \mathscr{H}$ and some real constant Q_0.

Example 14 Let $\mathscr{T} = \mathscr{Z}$. Let us consider the IVP

$$z^{\Delta}(x) + 2xz(x) = 0, \quad z(0) = 1.$$

Here

$$a(x) = 2x, \quad u(x) = 0, \quad z_0 = 1.$$

Then we get the integral equation

$$\phi(x) = -2x - 2x \int_0^x \phi(y)\Delta y.$$

Example 15 Let $\mathscr{T} = \mathscr{N}$. Let us consider the IVP

$$z^{\Delta}(x) + x^2 z(x) = x, \quad z(1) = 1.$$

Here

$$x_0 = 1, \quad a(x) = x^2, \quad u(x) = x, \quad z_0 = 1.$$

Then we obtain the following integral equation

$$\phi(x) = x - x^2 - \int_1^x x^2 \phi(y) \Delta y.$$

Example 16 Let $\mathscr{T} = 2^{\mathscr{N}_0} \cup \{0\}$. Let us consider the IVP

$$z^{\Delta}(x) + e_x(x, 1) z(x) = \sinh_x(x, 1), \quad z(0) = 1.$$

Here

$$x_0 = 0, \quad z_0 = 1, \quad a(x) = e_x(x, 1). \quad u(x) = \sinh_x(x, 1).$$

Then we get the following integral equation

$$\phi(x) = \sinh_x(x, 1) - e_x(x, 1) - \int_0^x e_x(x, 1)\phi(y) \Delta y.$$

Exercise 4 Convert the following IVPs to integral equations.

1.
$$\begin{cases} z^{\Delta}(x) + (x^2 + 2x - 1) z(x) = 3 \\ z(0) = 2, \quad \mathscr{T} = \mathscr{Z}, \end{cases}$$

2.
$$\begin{cases} z^{\Delta}(x) + e_x(x, 0) z(x) = 3x^2 \\ z(0) = 1, \quad \mathscr{T} = \mathscr{N}_0, \end{cases}$$

3.
$$\begin{cases} z^{\Delta}(x) - \cos_x(x, 1) z(x) = -x \\ z(0) = 0, \quad \mathscr{T} = 2^{\mathscr{N}_0} \cup \{0\}. \end{cases}$$

Answer

1. $\phi(x) = 3 - 2(x^2 + 2x - 1) - \int_0^x (x^2 + 2x - 1)\phi(y) \Delta y,$
2. $\phi(x) = 3x^2 - e_x(x, 0) - \int_0^x e_x(x, 0)\phi(y) \Delta y,$
3. $\phi(x) = -x + \int_0^x \cos_x(x, 1)\phi(y) \Delta y.$

Now we consider the problem (2.7), (2.8). Set

$$z^{\Delta^2}(x) = \phi(x). \tag{2.14}$$

Then, using (2.8), we get

$$z^{\Delta}(x) - z^{\Delta}(x_0) = \int_{x_0}^{x} \phi(y)\Delta y$$

or

$$z^{\Delta}(x) = z_0^{\Delta} + \int_{x_0}^{x} \phi(y)\Delta y, \tag{2.15}$$

whereupon

$$z(x) - z(x_0) = \int_{x_0}^{x} z_0^{\Delta}\Delta y + \int_{x_0}^{x}\int_{x_0}^{x_1} \phi(y)\Delta y\Delta x_1,$$

or

$$z(x) = z_0 + z_0^{\Delta}(x - x_0) + \int_{x_0}^{x}\int_{x_0}^{x_1} \phi(y)\Delta y\Delta x_1.$$

Hence, applying Theorem 1, we find

$$z(x) = z_0 + z_0^{\Delta}(x - x_0) + \int_{x_0}^{x} (x - \sigma(y))\phi(y)\Delta y. \tag{2.16}$$

Substituting (2.14), (2.15) and (2.16) in (2.7), we get

$$\begin{aligned}&\phi(x) + a(x)\left(z_0^{\Delta} + \textstyle\int_{x_0}^{x} \phi(y)\Delta y\right)\\ &+ b(x)\left(z_0 + z_0^{\Delta}(x - x_0) + \textstyle\int_{x_0}^{x}(x - \sigma(y))\phi(y)\Delta y\right) = u(x)\end{aligned}$$

or

$$\begin{aligned}\phi(x) &= u(x) - a(x)z_0^{\Delta} - b(x)z_0 - b(x)z_0^{\Delta}(x - x_0)\\ &- a(x)\textstyle\int_{x_0}^{x} \phi(y)\Delta y - b(x)\int_{x_0}^{x}(x - \sigma(y))\phi(y)\Delta y,\end{aligned}$$

i.e.,

$$\phi(x) = u(x) - a(x)z_0^{\Delta} - b(x)z_0 - b(x)z_0^{\Delta}(x - x_0) - \int_{x_0}^{x} \left[a(x) + b(x)(x - \sigma(y))\right]\phi(y)\Delta y. \tag{2.17}$$

Example 17 Let $\mathscr{T} = 2^{\mathscr{N}_0}$. Consider the IVP

$$z^{\Delta^2}(x) + x^2 z^{\Delta}(x) + xz(x) = x - 1,$$

$$z(1) = 1, \quad z^{\Delta}(1) = 2.$$

Here

$$\sigma(x) = 2x, \quad a(x) = x^2, \quad b(x) = x, \quad u(x) = x - 1,$$

$$z_0 = 1, \quad z_0^{\Delta} = 2, \quad x_0 = 1.$$

Then, using (2.17), we get the following integral equation

$$\phi(x) = x - 1 - 2x^2 - x - 2x(x-1) - \int_1^x \left[x^2 + x(x - 2y)\right] \phi(y) \Delta y$$

$$= -4x^2 + 2x - 1 - 2\int_1^x (x^2 - xy)\phi(y)\Delta y.$$

Example 18 Let $\mathscr{T} = 2\mathscr{Z}$. Consider the IVP

$$z^{\Delta^2}(x) + xz^{\Delta}(x) - x^2 z(x) = x,$$

$$z(0) = 0, \quad z^{\Delta}(0) = 1.$$

Here

$$\sigma(x) = x + 2, \quad a(x) = x, \quad b(x) = -x^2, \quad u(x) = x,$$

$$z_0 = 0, \quad z_0^{\Delta} = 1, \quad x_0 = 0.$$

Then, using (2.17), we get the following integral equation

$$\phi(x) = x - x + x^3 - \int_0^x \left[x - x^2(x - y - 2)\right] \phi(y) \Delta y$$

$$= x^3 - \int_0^x (-x^3 + 2x^2 + x + x^2 y)\phi(y)\Delta y.$$

Example 19 Let $\mathscr{T} = \mathscr{N}_0^3$. Consider the IVP

$$z^{\Delta^2}(x) - 2e_x(x, 0)z(x) = \sinh_x(x, 0),$$

$$z(0) = 0, \quad z^{\Delta}(0) = 1.$$

Here

$$\sigma(x) = (\sqrt[3]{x}+1)^3, \quad a(x) = 0, \quad b(x) = -2e_x(x,0), \quad u(x) = \sinh_x(x,0),$$

$$z_0 = 0, \quad z_0^{\Delta} = 1, \quad x_0 = 0.$$

Then, using (2.17), we get the following integral equation

$$\phi(x) = \sinh_x(x,0) + 2xe_x(x,0) + 2\int_0^x e_x(x,0)\left[x - (\sqrt[3]{y}+1)^3\right]\phi(y)\Delta y.$$

Exercise 5 Convert the following IVPs to integral equations.

1.
$$\begin{cases} z^{\Delta^2}(x) - x^2 z^{\Delta}(x) = 0, \\ z(0) = 0, \quad z^{\Delta}(0) = 0, \quad \mathscr{T} = \mathscr{N}_0, \end{cases}$$

2.
$$\begin{cases} z^{\Delta^2}(x) + z^{\Delta}(x) + z(x) = 1, \\ z(0) = 0, \quad z^{\Delta}(0) = 1, \quad \mathscr{T} = 2^{\mathscr{N}_0} \cup \{0\}, \end{cases}$$

3.
$$\begin{cases} z^{\Delta^2}(x) + e_x(x,1) z^{\Delta}(x) = 0, \\ z(1) = 1, \quad z^{\Delta}(1) = 2, \quad \mathscr{T} = \mathscr{N}. \end{cases}$$

Answer

1. $\phi(x) = \int_0^x x^2\phi(y)\Delta y,$
2. $\phi(x) = -x - \int_0^x (1+x-2y)\phi(y)\Delta y,$
3. $\phi(x) = -2e_x(x,1) - \int_1^x e_x(x,1)\phi(y)\Delta y.$

2.3 Converting Generalized Volterra Integral Equations to IVP

A method for solving generalized Volterra integral and Volterra integro-differential equation converts these equations to equivalent initial value problems. This method is achieved by differentiating both sides of generalized Volterra equations with respect to x as many times as we need to get rid of the integral sign and obtain a differential equation. The conversion of generalized Volterra equations requires to use Leibnitz rule for differentiating the integral at the right hand side. The initial conditions are obtained by substituting $x = a$ into $u(x)$ and its derivatives. For instance, after we differentiate (2.3) with respect to x we get

$$\phi^{\Delta}(x) = u^{\Delta}(x) + \lambda \int_a^x K_x^{\Delta}(x, y)\phi(y)\Delta y + \lambda K(\sigma(x), x)\phi(x) \tag{2.18}$$

and substituting $x = a$ in (2.3) we find

$$\phi(a) = u(a).$$

If there is an integral sign in (2.18), then we differentiate it with respect to x and so on.

Example 20 Let $\mathscr{T} = \mathscr{Z}$. Consider the equation

$$\phi(x) = x^2 + \int_0^x \phi(y)\Delta y. \tag{2.19}$$

We have

$$\sigma(x) = x + 1, \quad x \in \mathscr{T},$$

and

$$(x^2)^{\Delta} = \sigma(x) + x = x + 1 + x = 2x + 1.$$

Hence, differentiating (2.19) with respect to x, we get

$$\phi^{\Delta}(x) = (x^2)^{\Delta} + \phi(x)$$

$$= 2x + 1 + \phi(x).$$

Substituting $x = 0$ in (2.19), we find $\phi(0) = 0$.
In this way, we get the following IVP

$$\begin{cases} \phi^{\Delta}(x) - \phi(x) = 2x + 1 \\ \phi(0) = 0. \end{cases}$$

Example 21 Let $\mathscr{T} = 2^{\mathscr{N}_0}$. Consider the equation

$$\phi(x) = x^3 + \int_1^x (x - y)\phi(y)\Delta y. \tag{2.20}$$

We have

$$\sigma(x) = 2x, \quad x \in \mathscr{T},$$
$$(x^3)^{\Delta} = \sigma^2(x) + x\sigma(x) + x^2$$

$$
\begin{aligned}
&= 4x^2 + 2x^2 + x^2 \\
&= 7x^2, \\
(x-y)_x^{\Delta} &= 1.
\end{aligned}
$$

We differentiate with respect to x the Eq. (2.20) and we get

$$
\begin{aligned}
\phi^{\Delta}(x) &= (x^3)^{\Delta} + \left(\int_1^x (x-y)\phi(y)\Delta y \right)^{\Delta} \\
&= 7x^2 + \int_1^x \phi(y)\Delta y + (\sigma(x) - x)\phi(x) \\
&= 7x^2 + x\phi(x) + \int_1^x \phi(y)\Delta y,
\end{aligned}
$$

i.e.,

$$
\phi^{\Delta}(x) = 7x^2 + x\phi(x) + \int_1^x \phi(y)\Delta y. \tag{2.21}
$$

Now we differentiate (2.21) with respect to x and we find

$$
\begin{aligned}
\phi^{\Delta^2}(x) &= (7x^2)^{\Delta} + (x\phi(x))^{\Delta} + \left(\int_1^x \phi(y)\Delta y \right)^{\Delta} \\
&= 7(\sigma(x) + x) + \phi(x) + \sigma(x)\phi^{\Delta}(x) + \phi(x) \\
&= 21x + 2\phi(x) + 2x\phi^{\Delta}(x)
\end{aligned}
$$

or

$$
\phi^{\Delta^2}(x) - 2x\phi^{\Delta}(x) - 2\phi(x) = 21x.
$$

We put $x = 1$ in (2.20) and we get $\phi(1) = 1$.
We substitute $x = 1$ in (2.21) and we find

$$
\phi^{\Delta}(1) = 7 + \phi(1) = 8.
$$

In this way we go to the following IVP

$$
\begin{cases}
\phi^{\Delta^2}(x) - 2x\phi^{\Delta}(x) - 2\phi(x) = 21x, \\
\phi(1) = 1, \quad \phi^{\Delta}(1) = 8.
\end{cases}
$$

Example 22 Let $\mathscr{T} = 3^{\mathscr{Z}}$. Consider the equation

$$
\phi(x) = e_x(x, 1) + \int_1^x (x + 2y)\phi(y)\Delta y. \tag{2.22}
$$

Here

$$\sigma(x) = x + 3, \quad x \in \mathscr{T}.$$

Then, differentiating (2.22) with respect to x, we get

$$\begin{aligned}\phi^{\Delta}(x) &= e_x^{\Delta}(x, 1) + \left(\int_1^x (x + 2y)\phi(y)\Delta y\right)^{\Delta} \\ &= xe_x(x, 1) + \int_1^x \phi(y)\Delta y + (\sigma(x) + 2x)\phi(x) \\ &= xe_x(x, 1) + (3x + 3)\phi(x) + \int_1^x \phi(y)\Delta y,\end{aligned}$$

i.e.,

$$\phi^{\Delta}(x) = xe_x(x, 1) + 3(x + 1)\phi(x) + \int_1^x \phi(y)\Delta y. \tag{2.23}$$

Now we differentiate (2.23) with respect to x and we find

$$\begin{aligned}\phi^{\Delta^2}(x) &= (xe_x(x, 1))^{\Delta} + 3((x + 1)\phi(x))^{\Delta} + \left(\int_1^x \phi(y)\Delta y\right)^{\Delta} \\ &= e_x(x, 1) + \sigma(x)xe_x(x, 1) + 3\phi(x) + 3(\sigma(x) + 1)\phi^{\Delta}(x) + \phi(x) \\ &= (x^2 + 3x + 1)e_x(x, 1) + 4\phi(x) + 3(x + 4)\phi^{\Delta}(x),\end{aligned}$$

i.e.,

$$\phi^{\Delta^2}(x) - 3(x + 4)\phi^{\Delta}(x) - 4\phi(x) = (x^2 + 3x + 1)e_x(x, 1).$$

We put $x = 1$ in (2.22) and we find $\phi(1) = e_1(1, 1) = 1$.
We substitute $x = 1$ in (2.23) and we get

$$\phi^{\Delta}(1) = e_1(1, 1) + 6\phi(1) = 7e_1(1, 1) = 7.$$

In this way we get the following IVP

$$\begin{cases}\phi^{\Delta^2}(x) - 3(x + 4)\phi^{\Delta}(x) - 4\phi(x) = (x^2 + 3x + 1)e_x(x, 1), \\ \phi(1) = 1, \quad \phi^{\Delta}(1) = 7.\end{cases}$$

Exercise 6 Convert the following generalized Volterra integral equations to IVPs.

1. $\phi(x) = 2x^2 - 1 + \int_0^x \phi(y)\Delta y, \ \mathscr{T} = \mathscr{N}_0,$
2. $\phi(x) = x + 3 + x\int_1^x \phi(y)\Delta y, \ \mathscr{T} = 3^{\mathscr{N}_0},$
3. $\phi(x) = 2x + \int_0^x (x - y)\phi(y)\Delta y, \ \mathscr{T} = \mathscr{N}_0^2.$

Answer

1.
$$\begin{cases} \phi^{\Delta}(x) - \phi(x) = 4x + 2, \\ \phi(0) = -1, \end{cases}$$

2.
$$\begin{cases} \phi^{\Delta^2}(x) - 9x\phi^{\Delta}(x) - 4\phi(x) = 0, \\ \phi(1) = 4, \quad \phi^{\Delta}(1) = 13, \end{cases}$$

3.
$$\begin{cases} \phi^{\Delta^2}(x) - (3 + 2\sqrt{x})\phi^{\Delta}(x) - \frac{3+2\sqrt{x}}{1+2\sqrt{x}}\phi(x) = 0, \\ \phi(0) = 0, \quad \phi^{\Delta}(0) = 2. \end{cases}$$

Example 23 Let $\mathscr{T} = 2^{\mathscr{N}_0}$. Consider the generalized Volterra integro-differential equation

$$\phi(x) = \phi^{\Delta}(x) + x + \int_1^x (x + y)\phi(y)\Delta y. \tag{2.24}$$

We have $\sigma(x) = 2x$, $x \in \mathscr{T}$.
Then we differentiate with respect to x the Eq. (2.24) and we get

$$\begin{aligned} \phi^{\Delta}(x) &= \phi^{\Delta^2}(x) + 1 + \int_1^x \phi(y)\Delta y + (\sigma(x) + x)\phi(x) \\ &= \phi^{\Delta^2}(x) + 3x\phi(x) + 1 + \int_1^x \phi(y)\Delta y. \end{aligned}$$

i.e.,

$$\phi^{\Delta}(x) = \phi^{\Delta^2}(x) + 3x\phi(x) + 1 + \int_1^x \phi(y)\Delta y. \tag{2.25}$$

Now we differentiate (2.25) with respect to x and we find

$$\begin{aligned} \phi^{\Delta^2}(x) &= \phi^{\Delta^3}(x) + 3\phi(x) + 3\sigma(x)\phi^{\Delta}(x) + \phi(x) \\ &= \phi^{\Delta^3}(x) + 4\phi(x) + 6x\phi^{\Delta}(x) \end{aligned}$$

or

$$\phi^{\Delta^3}(x) - \phi^{\Delta^2}(x) + 6x\phi^{\Delta}(x) + 4\phi(x) = 0.$$

We put $x = 1$ in (2.24) and we get

$$\phi(1) = \phi^{\Delta}(1) + 1.$$

Now we substitute $x = 1$ in (2.25) and we find

$$\begin{aligned}\phi^{\Delta}(1) &= \phi^{\Delta^2}(1) + 3\phi(1) + 1\\ &= \phi^{\Delta^2}(1) + 3\phi^{\Delta}(1) + 3 + 1\\ &= \phi^{\Delta^2}(1) + 3\phi^{\Delta}(1) + 4\end{aligned}$$

or

$$\phi^{\Delta^2}(1) + 2\phi^{\Delta}(1) + 4 = 0.$$

In this way we go to the following problem

$$\begin{cases}\phi^{\Delta^3}(x) - \phi^{\Delta^2}(x) + 6x\phi^{\Delta}(x) + 4\phi(x) = 0,\\ \phi^{\Delta}(1) - \phi(1) + 1 = 0, \quad \phi^{\Delta^2}(1) + 2\phi^{\Delta}(1) + 4 = 0.\end{cases}$$

Example 24 Let $\mathscr{T} = 3\mathscr{N}_0$. Consider the generalized Volterra integro-differential equation

$$\phi(x) = \phi^{\Delta}(x) + \int_1^x y^2\phi(y)\Delta y. \tag{2.26}$$

Then, differentiating (2.26) with respect to x, we get

$$\phi^{\Delta}(x) = \phi^{\Delta^2}(x) + x^2\phi(x)$$

or

$$\phi^{\Delta^2}(x) - \phi^{\Delta}(x) + x^2\phi(x) = 0.$$

We put $x = 1$ in (2.26) and we find

$$\phi(1) = \phi^{\Delta}(1).$$

Therefore we obtain the following problem

$$\begin{cases}\phi^{\Delta^2}(x) - \phi^{\Delta}(x) + x^2\phi(x) = 0\\ \phi(1) = \phi^{\Delta}(1).\end{cases}$$

Example 25 Let $\mathscr{T} = 5^{\mathscr{N}_0} \cup \{0\}$. Consider the generalized Volterra integro-differential equation

$$\phi^{\Delta}(x) = \phi(x) - x\int_0^x \phi(y)\Delta y. \tag{2.27}$$

Here $\sigma(x) = 5x$, $x \in \mathscr{T}$. Then, differentiating (2.27) with respect to x, we get

$$\begin{aligned}\phi^{\Delta^2}(x) &= \phi^{\Delta}(x) - \int_0^x \phi(y)\Delta y - \sigma(x)\phi(x)\\ &= \phi^{\Delta}(x) - \int_0^x \phi(y)\Delta y - 5x\phi(x),\end{aligned}$$

i.e.,

$$\phi^{\Delta^2}(x) - \phi^{\Delta}(x) + \int_0^x \phi(y)\Delta y + 5x\phi(x) = 0. \tag{2.28}$$

Hence,

$$\phi^{\Delta^3}(x) - \phi^{\Delta^2}(x) + \phi(x) + 5\phi(x) + 5\sigma(x)\phi^{\Delta}(x) = 0$$

or

$$\phi^{\Delta^3}(x) - \phi^{\Delta^2}(x) + 25x\phi^{\Delta}(x) + 6\phi(x) = 0.$$

We substitute $x = 0$ in (2.27) and (2.28) and we obtain

$$\phi(0) = \phi^{\Delta}(0) = \phi^{\Delta^2}(0).$$

Consequently we get the following problem

$$\begin{cases}\phi^{\Delta^3}(x) - \phi^{\Delta^2}(x) + 25x\phi^{\Delta}(x) + 6\phi(x) = 0\\ \phi(0) = \phi^{\Delta}(0) = \phi^{\Delta^2}(0).\end{cases}$$

Exercise 7 Convert the following generalized Volterra integro-differential equations to IVPs.

1. $\phi^{\Delta^2}(x) = \phi(x) + x^2 + \int_1^x \sinh_y(y, 1)\phi(y)\Delta y, \quad \mathscr{T} = \mathscr{N}_0,$
2. $\phi^{\Delta}(x) = \phi(x) + x^3 + \int_1^x (x+y)e_y(y, 1)\phi(y)\Delta y, \quad \mathscr{T} = 2^{\mathscr{N}_0},$
3. $\phi(x) = \phi^{\Delta}(x) + x^4 - \int_0^x (2x+y)\phi(y)\Delta y, \quad \mathscr{T} = 2\mathscr{Z}.$

Answer

1.

$$\begin{cases} \phi^{\Delta^3}(x) - \phi^{\Delta}(x) - \sinh_x(x,1)\phi(x) = 2x+1, \\ \phi^{\Delta^2}(1) = \phi(1)+1, \end{cases}$$

2.

$$\begin{cases} \phi^{\Delta^3}(x) - \phi^{\Delta^2}(x) - 6xe_{2x}(2x,1)\phi^{\Delta}(x) - (4+6x^2)e_x(x,1)\phi(x) = 21x, \\ \phi^{\Delta}(1) = \phi(1)+1, \quad \phi^{\Delta^2}(1) = \phi^{\Delta}(1) + 3\phi(1) + 7, \end{cases}$$

3.

$$\begin{cases} \phi^{\Delta^3}(x) - \phi^{\Delta^2}(x) - (3x+10)\phi^{\Delta}(x) - 5\phi(x) = -12x^2 - 48x - 56, \\ \phi(0) = \phi^{\Delta}(0), \quad \phi^{\Delta}(0) = \phi^{\Delta^2}(0) - 4\phi(0) + 8. \end{cases}$$

2.4 Converting BVP to Generalized Fredholm Integral Equation

In this section we will represent a method for converting boundary value problems to generalized Fredholm integral equation. This method is similar to the method for converting of IVP to generalized Volterra integral equation. Here boundary conditions will be used instead of initial conditions. In this case we will determine another initial condition that is not given in the problem.
We consider the following boundary value problem

$$z^{\Delta^2}(x) + f(x)z(x) = g(x), \quad x_0 < x < x_1, \quad x_0, x_1 \in \mathscr{T}, \tag{2.29}$$

$$z(x_0) = z_0, \quad z(x_1) = z_1. \tag{2.30}$$

We set

$$\phi(x) = z^{\Delta^2}(x). \tag{2.31}$$

Integrating both sides of (2.31) from x_0 to x we obtain

$$\int_{x_0}^{x} z^{\Delta^2}(t)\Delta t = \int_{x_0}^{x} \phi(t)\Delta t$$

or

$$z^{\Delta}(x) - z^{\Delta}(x_0) = \int_{x_0}^{x} \phi(t)\Delta t,$$

or

$$z^{\Delta}(x) = z^{\Delta}(x_0) + \int_{x_0}^{x} \phi(t)\Delta t.$$

We integrate the last equation from x_0 to x and we find

$$\int_{x_0}^{x} z^{\Delta}(y)\Delta y = \int_{x_0}^{x} z^{\Delta}(x_0)\Delta y + \int_{x_0}^{x}\int_{x_0}^{x_2} \phi(t)\Delta t \Delta x_2$$

or

$$z(x) - z(x_0) = z^{\Delta}(x_0)(x - x_0) + \int_{x_0}^{x}\int_{x_0}^{x_2} \phi(t)\Delta t \Delta x_2,$$

or

$$z(x) = z_0 + z^{\Delta}(x_0)(x - x_0) + \int_{x_0}^{x}\int_{x_0}^{x_2} \phi(t)\Delta t \Delta x_2.$$

Applying Theorem 1 we find

$$z(x) = z_0 + z^{\Delta}(x_0)(x - x_0) + \int_{x_0}^{x} (x - \sigma(t))\Delta t. \tag{2.32}$$

We substitute $x = x_1$ in the last equation and using that $z(x_1) = z_1$ we go to

$$z_1 = z_0 + z^{\Delta}(x_0)(x_1 - x_0) + \int_{x_0}^{x_1} (x_1 - \sigma(t))\phi(t)\Delta t$$

or

$$z^{\Delta}(x_0)(x_1 - x_0) = z_1 - z_0 - \int_{x_0}^{x_1} (x_1 - \sigma(t))\phi(t)\Delta t,$$

or

$$z^{\Delta}(x_0) = \frac{z_1 - z_0}{x_1 - x_0} - \frac{1}{x_1 - x_0}\int_{x_0}^{x_1} (x_1 - \sigma(t))\phi(t)\Delta t.$$

We substitute the last expression in (2.32) and we find

$$z(x) = z_0 + \frac{z_1 - z_0}{x_1 - x_0}(x - x_0) - \frac{x - x_0}{x_1 - x_0}\int_{x_0}^{x_1} (x_1 - \sigma(t))\phi(t)\Delta t$$
$$+ \int_{x_0}^{x} (x - \sigma(t))\phi(t)\Delta t.$$

The last expression and (2.31) we put in (2.29). Then

$$\phi(x) + f(x)z_0 + \frac{z_1 - z_0}{x_1 - x_0} f(x)(x - x_0) - f(x)\frac{x - x_0}{x_1 - x_0}\int_{x_0}^{x_1} (x_1 - \sigma(t))\phi(t)\Delta t$$
$$+ f(x)\int_{x_0}^{x} (x - \sigma(t))\phi(t)\Delta t = g(x)$$

or

$$\phi(x) = g(x) - f(x)z_0 - \frac{z_1 - z_0}{x_1 - x_0} f(x)(x - x_0) + f(x)\frac{x - x_0}{x_1 - x_0}\int_{x_0}^{x_1} (x_1 - \sigma(t))\phi(t)\Delta t$$
$$- f(x)\int_{x_0}^{x} (x - \sigma(t))\phi(t)\Delta(t)$$
$$= g(x) - f(x)z_0 - \frac{z_1 - z_0}{x_1 - x_0} f(x)(x - x_0) + f(x)\frac{x - x_0}{x_1 - x_0}\int_{x_0}^{x} (x_1 - \sigma(t))\phi(t)\Delta t$$
$$+ f(x)\frac{x - x_0}{x_1 - x_0}\int_{x}^{x_1} (x_1 - \sigma(t))\phi(t)\Delta t - f(x)\int_{x_0}^{x} (x - \sigma(t))\phi(t)\Delta t$$
$$= g(x) - f(x)z_0 - \frac{z_1 - z_0}{x_1 - x_0} f(x)(x - x_0)$$
$$+ \int_{x_0}^{x} f(x)\left(x_1\frac{x - x_0}{x_1 - x_0} - \sigma(t)\frac{x - x_0}{x_1 - x_0} - x + \sigma(t)\right)\phi(t)\Delta t$$
$$+ f(x)\frac{x - x_0}{x_1 - x_0}\int_{x}^{x_1} (x_1 - \sigma(t))\phi(t)\Delta t$$
$$= g(x) - f(x)z_0 - \frac{z_1 - z_0}{x_1 - x_0} f(x)(x - x_0)$$
$$+ \int_{x_0}^{x} f(x)\left(-\frac{(x_1 - x)x_0}{x_1 - x_0} - \frac{x - x_1}{x_1 - x_0}\sigma(t)\right)\phi(t)\Delta t$$
$$+ f(x)\frac{x - x_0}{x_1 - x_0}\int_{x}^{x_1} (x_1 - \sigma(t))\phi(t)\Delta t,$$

i.e.,

$$\phi(x) = g(x) - f(x)z_0 - \frac{z_1 - z_0}{x_1 - x_0} f(x)(x - x_0)$$
$$+ \int_{x_0}^{x} f(x)\left(-\frac{(x_1 - x)x_0}{x_1 - x_0} - \frac{x - x_1}{x_1 - x_0}\sigma(t)\right)\phi(t)\Delta t$$

$$+f(x)\frac{x-x_0}{x_1-x_0}\int_x^{x_1}(x_1-\sigma(t))\phi(t)\Delta t.$$

Let

$$K(x,t)=\begin{cases} f(x)\left(-\frac{(x_1-x)x_0}{x_1-x_0}-\frac{x-x_1}{x_1-x_0}\sigma(t)\right) & for \quad x_0\le t\le x \\ f(x)(x_1-\sigma(t))\frac{x-x_0}{x_1-x_0} & for \quad x\le t\le x_1 \end{cases}$$

and

$$h(x)=g(x)-f(x)z_0-\frac{z_1-z_0}{x_1-x_0}f(x)(x-x_0) \quad for \quad x_0\le x\le x_1.$$

Consequently we obtain the following generalized Fredholm integral equation

$$\phi(x)=h(x)+\int_{x_0}^{x_1}K(x,t)\phi(t)\Delta t. \tag{2.33}$$

Example 26 Let $\mathscr{T}=\mathscr{N}_0^3$. Consider the following BVP

$$z^{\Delta^2}(x)+x^2z(x)=x, \quad 0<x<8,$$

$$z(0)=0, \quad z(8)=2.$$

Here

$$\sigma(t)=(\sqrt[3]{t}+1)^3, \quad t\in\mathscr{T}, \quad f(x)=x^2, \quad g(x)=x, \quad x\in[0,8],$$

$$x_0=0, \quad x_1=8, \quad z_0=0, \quad z_1=2.$$

From here we obtain

$$h(x)=x-2x^3.$$

Substituting this in (2.33) gives the following generalized Fredholm integral equation

$$\phi(x)=x-\frac{1}{4}x^3+\int_0^8K(x,t)\phi(t)\Delta t,$$

where

$$K(x,t)=\begin{cases} -\frac{1}{8}x^2(x-8)(\sqrt[3]{t}+1)^3 & for \quad 0\le t\le x \\ \frac{1}{8}x^3\left(8-(\sqrt[3]{t}+1)^3\right) & for \quad x\le t\le 8. \end{cases}$$

Example 27 Let $\mathscr{T} = 4^{\mathscr{N}_0} \cup \{0\}$. Consider the following BVP

$$z^{\Delta^2}(x) + \cosh_x(x, 1)z(x) = x, \quad 1 < x < 16,$$

$$z(1) = 0, \quad z(16) = -3.$$

Here

$$\sigma(x) = 4x, \quad x \in \mathscr{T}, \quad f(x) = \cosh_x(x, 1), \quad g(x) = x,$$

$$x_0 = 1, \quad x_1 = 16, \quad z_0 = 0, \quad z_1 = -3.$$

Hence, we find

$$h(x) = x + \frac{1}{15}\cosh_x(x, 1)(x - 1).$$

Substituting this in (2.33) gives the following generalized Fredholm integral equation

$$\phi(x) = x + \frac{1}{15}\cosh_x(x, 1)(x - 1) + \int_1^{16} K(x, t)\phi(t)\Delta t,$$

where

$$K(x, t) = \begin{cases} -\frac{1}{15}\cosh_x(x, 1)(16 - x)(1 - 4t) & \textit{for} \quad 1 \le x \le t \\ \frac{4}{15}(x - 1)(4 - t)\cosh_x(x, 1) & \textit{for} \quad t \le x \le 16. \end{cases}$$

Example 28 Let $\mathscr{T} = 3\mathscr{N}_0$. Consider the following BVP

$$z^{\Delta^2}(x) + e_x(x, 1)z(x) = x^2, \quad 0 < x < 6,$$

$$z(0) = 1, \quad z(6) = 1.$$

Here

$$\sigma(x) = x + 3, \quad x \in \mathscr{T}, \quad f(x) = e_x(x, 1), \quad g(x) = x^2,$$

$$x_0 = 0, \quad x_1 = 6, \quad z_0 = z_1 = 1.$$

Then

$$h(x) = x^2 - e_x(x, 1).$$

Substituting this in (2.33) we get

$$\phi(x) = x^2 - e_x(x, 1) + \int_0^6 K(x, t)\phi(t)\Delta t,$$

where

$$K(x,t) = \begin{cases} -\frac{1}{6}e_x(x,1)(x-6)(t+3) & \textit{for} \quad 0 \le t \le x \\ \frac{1}{6}e_x(x,1)x(3-t) & \textit{for} \quad x \le t \le 6. \end{cases}$$

Exercise 8 Convert the following BVPs to generalized Fredholm integral equations.

1.
$$\begin{cases} z^{\Delta^2}(x) - z(x) = x^2, & 0 < x < 2, \\ z(0) = z(2) = 0, & \mathscr{T} = \mathscr{Z}, \end{cases}$$

2.
$$\begin{cases} z^{\Delta^2}(x) + 2z(x) = x, & 1 < x < 5, \\ z(1) = 0, \quad z(5) = 1, & \mathscr{T} = \mathscr{N}, \end{cases}$$

3.
$$\begin{cases} z^{\Delta^2}(x) - (2x+1)z(x) = 1, & 0 < x < 9, \\ z(0) = 0, \quad z(9) = 1, & \mathscr{T} = \mathscr{N}_0^2. \end{cases}$$

Answer

1. $h(x) = x^2$,

$$K(x,t) = \begin{cases} \frac{1}{2}(x-2)(t+1) & \textit{for} \quad 0 \le x \le t \\ -\frac{1}{2}x(1-t) & \textit{for} \quad t \le x \le 2, \end{cases}$$

2. $h(x) = \dfrac{1}{2}(x+1)$,

$$K(x,t) = \begin{cases} -\frac{1}{2}t(x-5) & \textit{for} \quad 1 \le x \le t \\ \frac{1}{2}(x-1)(4-t) & \textit{for} \quad t \le x \le 5, \end{cases}$$

3. $h(x) = 1 + \dfrac{1}{9}x(2x+1)$,

$$K(x,t) = \begin{cases} \frac{1}{9}(2x+1)(x-9)(\sqrt{t}+1)^2 & \textit{for} \quad 0 \le x \le t \\ -\frac{1}{9}x(2x+1)\left(9-(\sqrt{t}+1)^2\right) & \textit{for} \quad x \le t \le 9. \end{cases}$$

We next consider the following boundary value problem for the Eq. (2.29) with boundary conditions

$$z(x_0) = z_2, \quad z^{\Delta}(x_1) = z_3. \tag{2.34}$$

Again we set $z^{\Delta^2}(x) = \phi(x)$. Integrating both sides of (2.31) from x_0 to x we get

$$z^{\Delta}(x) = z^{\Delta}(x_0) + \int_{x_0}^{x} \phi(t)\Delta t. \tag{2.35}$$

We put $x = x_1$ in the last expression and we find

$$z^{\Delta}(x_1) = z^{\Delta}(x_0) + \int_{x_0}^{x_1} \phi(t)\Delta t$$

or

$$z_3 = z^{\Delta}(x_0) + \int_{x_0}^{x_1} \phi(t)\Delta t,$$

whereupon

$$z^{\Delta}(x_0) = z_3 - \int_{x_0}^{x_1} \phi(t)\Delta t.$$

We substitute the last expression in (2.35) and we obtain

$$z^{\Delta}(x) = z_3 - \int_{x_0}^{x_1} \phi(t)\Delta t + \int_{x_0}^{x} \phi(t)\Delta t,$$

which we integrate from x_0 to x_1 and we get

$$\int_{x_0}^{x} z^{\Delta}(t)\Delta t = \int_{x_0}^{x} \left(z_3 - \int_{x_0}^{x_1} \phi(t)\Delta t\right) \Delta y + \int_{x_0}^{x} \int_{x_0}^{x_2} \phi(t)\Delta t \Delta x_2,$$

or

$$z(x) - z(x_0) = \left(z_3 - \int_{x_0}^{x_1} \phi(t)\Delta t\right)(x - x_0) + \int_{x_0}^{x} \int_{x_0}^{x_2} \phi(t)\Delta t \Delta x_2,$$

or

$$z(x) = z_2 + \left(z_3 - \int_{x_0}^{x_1} \phi(t)\Delta t\right)(x - x_0) + \int_{x_0}^{x} \int_{x_0}^{x_2} \phi(t)\Delta t \Delta x_2.$$

Applying Theorem 1 we get

$$z(x) = z_2 + \left(z_3 - \int_{x_0}^{x_1} \phi(t)\Delta t\right)(x - x_0) + \int_{x_0}^{x} (x - \sigma(t))\phi(t)\Delta t. \tag{2.36}$$

We substitute (2.31) and (2.36) in (2.29) and we find

$$\phi(x) + z_2 f(x) + f(x)\left(z_3 - \int_{x_0}^{x_1} \phi(t)\Delta t\right)(x - x_0)$$

$$+f(x)\int_{x_0}^{x}(x - \sigma(t))\Delta t = g(x)$$

or

$$\phi(x) = g(x) - z_2 f(x) - z_3(x - x_0)f(x) + f(x)(x - x_0)\int_{x_0}^{x_1} \phi(t)\Delta t$$

$$-f(x)\int_{x_0}^{x}(x - \sigma(t))\phi(t)\Delta t$$

$$= g(x) - z_2 f(x) - z_3(x - x_0)f(x) + f(x)(x - x_0)\int_{x_0}^{x} \phi(t)\Delta t$$

$$+f(x)(x - x_0)\int_{x}^{x_1} \phi(t)\Delta t - f(x)\int_{x_0}^{x}(x - \sigma(t))\phi(t)\Delta t$$

$$= g(x) - z_2 f(x) - z_3(x - x_0)f(x)$$

$$+\int_{x_0}^{x}\left(f(x)(x - x_0) - f(x)(x - \sigma(t))\right)\phi(t)\Delta t$$

$$+f(x)(x - x_0)\int_{x}^{x_1} \phi(t)\Delta t$$

$$= g(x) - z_2 f(x) - z_3(x - x_0)f(x) + \int_{x_0}^{x} f(x)(-x_0 + \sigma(t))\phi(t)\Delta t$$

$$+\int_{x}^{x_1} f(x)(x - x_0)\phi(t)\Delta t.$$

Let

$$h_1(x) = g(x) - z_2 f(x) - z_3(x - x_0)f(x),$$

$$K_1(x,t) = \begin{cases} f(x)(-x_0 + \sigma(t)) & \textit{for} \quad x_0 \le t \le x \\ f(x)(x - x_0) & \textit{for} \quad x \le t \le x_1. \end{cases}$$

Then we get the following generalized Fredholm integral equation

$$\phi(x) = h_1(x) + \int_{x_0}^{x_1} K_1(x,t)\phi(t)\Delta t. \tag{2.37}$$

Example 29 Let $\mathcal{T} = \mathcal{N}_0^4$. Consider the following BVP

$$z^{\Delta^2}(x) + x^2 z(x) = \cos_x(x, 1) + \sinh_x(x, 2), \quad 0 < x < 81,$$

$$z(0) = 0, \quad z^{\Delta}(81) = 1.$$

Here

$$\sigma(x) = (\sqrt[4]{x} + 1)^4, \quad x \in \mathcal{T},$$

$$f(x) = x^2, \quad g(x) = \cos_x(x, 1) + \sinh_x(x, 2),$$

$$x_0 = 0, \quad x_1 = 81, \quad z_2 = 0, \quad z_3 = 1.$$

Then

$$h_1(x) = -x^3 + \cos_x(x, 1) + \sinh_x(x, 2).$$

Substituting this in (2.37) gives the following generalized Fredholm integral equation

$$\phi(x) = -x^3 + \cos_x(x, 1) + \sinh_x(x, 2) + \int_0^{81} K_1(x, t)\phi(t)\Delta t,$$

where

$$K_1(x, t) = \begin{cases} x^2(\sqrt[4]{t} + 1)^4 & for \quad 0 \le t \le x \\ x^3 \quad for \quad x \le t \le 81. \end{cases}$$

Example 30 Let $\mathcal{T} = 3^{\mathcal{N}_0} \cup \{0\}$. Consider the BVP

$$z^{\Delta^2}(x) + (2x - 3)z^{\Delta}(x) = x^2 - 1, \quad 0 < x < 27,$$

$$z(0) = 1, \quad z^{\Delta}(1) = -3.$$

Here

$$\sigma(x) = 3x, \quad x \in \mathcal{T},$$
$$f(x) = 2x - 3, \quad g(x) = x^2 - 1,$$
$$x_0 = 0, \quad x_1 = 1, \quad z_2 = 1, \quad z_3 = -3.$$

Then

$$\begin{aligned} h_1(x) &= g(x) - z_2 f(x) - z_3(x - x_0)f(x) \\ &= x^2 - 1 - (2x - 3) - (-3)x(2x - 3) \\ &= x^2 - 1 - 2x + 3 + 6x^2 - 9x \\ &= 7x^2 - 11x + 2. \end{aligned}$$

Substituting this is (2.37) gives the following generalized Fredholm integral equation

$$\phi(x) = 7x^2 - 11x + 2 + \int_0^{27} K_1(x,t)\phi(t)\Delta t,$$

where

$$K_1(x,t) = \begin{cases} 3(2x-3)t & \textit{for} \quad 0 \le t \le x \\ \\ 2x^2 - 3x & \textit{for} \quad x \le t \le 27. \end{cases}$$

Example 31 Let $\mathscr{T} = 2\mathscr{Z}$. Consider the BVP

$$z^{\Delta^2}(x) - xz^{\Delta}(x) = 1 + x, \quad -2 < x < 6,$$

$$z(-2) = 0, \quad z^{\Delta}(6) = 1.$$

Here

$$\sigma(x) = x + 2, \quad x \in \mathscr{T},$$

$$f(x) = -x, \quad g(x) = 1 + x,$$

$$x_0 = -2, \quad x_1 = 6, \quad z_2 = 0, \quad z_3 = 1.$$

Then

$$\begin{aligned} h_1(x) &= g(x) - z_2 f(x) - z_3(x - x_0) f(x) \\ &= 1 + x - (x+2)(-x) \\ &= 1 + x + x^2 + 2x \\ &= x^2 + 3x + 1. \end{aligned}$$

Substituting this in (2.37) gives the following generalized Fredholm integral equation

$$\phi(x) = x^2 + 3x + 1 + \int_{-2}^{6} K_1(x,t)\phi(t)\Delta t,$$

where

$$K_1(x,t) = \begin{cases} -x(t+4) & \textit{for} \quad -2 \le t \le x \\ \\ -x^2 - 2x & \textit{for} \quad x \le t \le 6. \end{cases}$$

Exercise 9 Convert the following BVPs to generalized Fredholm integral equations.

1.
$$\begin{cases} z^{\Delta^2}(x) + \sin x z(x) = \cos x, \quad 0 < x < 4, \\ z(0) = 0, \quad z^{\Delta}(4) = 3, \quad \mathcal{T} = \mathcal{R}, \end{cases}$$

2.
$$\begin{cases} z^{\Delta^2}(x) - 3z(x) = x^2 + 1, \quad -1 < x < 4, \\ z(-1) = 1, \quad z(4) = 0, \quad \mathcal{T} = \mathcal{R}, \end{cases}$$

3.
$$\begin{cases} z^{\Delta^2}(x) + \frac{1}{1+x^2} z(x) = 1, \quad 1 < x < 27, \\ z(1) = 0, \quad z^{\Delta}(27) = 1, \quad \mathcal{T} = 3^{\mathcal{N}_0}. \end{cases}$$

Answer

1. $h_1(x) = \cos x - 3x \sin x,$

$$K_1(x,t) = \begin{cases} (t+1)\sin x & for \quad 0 \le t \le x, \\ x \sin x & for \quad x \le t \le 4, \end{cases}$$

2. $h_1(x) = x^2 + 4,$

$$K_1(x,t) = \begin{cases} -3(t+2) & for \quad -1 \le t \le x \\ -3(x+1) & for \quad x \le t \le 4, \end{cases}$$

3. $h_1(x) = \dfrac{x^2 - x + 2}{x^2 + 1},$

$$K_1(x,t) = \begin{cases} \frac{1}{1+x^2}(-1+3t) & for \quad 1 \le t \le x \\ \frac{x-1}{1+x^2} & for \quad x \le t \le 27. \end{cases}$$

2.5 Converting Generalized Fredholm Integral Equation to BVP

In the previous sections, we have represented a technique to convert Volterra integral equations to equivalent initial value problems. In a similar manner, we will represent a technique that converts Fredholm integral equations to equivalent boundary value problems.

We first consider the generalized Fredholm integral equation

$$\phi(x) = g(x) + \int_{x_0}^{x_1} K(x,t)\phi(t)\Delta t, \tag{2.38}$$

where

$$K(x,t) = \begin{cases} f(x)(x_1 - x)(\sigma(t) - x_0) & for \quad x_0 \le t \le x \\ f(x)(x_1 - \sigma(t))(x - x_0) & for \quad x \le t \le x_1. \end{cases}$$

The Eq. (2.38) we can rewrite in the following form

$$\begin{aligned} \phi(x) &= g(x) + \textstyle\int_{x_0}^{x} f(x)(x_1 - x)(\sigma(t) - x_0)\phi(t)\Delta t \\ &+ \textstyle\int_{x}^{x_1} f(x)(x_1 - \sigma(t))(x - x_0)\phi(t)\Delta t. \end{aligned} \tag{2.39}$$

For simplicity reason, we may assume that $f(x) = a$, where a is a constant. Then (2.39) takes the form

$$\phi(x) = g(x) + a(x_1 - x)\int_{x_0}^{x} (\sigma(t) - x_0)\phi(t)\Delta t + a(x - x_0)\int_{x}^{x_1} (x_1 - \sigma(t))\phi(t)\Delta t. \tag{2.40}$$

We differentiate (2.40) with respect to x and we get

$$\begin{aligned} \phi^{\Delta}(x) &= g^{\Delta}(x) - a\int_{x_0}^{x} (\sigma(t) - x_0)\phi(t)\Delta t \\ &\quad + a(x_1 - \sigma(x))(\sigma(x) - x_0)\phi(x) + a\int_{x}^{x_1} (x_1 - \sigma(t))\phi(t)\Delta t \\ &\quad - a(\sigma(x) - x_0)(x_1 - \sigma(x))\phi(x) \\ &= g^{\Delta}(x) - a\int_{x_0}^{x} (\sigma(t) - x_0)\phi(t)\Delta t + a\int_{x}^{x_1} (x_1 - \sigma(t))\phi(t)\Delta t. \end{aligned}$$

Again we differentiate with respect to x and we find

$$\begin{aligned} \phi^{\Delta^2}(x) &= g^{\Delta^2}(x) - a(\sigma(x) - x_0)\phi(x) - a(x_1 - \sigma(x))\phi(x) \\ &= g^{\Delta^2}(x) - a(x_1 - x_0)\phi(x), \end{aligned}$$

i.e.,

$$\phi^{\Delta^2}(x) + a(x_1 - x_0)\phi(x) = g^{\Delta^2}(x). \tag{2.41}$$

By substituting $x = x_0$ and $x = x_1$ in (2.40) we find that

$$\phi(x_0) = g(x_0) \quad and \quad \phi(x_1) = g(x_1). \tag{2.42}$$

Combining (2.41) and (2.42) gives the following boundary value problem

$$\begin{cases} \phi^{\Delta^2}(x) + a(x_1 - x_0)\phi(x) = g^{\Delta^2}(x), \quad x_0 < x < x_1, \\ \phi(x_0) = g(x_0), \quad \phi(x_1) = g(x_1). \end{cases} \tag{2.43}$$

Example 32 Let $\mathscr{T} = 2^{\mathscr{N}_0} \cup \{0\}$. We consider the following generalized Fredholm integral equation

$$\phi(x) = 2x + 3 + \int_0^4 K(x,t)\phi(t)\Delta t,$$

where

$$K(x,t) = \begin{cases} 2t(4-x) \quad for \quad 0 \le t \le x \\ 2x(2-t) \quad for \quad x \le t \le 4. \end{cases}$$

Here $\sigma(x) = 2x,\ x \in \mathscr{T},\ a = 1,\ g(x) = 2x + 3.$ Then $g^{\Delta}(x) = 2,\ g^{\Delta^2}(x) = 0.$ Hence, using (2.43), we get the following BVP

$$\begin{cases} \phi^{\Delta^2}(x) + 4\phi(x) = 0, \quad 0 < x < 4, \\ \phi(0) = 3, \quad \phi(4) = 11. \end{cases}$$

Example 33 Let $\mathscr{T} = \mathscr{N}_0^3 \cup \{0\}$. Consider the following generalized Fredholm integral equation

$$\phi(x) = x^2 + 2x + \int_0^{27} K(x,t)\phi(t)\Delta t,$$

where

$$K(x,t) = \begin{cases} 2(27-x)(\sqrt[3]{t}+1)^3 \quad for \quad 0 \le t \le x \\ 2x(27-(\sqrt[3]{t}+1)^3) \quad for \quad x \le t \le 27. \end{cases}$$

Here

$$\sigma(x) = (\sqrt[3]{x}+1)^3, \quad x \in \mathscr{T},$$

$$a = 2, \quad g(x) = x^2 + 2x, \quad x_0 = 0, \quad x_1 = 27.$$

Then

$$\begin{aligned} g^{\Delta}(x) &= \sigma(x) + x + 2 \\ &= (\sqrt[3]{x}+1)^3 + x + 2 \\ &= x + 3\sqrt[3]{x^2} + 3\sqrt[3]{x} + 1 + x + 2 \\ &= 2x + 3\sqrt[3]{x^2} + 3\sqrt[3]{x} + 3, \end{aligned}$$

$$\begin{aligned} g^{\Delta^2}(x) &= 2 + 3\frac{\sqrt[3]{\sigma^2(x)} - \sqrt[3]{x^2}}{\sigma(x) - x} + 3\frac{\sqrt[3]{\sigma(x)} - \sqrt[3]{x}}{\sigma(x) - x} \\ &= 2 + 3\frac{(\sqrt[3]{\sigma(x)} - \sqrt[3]{x})(\sqrt[3]{\sigma(x)} + \sqrt[3]{x})}{(\sqrt[3]{\sigma(x)} - \sqrt[3]{x})(\sqrt[3]{\sigma^2(x)} + \sqrt[3]{x}\sqrt[3]{\sigma(x)} + \sqrt[3]{x^2})} \\ &\quad + 3\frac{\sqrt[3]{\sigma(x)} - \sqrt[3]{x}}{(\sqrt[3]{\sigma(x)} - \sqrt[3]{x})(\sqrt[3]{\sigma^2(x)} + \sqrt[3]{x}\sqrt[3]{\sigma(x)} + \sqrt[3]{x^2})} \\ &= 2 + 3\frac{\sqrt[3]{\sigma(x)} + \sqrt[3]{x} + 1}{\sqrt[3]{\sigma^2(x)} + \sqrt[3]{x}\sqrt[3]{\sigma(x)} + \sqrt[3]{x^2}} \\ &= 2 + 3\frac{\sqrt[3]{x} + 1 + \sqrt[3]{x} + 1}{(\sqrt[3]{x}+1)^2 + \sqrt[3]{x}(\sqrt[3]{x}+1) + \sqrt[3]{x^2}} \\ &= 2 + 6\frac{\sqrt[3]{x} + 1}{\sqrt[3]{x^2} + 2\sqrt[3]{x} + 1 + \sqrt[3]{x^2} + \sqrt[3]{x} + \sqrt[3]{x^2}} \\ &= 2 + 6\frac{\sqrt[3]{x} + 1}{3\sqrt[3]{x^2} + 3\sqrt[3]{x} + 1}. \end{aligned}$$

Also,

$$g(0) = 0, \quad g(27) = 783.$$

Hence, using (2.43), we get the following boundary value problem

$$\begin{cases} \phi^{\Delta^2}(x) + 54\phi(x) = 2 + 6\frac{\sqrt[3]{x}+1}{3\sqrt[3]{x^2}+3\sqrt[3]{x}+1}, & 0 < x < 27, \\ \phi(0) = 0, \quad \phi(27) = 783. \end{cases}$$

Example 34 Let $\mathscr{T} = 3^{\mathscr{N}_0}$. Consider the following generalized Fredholm integral equation

$$\phi(x) = x^3 - 2x^2 + \int_1^9 K(x,t)\phi(t)\Delta t,$$

where

$$K(x,t) = \begin{cases} (9-x)(3t-1) & \textit{for} \quad 1 \le t \le x \\ (9-3t)(x-1) & \textit{for} \quad x \le t \le 9. \end{cases}$$

Here

$$a = 1, \quad \sigma(x) = 3x, \quad x \in \mathscr{T},$$

$$x_0 = 1, \quad x_1 = 9, \quad g(x) = x^3 - 2x^2.$$

Then

$$g^{\Delta}(x) = \sigma^2(x) + x\sigma(x) + x^2 - 2(x + \sigma(x))$$

$$= 9x^2 + 3x^2 + x^2 - 8x$$

$$= 13x^2 - 8x,$$

$$g^{\Delta^2}(x) = 13(\sigma(x) + x) - 8$$

$$= 52x - 8,$$

$$g(1) = -1, \quad g(9) = 567.$$

Hence, using (2.43), we get the following boundary value problem

$$\begin{cases} \phi^{\Delta^2}(x) + 8\phi(x) = 52x - 8, & 1 < x < 9, \\ \phi(1) = -1, \quad \phi(9) = 567. \end{cases}$$

Exercise 10 Convert the following generalized Fredholm integral equations to BVPs.

1.

$$\phi(x) = x - 2 + \int_0^{10} K(x,t)\phi(t)\Delta t,$$

where

$$K(x,t) = \begin{cases} (10-x)(t+1) & \textit{for} \quad 0 \le t \le x \\ (9-t)x & \textit{for} \quad x \le t \le 10, \end{cases}$$

$\mathscr{T} = \mathscr{Z}$,

2.

$$\phi(x) = x^2 + \int_{-1}^{2} K(x,t)\phi(t)\Delta t,$$

where

$$K(x,t) = \begin{cases} 2(2-x)(t+2) & \textit{for} \quad -1 \le t \le x \\ 2(1-t)(x+1) & \textit{for} \quad x \le t \le 2, \end{cases}$$

$\mathscr{T} = \mathscr{Z}$,

3.

$$\phi(x) = x^3 + \int_{1}^{4} K(x,t)\phi(t)\Delta t,$$

where

$$K(x,t) = \begin{cases} (4-x)(2t-1) & \textit{for} \quad 1 \le t \le x \\ 2(2-t)(x-1) & \textit{for} \quad x \le t \le 4, \end{cases}$$

$\mathscr{T} = 2^{\mathscr{N}_0}$.

Answer

1.

$$\begin{cases} \phi^{\Delta^2}(x) + 10\phi(x) = 0, \quad 0 < x < 10, \\ \phi(0) = -2, \quad \phi(10) = 8, \end{cases}$$

2.

$$\begin{cases} \phi^{\Delta^2}(x) + 6\phi(x) = 2, \quad -1 < x < 2, \\ \phi(-1) = 1, \quad \phi(2) = 4, \end{cases}$$

3.

$$\begin{cases} \phi^{\Delta^2}(x) + 3\phi(x) = 21x, \quad 1 < x < 4, \\ \phi(1) = 1, \quad \phi(4) = 64. \end{cases}$$

Next we consider the following generalized Fredholm integral equation

$$\phi(x) = g(x) + \int_{x_0}^{x_1} K(x,t)\phi(t)\Delta t, \tag{2.44}$$

where

$$K(x,t) = \begin{cases} f(x)(-x_0+\sigma(t)) & for \quad x_0 \le t \le x \\ f(x)(x-x_0) & for \quad x \le t \le x_1. \end{cases}$$

The Eq. (2.44) we can rewrite in the following form

$$\begin{aligned}\phi(x) &= g(x) + \int_{x_0}^{x} K(x,t)\phi(t)\Delta t + \int_{x}^{x_1} K(x,t)\phi(t)\Delta t \\ &= g(x) + \int_{x_0}^{x} f(x)(-x_0+\sigma(t))\phi(t)\Delta t \\ &+ \int_{x}^{x_1} f(x)(x-x_0)\phi(t)\Delta t.\end{aligned}$$

For simplicity reasons, we may assume that $f(x) = b$, where b is a real constant. Then the Eq. (2.44) takes the form

$$\phi(x) = g(x) + b\int_{x_0}^{x} (-x_0+\sigma(t))\phi(t)\Delta t + b\int_{x}^{x_1} (x-x_0)\phi(t)\Delta t. \tag{2.45}$$

We differentiate the last equation with respect to x and we find

$$\begin{aligned}\phi^{\Delta}(x) &= g^{\Delta}(x) + b(-x_0+\sigma(x))\phi(x) + b\int_{x}^{x_1} \phi(t)\Delta t \\ &- b(\sigma(x)-x_0)\phi(x) \\ &= g^{\Delta}(x) + b\int_{x}^{x_1} \phi(t)\Delta t,\end{aligned}$$

i.e.,

$$\phi^{\Delta}(x) = g^{\Delta}(x) + b\int_{x}^{x_1} \phi(t)\Delta t. \tag{2.46}$$

We differentiate with respect to x the Eq. (2.46) and we find

$$\phi^{\Delta^2}(x) = g^{\Delta^2}(x) - b\phi(x)$$

or

$$\phi^{\Delta^2}(x) + b\phi(x) = g^{\Delta^2}(x). \tag{2.47}$$

We substitute $x = x_0$ and $x = x_1$ in (2.45) and (2.46), respectively. We find

$$\phi(x_0) = g(x_0), \quad \phi^{\Delta}(x_1) = g^{\Delta}(x_1). \tag{2.48}$$

Combining (2.47) and (2.48) we get the following boundary value problem

$$\begin{cases} \phi^{\Delta^2}(x) + b\phi(x) = g^{\Delta^2}(x), \quad x_0 < x < x_1, \\ \phi(x_0) = g(x_0), \quad \phi^{\Delta}(x_1) = g^{\Delta}(x_1). \end{cases} \tag{2.49}$$

Example 35 Let $\mathcal{T} = \mathcal{Z}$. Consider the generalized Fredholm integral equation

$$\phi(x) = x^4 + 3x^2 + 3x + \int_0^4 K(x, t)\phi(t)\Delta t,$$

where

$$K(x, t) = \begin{cases} t + 1 & for \quad 0 \le x \le t \\ x & for \quad x \le t \le 4. \end{cases}$$

Here

$$\sigma(x) = x + 1, \quad x \in \mathcal{T}, \quad b = 1, \quad x_0 = 0, \quad x_1 = 4,$$
$$g(x) = x^4 + 3x^2 + 3x.$$

Then

$$\begin{aligned} g^{\Delta}(x) &= \sigma^3(x) + x\sigma^2(x) + x^2\sigma(x) + x^3 + 3(\sigma(x) + x) + 3 \\ &= (x+1)^3 + x(x+1)^2 + x^2(x+1) + x^3 + 3(x+1+x) + 3 \\ &= x^3 + 3x^2 + 3x + 1 + x^3 + 2x^2 + x + x^3 + x^2 \\ &\quad + x^3 + 6x + 6 \\ &= 4x^3 + 6x^2 + 10x + 7, \\ g^{\Delta^2}(x) &= 4(\sigma^2(x) + x\sigma(x) + x^2) + 6(\sigma(x) + x) + 10 \\ &= 4(x+1)^2 + 4x(x+1) + 4x^2 + 6(x+1+x) + 10 \\ &= 4x^2 + 8x + 4 + 4x^2 + 4x + 4x^2 + 12x + 16 \\ &= 12x^2 + 24x + 20, \\ g(0) &= 0, \quad g^{\Delta}(4) = 399. \end{aligned}$$

Hence and (2.49) we get the following BVP

$$\begin{cases} \phi^{\Delta^2}(x) + \phi(x) = 12x^2 + 24x + 20, & 0 < x < 4, \\ \phi(0) = 0, \quad \phi^{\Delta}(4) = 399. \end{cases}$$

Example 36 Let $\mathscr{T} = 2^{\mathscr{N}_0} \cup \{0\}$. Consider the following generalized Fredholm integral equation

$$\phi(x) = x^3 - 7x^2 + 2x + \int_0^4 K(x, t)\phi(t)\Delta t,$$

where

$$K(x, t) = \begin{cases} 4t & for \quad 0 \le x \le t \\ 2x & for \quad x \le t \le 4. \end{cases}$$

Here

$$\sigma(x) = 2x, \quad x \in \mathscr{T}, \quad x_0 = 0, \quad x_1 = 4, \quad b = 2,$$

$$g(x) = x^3 - 7x^2 + 2x.$$

Then

$$g^{\Delta}(x) = \sigma^2(x) + x\sigma(x) + x^2 - 7(\sigma(x) + x) + 2$$

$$= 4x^2 + 2x^2 + x^2 - 21x + 2$$

$$= 7x^2 - 21x + 2,$$

$$g^{\Delta^2}(x) = 7(\sigma(x) + x) - 21$$

$$= 21x - 21,$$

$$g(0) = 0, \quad g^{\Delta}(4) = 30.$$

Hence, using (2.49), we get the following BVP

$$\begin{cases} \phi^{\Delta^2}(x) + 2\phi(x) = 21x - 21, & 0 < x < 4, \\ \phi(0) = 0, \quad \phi^{\Delta}(4) = 30. \end{cases}$$

Example 37 Let $\mathscr{T} = 4\mathscr{Z}$. Consider the following generalized Fredholm integral equation

$$\phi(x) = 2x^3 - x^2 + 4x + 2 + \int_{-4}^{8} K(x,t)\phi(t)\Delta t,$$

where

$$K(x,t) = \begin{cases} -t-8 & \mathit{for} \quad -4 \le t \le x \\ -x-4 & \mathit{for} \quad x \le t \le 8. \end{cases}$$

Here

$$\sigma(x) = x+4, \quad x \in \mathscr{T}, \quad x_0 = -4, \quad x_1 = 8, \quad b = -1,$$

$$g(x) = 2x^3 - x^2 + 4x + 2.$$

Then

$$\begin{aligned} g^{\Delta}(x) &= 2(\sigma^2(x) + x\sigma(x) + x^2) - (\sigma(x) + x) + 4 \\ &= 2((x+4)^2 + x(x+4) + x^2) - (x+4+x) + 4 \\ &= 2(x^2 + 8x + 16 + x^2 + 4x + x^2) - 2x \\ &= 2(3x^2 + 12x + 16) - 2x \\ &= 6x^2 + 24x + 32 - 2x \\ &= 6x^2 + 22x + 32, \\ g^{\Delta^2}(x) &= 6(\sigma(x) + x) + 22 \\ &= 6(x+4+x) + 22 \\ &= 6(2x+4) + 22 \\ &= 12x + 24 + 22 \\ &= 12x + 46, \end{aligned}$$

$$g(-4) = -158, \quad g^{\Delta}(8) = 592.$$

Hence, using (2.49), we get the following BVP

$$\begin{cases} \phi^{\Delta^2}(x) - \phi(x) = 12x + 46, \quad -4 < x < 8, \\ \phi(-4) = -158, \quad \phi^{\Delta}(8) = 592. \end{cases}$$

Exercise 11 Convert the following generalized Fredholm integral equations to BVPs.

1. $\phi(x) = x + 10 + \int_{-2}^{4} K(x,t)\phi(t)\Delta t$, $\mathscr{T} = \mathscr{Z}$, where

$$K(x,t) = \begin{cases} t + 3 \quad for \quad -2 \le t \le x \\ x + 2 \quad for \quad x \le t \le 4, \end{cases}$$

2. $\phi(x) = x^3 + 1 + \int_{0}^{6} K(x,t)\phi(t)\Delta t$, $\mathscr{T} = 2\mathscr{Z}$, where

$$K(x,t) = \begin{cases} -2(t+2) \quad for \quad 0 \le t \le x \\ -2x \quad for \quad x \le t \le 6, \end{cases}$$

3. $\phi(x) = x^2 + 3x + \int_{0}^{4} K(x,t)\phi(t)\Delta t$, $\mathscr{T} = 2^{\mathscr{N}_0} \cup \{0\}$, where

$$K(x,t) = \begin{cases} 2t \quad for \quad 0 \le t \le x \\ x \quad for \quad x \le t \le 4. \end{cases}$$

Answer

1.
$$\begin{cases} \phi^{\Delta^2}(x) + \phi(x) = 0, \quad -2 < x < 4, \\ \phi(-2) = 8, \quad \phi^{\Delta}(4) = 1, \end{cases}$$

2.
$$\begin{cases} \phi^{\Delta^2}(x) - 2\phi(x) = 6x + 12, \quad 0 < x < 6, \\ \phi(0) = 1, \quad \phi^{\Delta}(6) = 148, \end{cases}$$

3.
$$\begin{cases} \phi^{\Delta^2}(x) + \phi(x) = 3, \quad 0 < x < 4, \\ \phi(0) = 0, \quad \phi^{\Delta}(4) = 15. \end{cases}$$

2.6 Solutions of Generalized Integral Equations and Generalized Integro-Differential Equations

Definition 1 A solution of a generalized integral equation or generalized integro-differential equation is a function $\phi(x)$ that satisfies the given equation. In other words, the solution $\phi(x)$ must satisfy both sides of the examined equation.

Definition 2 The solution is called exact if it can be represented in a closed form, such as a polynomial, exponential function, trigonometric function or the combination of two or more of these elementary functions.

The following examples will be examined to explain the meaning of a solution.

Example 38 Some examples of exact solutions are as follows:

$$\phi(x) = x^2 + x + e^x + \cos x,$$

$$\phi(x) = x - 2e_x(x, 3),$$

$$\phi(x) = 1 + \cosh_x(x, 1) + \cos x.$$

Example 39 Let $\mathscr{T} = \mathscr{Z}$. Consider the equation

$$\phi(x) = x^2 + 4x + 4 - \int_{-1}^{x} \phi(t)\Delta t.$$

We will prove that $\phi(x) = 2x + 3$ is its solution. Indeed,

$$\begin{aligned}
\int_{-1}^{x} \phi(t)\Delta t &= \int_{-1}^{x} (2t + 3)\Delta t \\
&= 2\int_{-1}^{x} t\Delta t + 3\int_{-1}^{x} \Delta t \\
&= 2\int_{-1}^{x} \left(\frac{1}{2}(t^2)^{\Delta} - \frac{1}{2}\right) \Delta t + 3(x + 1) \\
&= \int_{-1}^{x} (t^2)^{\Delta}\Delta t - \int_{-1}^{x} \Delta t + 3x + 3 \\
&= x^2 - 1 - (x + 1) + 3x + 3 \\
&= x^2 - 1 - x - 1 + 3x + 3 \\
&= x^2 + 2x + 1.
\end{aligned}$$

Hence,

$$x^2 + 4x + 4 - \int_{-1}^{x} \phi(t)\Delta t = x^2 + 4x + 4 - (x^2 + 2x + 1)$$

$$= 2x + 3$$

$$= \phi(x).$$

Example 40 Let $\mathcal{T} = 2\mathcal{Z}$. Consider the equation

$$\phi(x) = \frac{1}{3}x^4 - \frac{1}{2}x^3 - \frac{1}{3}x^2 + x + 1 - x\int_0^x t\phi(t)\Delta t.$$

We will prove that $\phi(x) = x + 1$ is its solution. Indeed, we have that $\sigma(x) = x + 2$ and

$$\begin{aligned}\int_0^x t\phi(t)\Delta t &= \int_0^x t(t+1)\Delta t\\ &= \int_0^x (t^2 + t)\Delta t\\ &= \int_0^x \left(\frac{1}{3}(t^3)^\Delta - (t^2)^\Delta + \frac{2}{3} + \frac{1}{2}(t^2)^\Delta - 1\right)\Delta t\\ &= \int_0^x \left(\frac{1}{3}(t^3)^\Delta - \frac{1}{2}(t^2)^\Delta - \frac{1}{3}\right)\Delta t\\ &= \frac{1}{3}\int_0^x (t^3)^\Delta \Delta t - \frac{1}{2}\int_0^x (t^2)^\Delta \Delta t - \frac{1}{3}\int_0^x \Delta t\\ &= \frac{1}{3}x^3 - \frac{1}{2}x^2 - \frac{1}{3}x.\end{aligned}$$

Hence,

$$\begin{aligned}&\frac{1}{3}x^4 - \frac{1}{2}x^3 - \frac{1}{3}x^2 + x + 1 - x\int_0^x t\phi(t)\Delta t\\ &=\frac{1}{3}x^4 - \frac{1}{2}x^3 - \frac{1}{3}x^2 + x + 1 - x\left(\frac{1}{3}x^3 - \frac{1}{2}x^2 - \frac{1}{3}x\right)\\ &=\frac{1}{3}x^4 - \frac{1}{2}x^3 - \frac{1}{3}x^2 + x + 1 - \frac{1}{3}x^4 + \frac{1}{2}x^3 + \frac{1}{3}x^2\\ &=x + 1\\ &=\phi(x).\end{aligned}$$

Example 41 Let $\mathscr{T} = 2^{\mathscr{N}_0}$. Consider the integral equation

$$\phi(x) = -\frac{1}{3}x^2 + x + \frac{4}{3} + (x-1)e_x(x, 1) + \int_1^x \phi(t)\Delta t.$$

We will prove that $\phi(x) = e_x(x, 1) + x$ is its solution. Really, we have that $\sigma(x) = 2x$ and

$$\begin{aligned}
\int_1^x \phi(t)\Delta t &= \int_1^x (te_t(t, 1) + t)\Delta t \\
&= \int_1^x te_t(t, 1)\Delta t + \int_1^x t\Delta t \\
&= e_t(t, 1)\Big|_{t=1}^{t=x} + \frac{1}{3}\int_1^x (t^2)^\Delta \Delta t \\
&= e_x(x, 1) - e_1(1, 1) + \frac{1}{3}t^2\Big|_{t=1}^{t=x} \\
&= e_x(x, 1) + \frac{1}{3}x^2 - \frac{4}{3}.
\end{aligned}$$

Hence,

$$\begin{aligned}
&-\frac{1}{3}x^2 + x + \frac{4}{3} + (x-1)e_x(x, 1) + \int_1^x \phi(t)\Delta t \\
&= -\frac{1}{3}x^2 + x + \frac{4}{3} + (x-1)e_x(x, 1) + e_x(x, 1) - e_1(1, 1) + \frac{1}{3}x^2 - \frac{1}{3} \\
&= xe_x(x, 1) + x \\
&= \phi(x).
\end{aligned}$$

Exercise 12 Show that the given function is a solution of the corresponding generalized Volterra integral equation.

1. $\phi(x) = x^2, \quad \mathscr{T} = 2\mathscr{Z}$,

$$\phi(x) = -\frac{1}{5}x^6 + x^5 - \frac{4}{3}x^4 + x^2 + \frac{8}{15}x + x\int_0^x t^2\phi(t)\Delta t,$$

2. $\phi(x) = x + 2, \quad \mathscr{T} = \mathscr{Z}$,

$$\phi(x) = -\frac{1}{3}x^4 - \frac{1}{6}x^3 + \frac{4}{3}x^2 + \frac{1}{6}x + 2 + (x-1)\int_0^x t\phi(t)\Delta t,$$

3. $\phi(x) = 2\sqrt{x} + 2x + 1, \quad \mathcal{T} = \mathcal{N}_0^2,$

$$\phi(x) = -x^3 + 2x + 2\sqrt{x} + 1 + x\int_1^x \phi(t)\Delta t.$$

Example 42 Let $\mathcal{T} = \mathcal{Z}$. Consider the equation

$$\phi(x) = x^2 - 4x + x\int_0^4 \phi(t)\Delta t.$$

We will prove that $\phi(x) = x^2 - 2x$ is its solution. Indeed, we have that $\sigma(x) = x + 1$ and

$$x^2 - 4x + x\int_0^4 \phi(t)\Delta t = x^2 - 4x + x\int_0^4 (t^2 - 2t)\Delta t$$

$$= x^2 - 4x + x\int_0^4 \left(\frac{1}{3}(t^3)^\Delta - \frac{1}{2}(t^2)^\Delta + \frac{1}{6} - 2\left(\frac{1}{2}(t^2)^\Delta - \frac{1}{2}\right)\right)\Delta t$$

$$= x^2 - 4x + x\int_0^4 \left(\frac{1}{3}(t^3)^\Delta - \frac{1}{2}(t^2)^\Delta + \frac{1}{6} - (t^2)^\Delta + 1\right)\Delta t$$

$$= x^2 - 4x + x\int_0^4 \left(\frac{1}{3}(t^3)^\Delta - \frac{3}{2}(t^2)^\Delta + \frac{7}{6}\right)\Delta t$$

$$= x^2 - 4x + x\left(\frac{1}{3}t^3\Big|_{t=0}^{t=4} - \frac{3}{2}t^2\Big|_{t=0}^{t=4} + \frac{7}{6}t\Big|_{t=0}^{t=4}\right)$$

$$= x^2 - 4x + x\left(\frac{64}{3} - 24 + \frac{14}{3}\right)$$

$$= x^2 - 4x + 2x$$

$$= x^2 - 2x$$

$$= \phi(x).$$

Example 43 Let $\mathscr{T} = 3^{\mathscr{N}_0}$. Consider the equation

$$\phi(x) = \frac{1}{164}x^2 \int_1^9 t\phi(t)\Delta t.$$

We will prove that $\phi(x) = x^2$ is its solution. Really, we have that $\sigma(x) = 3x$ and

$$\begin{aligned}
\frac{1}{164}x^2 \int_1^9 t\phi(t)\Delta t &= \frac{1}{164}x^2 \int_1^9 t^3 \Delta t \\
&= \frac{1}{164}x^2 \frac{1}{40} \int_1^9 (t^4)^{\Delta} \Delta t \\
&= \frac{1}{6560}x^2 t^4 \Big|_{t=1}^{t=9} \\
&= x^2 \\
&= \phi(x).
\end{aligned}$$

Example 44 Let $\mathscr{T} = \mathscr{N}_0^2 \cup \{0\}$. Consider the equation

$$\phi(x) = 12x\sqrt{x} + 6x - 2 + \frac{3}{32}x^2 \int_0^4 \phi(t)\Delta t.$$

We will prove that

$$\phi(x) = 6x^2 + 12x\sqrt{x} + 6x - 2$$

is its solution. Really, we have that $\sigma(x) = (\sqrt{x} + 1)^2$ and

$$\begin{aligned}
&12x\sqrt{x} + 6x - 2 + \frac{3}{32}x^2 \int_0^4 \phi(t)\Delta t \\
&= 12x\sqrt{x} + 6x - 2 + \frac{3}{32}x^2 \int_0^4 (6x^2 + 12x\sqrt{x} + 6x - 2)\Delta x \\
&= 12x\sqrt{x} + 6x - 2 \\
&+ \frac{3}{32}x^2 \int_0^4 \left(2(3x^2 + 6x\sqrt{x} + 7x + 4\sqrt{x} + 1) - 4(2x + 2\sqrt{x} + 1)\right) \Delta x
\end{aligned}$$

$$= 12x\sqrt{x} + 6x - 2 + \frac{3}{32}x^2 \int_0^4 \left(2(x^3)^\Delta - 4(x^2)^\Delta\right) \Delta x$$

$$= 12x\sqrt{x} + 6x - 2 + \frac{3}{32}\left(2x^3\Big|_{x=0}^{x=4} - 4x^2\Big|_{x=0}^{x=4}\right)$$

$$= 12x\sqrt{x} + 6x - 2 + \frac{3}{32}x^2(128 - 64)$$

$$= 12x\sqrt{x} + 6x - 2 + 6x^2$$

$$= \phi(x).$$

Exercise 13 Show that the given function is a solution of the corresponding generalized Fredholm integral equation.

1. $\phi(x) = x^2 + 2x + 1, \quad \mathscr{T} = \mathscr{Z},$

$$\phi(x) = x^2 - 4x + 1 + x\int_{-2}^{2} \phi(t)\Delta t,$$

2. $\phi(x) = x - 4, \quad \mathscr{T} = 3\mathscr{Z},$

$$\phi(x) = 16x - 4 + x\int_0^6 \phi(t)\Delta t,$$

3. $\phi(x) = x - 1, \quad \mathscr{T} = 2^{\mathscr{N}_0} \cup \{0\},$

$$\phi(x) = -\frac{1}{3}x - 1 + x\int_0^4 \phi(t)\Delta t.$$

2.7 Advanced Practical Exercises

Problem 1 Convert the following multiple integrals to single integrals.

1. $\int_0^x \int_0^{x_1} e_{t^2}(t, 1)\Delta t \Delta x_1, \ \mathscr{T} = 3\mathscr{Z}.$
2. $\int_0^x \int_0^{x_1} e_{t \ominus t^2}(t, 1)\Delta t \Delta x_1, \ \mathscr{T} = 3^{\mathscr{N}_0} \cup \{0\}.$
3. $\int_0^x \int_0^{x_1} \sinh_t(t, 2)\Delta t \Delta x_1, \ \mathscr{T} = 2^{\mathscr{N}_0} \cup \{0\}.$

Answer

1. $\int_0^x (x-t-3)e_{t^2}(t,1)\Delta t$,
2. $\int_0^x (x-3t)e_{t\ominus t^2}(t,1)\Delta t$,
3. $\int_0^x (x-2t)\sinh_t(t,2)\Delta t$.

Problem 2 Convert the following multiple integrals to single integrals.

1. $\int_0^x \int_0^{x_1} (x_1-t)t^4\Delta t\Delta x_1,\ \mathcal{T}=\mathcal{R}$,
2. $\int_0^x \int_0^{x_1} (x_1-3t)e_t(t,1)\Delta t\Delta x_1,\ \mathcal{T}=3^{\mathcal{N}_0}\cup\{0\}$,
3. $\int_2^x \int_2^{x_1} (x_1-2t)\sinh_t(t,2)\Delta t\Delta x_1,\ \mathcal{T}=2^{\mathcal{N}_0}$.

Answer

1. $\int_0^x (x-t)^2 t^4\Delta t$,
2. $\int_0^x (x-3t)^2 e_t(t,1)\Delta t$,
3. $\int_2^x (x-2t)^2\sinh_t(t,2)\Delta t$.

Problem 3 Convert the following IVPs to integral equations.

1.
$$\begin{cases} z^{\Delta}(x)-\frac{x^2+1}{x^2+3}z(x)=-1 \\ z(0)=0, \quad \mathcal{T}=\mathcal{Z}, \end{cases}$$

2.
$$\begin{cases} z^{\Delta}(x)-(2x^2+1)z(x)=2 \\ z(0)=1, \quad \mathcal{T}=\mathcal{N}_0, \end{cases}$$

3.
$$\begin{cases} z^{\Delta}(x)-z(x)=-1 \\ z(2)=10, \quad \mathcal{T}=\mathcal{N}. \end{cases}$$

Answer

1. $\phi(x)=-x+\int_0^x \frac{y^2+1}{y^2+3}\phi(y)\Delta y$,
2. $\phi(x)=2x+1+\int_0^x (2y^2+1)\phi(y)\Delta y$,

3. $\phi(x) = 12 - x + \int_2^x \phi(y)\Delta y.$

Problem 4 Convert the following IVPs to integral equations.

1.
$$\begin{cases} z^{\Delta^2}(x) - \sin_x(x,0)z(x) = 1, \\ z(0) = 0, \quad z^{\Delta}(0) = 1, \quad \mathscr{T} = 2\mathscr{Z}, \end{cases}$$

2.
$$\begin{cases} z^{\Delta^2}(x) + 2z^{\Delta}(x) + z(x) = 0, \\ z(1) = 1, \quad z^{\Delta}(1) = 2, \quad \mathscr{T} = \mathscr{N}, \end{cases}$$

3.
$$\begin{cases} z^{\Delta^2}(x) + z^{\Delta}(x) + z(x) = e_x(x,2), \\ z(0) = 1, \quad z^{\Delta}(0) = 2, \quad \mathscr{T} = \mathscr{N}_0. \end{cases}$$

Answer

1. $\phi(x) = \frac{1}{2}x^2 + \int_0^x \sin_y(y,0)(x-2-y)\phi(y)\Delta y,$
2. $\phi(x) = -3 + 4x - \int_1^x (1+x-y)\phi(y)\Delta y,$
3. $\phi(x) = 3x + 1 - \int_0^x (x-y)\phi(y)\Delta y + \int_0^x (x-y-1)e_y(y,2)\Delta y.$

Problem 5 Convert the following generalized Volterra integral equations to IVPs.

1. $\phi(x) = \sinh_x(x,2) + \int_2^x (1-2y)\phi(y)\Delta y,\ \mathscr{T} = \mathscr{Z},$
2. $\phi(x) = x^2 + x + \int_1^x y^2\phi(y)\Delta y,\ \mathscr{T} = \mathscr{N}_0^3,$
3. $\phi(x) = x^2 - 2x + 2 + \int_1^x (x+y)\phi(y)\Delta y,\ \mathscr{T} = 4^{\mathscr{N}_0}.$

Answer

1.
$$\begin{cases} \phi^{\Delta}(x) - (1-2x)\phi(x) = x\cosh_x(x,2), \\ \phi(2) = 0, \end{cases}$$

2.
$$\begin{cases} \phi^{\Delta}(x) - x^2\phi(x) = 2x + 3\sqrt[3]{x^2} + 3\sqrt[3]{x} + 2, \\ \phi(1) = 2, \end{cases}$$

3.

$$\begin{cases} \phi^{\Delta^2}(x) - 20x\phi^{\Delta}(x) - 6\phi(x) = 5, \\ \phi(1) = 1, \quad \phi^{\Delta}(1) = 8. \end{cases}$$

Problem 6 Convert the following generalized Volterra integro-differential equations to IVPs.

1. $\phi(x) = \phi^{\Delta}(x) + \int_0^x y\phi(y)\Delta y, \quad \mathscr{T} = \mathscr{N}_0^2,$
2. $\phi^{\Delta}(x) = \phi(x) + x^2 + \int_0^x \phi(y)\Delta y, \quad \mathscr{T} = \mathscr{N}_0^2,$
3. $\phi^{\Delta}(x) = \phi(x) + x^3 - 2x + \int_0^x x\phi(y)\Delta y, \quad \mathscr{T} = 3^{\mathscr{N}_0} \cup \{0\}.$

Answer

1.

$$\begin{cases} \phi^{\Delta^2}(x) - \phi^{\Delta}(x) + x\phi(x) = 0, \\ \phi(0) = \phi^{\Delta}(0), \end{cases}$$

2.

$$\begin{cases} \phi^{\Delta^2}(x) - \phi^{\Delta}(x) - \phi(x) = 2x + 2\sqrt{x} + 1, \\ \phi^{\Delta}(0) = \phi(0), \end{cases}$$

3.

$$\begin{cases} \phi^{\Delta^3}(x) - \phi^{\Delta^2}(x) - 9x\phi^{\Delta}(x) - 4\phi(x) = 52x, \\ \phi^{\Delta}(0) = \phi(0), \quad \phi^{\Delta^2}(0) = \phi^{\Delta}(0) - 2. \end{cases}$$

Problem 7 Convert the following BVPs to generalized Fredholm integral equations.

1.

$$\begin{cases} z^{\Delta^2}(x) + z(x) = 1, \quad 0 < x < 9, \\ z(0) = z(9) = 0, \quad \mathscr{T} = \mathscr{Z}, \end{cases}$$

2.

$$\begin{cases} z^{\Delta^2}(x) + xz(x) = 1, \quad 1 < x < 16, \\ z(1) = z(16) = 0, \quad \mathscr{T} = \mathscr{N}_0^2, \end{cases}$$

3.

$$\begin{cases} z^{\Delta^2}(x) + x^2 z(x) = 1 - x, \quad 1 < x < 8, \\ z(1) = 0, \quad z(8) = 1, \quad \mathscr{T} = 2^{\mathscr{N}_0}. \end{cases}$$

Answer

1. $h(x) = 1$,

$$K(x,t) = \begin{cases} -\frac{1}{9}(x-9)(t+1) & \text{for} \quad 0 \le t \le x \\ \frac{x}{9}(8-t) & \text{for} \quad x \le t \le 9, \end{cases}$$

2. $h(x) = 1$,

$$K(x,t) = \begin{cases} -\frac{1}{15}x(x-16)(t+2\sqrt{t}) & \text{for} \quad 1 \le t \le x \\ \frac{1}{15}x(x-1)(15-t-2\sqrt{t}) & \text{for} \quad x \le t \le 16, \end{cases}$$

3. $h(x) = -(x-1)\left(\frac{1}{7}x^2+1\right)$,

$$K(x,t) = \begin{cases} -\frac{1}{7}(x-8)x^2(-1+2t) & \text{for} \quad 1 \le t \le x \\ \frac{2}{7}x^2(x-1)(4-t) & \text{for} \quad x \le t \le 8. \end{cases}$$

Problem 8 Convert the following BVPs to generalized Fredholm integral equations.

1.

$$\begin{cases} z^{\Delta^2}(x) - (3x+7)z(x) = 1+\cos x, \quad 0 < x < 4, \\ z(0) = 0, \quad z^{\Delta}(4) = 1, \quad \mathcal{T} = \mathcal{Z}, \end{cases}$$

2.

$$\begin{cases} z^{\Delta^2}(x) - z(x) = 1, \quad -1 < x < 9, \\ z(-1) = z^{\Delta}(9) = 0, \quad \mathcal{T} = \mathcal{Z}, \end{cases}$$

3.

$$\begin{cases} z^{\Delta^2}(x) + xz(x) = x^2 + 2x, \quad 1 < x < 27, \\ z(1) = 0, \quad z^{\Delta}(27) = 1, \quad \mathcal{T} = 3^{\mathcal{N}_0}. \end{cases}$$

Answer

1. $h_1(x) = 1 + \cos x + 3x^2 + 7x$,

$$K_1(x,t) = \begin{cases} -(3x+7)(t+1) & \text{for} \quad 0 \le t \le x \\ -x(3x+7) & \text{for} \quad x \le t \le 4, \end{cases}$$

2. $h_1(x) = 1$,

$$K_1(x,t) = \begin{cases} -t-2 & for \quad -1 \le t \le x \\ -x-1 & for \quad x \le t \le 9, \end{cases}$$

3. $h_1(x) = 3x$,

$$K_1(x,t) = \begin{cases} x(-1+3t) & for \quad 1 \le t \le x \\ x(x-1) & for \quad x \le t \le 27. \end{cases}$$

Problem 9 Convert the following generalized Fredholm integral equations to BVPs

1.
$$\phi(x) = x^2 + 2x + 4 + \int_0^4 K(x,t)\phi(t)\Delta t,$$

where
$$K(x,t) = \begin{cases} 3(4-x)(t+1) & for \quad 0 \le t \le x \\ 3x(3-t) & for \quad x \le t \le 4, \end{cases}$$

$\mathcal{T} = \mathcal{Z}$,

2.
$$\phi(x) = x^3 - 3x^2 + \int_{-1}^2 K(x,t)\phi(t)\Delta t,$$

where
$$K(x,t) = \begin{cases} 2(2-x)(t+2) & for \quad -1 \le t \le x \\ 2(1-t)(x+1) & for \quad x \le t \le 2, \end{cases}$$

$\mathcal{T} = \mathcal{Z}$,

3.
$$\phi(x) = x^4 + x + \int_0^3 K(x,t)\phi(t)\Delta t,$$

where
$$K(x,t) = \begin{cases} 3t(3-x) & for \quad 0 \le t \le x \\ 3x(1-t) & for \quad x \le t \le 3, \end{cases}$$

$\mathcal{T} = 3^{\mathcal{N}_0} \cup \{0\}$.

Answer

1.
$$\begin{cases} \phi^{\Delta^2}(x) + 12\phi(x) = 2 \\ \phi(0) = 4, \quad \phi(4) = 28, \end{cases}$$

2.
$$\begin{cases} \phi^{\Delta^2}(x) + 6\phi(x) = 6x \\ \phi(-1) = -4, \quad \phi(2) = -4, \end{cases}$$

3.
$$\begin{cases} \phi^{\Delta^2}(x) + 3\phi(x) = 520x^2 \\ \phi(0) = 0, \quad \phi(3) = 84. \end{cases}$$

Problem 10 Convert the following generalized Fredholm integral equations to BVPs.

1. $\phi(x) = x^2 - 10x + 5 + \int_0^8 K(x,t)\phi(t)\Delta t$, $\mathscr{T} = \mathscr{Z}$, where
$$K(x,t) = \begin{cases} t+1 & for \quad 0 \le t \le x \\ x+2 & for \quad x \le t \le 8, \end{cases}$$

2. $\phi(x) = x^4 + \int_{-2}^8 K(x,t)\phi(t)\Delta t$, $\mathscr{T} = 2\mathscr{Z}$, where
$$K(x,t) = \begin{cases} t+4 & for \quad -2 \le t \le x \\ x+2 & for \quad x \le t \le 8, \end{cases}$$

3. $\phi(x) = x^2 + \int_0^4 K(x,t)\phi(t)\Delta t$, $\mathscr{T} = \mathscr{N}_0^2 \cup \{0\}$, where
$$K(x,t) = \begin{cases} 2(\sqrt{t}+1)^2 & for \quad 0 \le t \le x \\ 2x & for \quad x \le t \le 4. \end{cases}$$

Answer

1.

$$\begin{cases} \phi^{\Delta^2}(x) + \phi(x) = 2, \quad 0 < x < 8, \\ \phi(0) = 5, \quad \phi^{\Delta}(8) = 7, \end{cases}$$

2.

$$\begin{cases} \phi^{\Delta^2}(x) + \phi(x) = 12x^2 + 48x + 56, \quad -2 < x < 8, \\ \phi(-2) = 16, \quad \phi^{\Delta}(8) = 2952, \end{cases}$$

3.

$$\begin{cases} \phi^{\Delta^2}(x) + 2\phi(x) = 2 + \frac{2}{1+2\sqrt{x}}, \quad 0 < x < 4, \\ \phi(0) = 0, \quad \phi^{\Delta}(4) = 13. \end{cases}$$

Problem 11 Show that the given function is a solution of corresponding generalized Volterra integral equation.

1. $\phi(x) = x^2, \quad \mathscr{T} = \mathscr{Z},$

$$\phi(x) = -\frac{1}{5}x^6 + \frac{1}{4}x^5 + \frac{1}{6}x^4 - \frac{1}{4}x^3 + \frac{1}{30}x + x^2 + x\int_0^x (t^2 + t)\phi(t)\Delta t,$$

2. $\phi(x) = x^2 + \sin x, \quad \mathscr{T} = \mathscr{R},$

$$\phi(x) = -\frac{x^5}{4} + \frac{x^4}{3} + \frac{7}{12}x - (x-1)\sin x + (x^2 - x)\cos x + x^2$$
$$+ 2x\cos 1 - x\sin 1 + x\int_{-1}^x (t-1)\phi(t)\Delta t,$$

3. $\phi(x) = x, \quad \mathscr{T} = 3^{\mathscr{N}_0},$

$$\phi(x) = -\frac{1}{13}x(x^3 - 14) + x\int_1^x t\phi(t)\Delta t.$$

Problem 12 Show that the given function is a solution of the corresponding generalized Fredholm integral equation.

1. $\phi(x) = x^3 + x, \quad \mathscr{T} = \mathscr{Z},$

$$\phi(x) = x^3 - 23x + 2x\int_0^3 \phi(t)\Delta t,$$

2. $\phi(x) = x^2 + 2x - 4, \quad \mathcal{T} = 2\mathcal{Z},$

$$\phi(x) = x^2 - 13x - 4 + x \int_0^5 \phi(t) \Delta t,$$

3. $\phi(x) = 2x + 2\sqrt{x} + 1, \quad \mathcal{T} = \mathcal{N}_0^2,$

$$\phi(x) = 2x + 1 + \frac{1}{8}\sqrt{x} \int_0^4 \phi(t) \Delta t.$$

Chapter 3
Generalized Volterra Integral Equations

In this chapter we investigate generalized Volterra integral equations. They are described different methods for finding a solution as an infinite series such as: the Adomian decomposition method, the modified decomposition method, the noise terms phenomenon, the differential equations method and the successive iterations method. It is given a procedure for conversion of generalized Volterra integral equations of the first kind to generalized Volterra integral equations of the second kind. They are provided existence and uniqueness of the solution.

The generalized Volterra integral equations arise in many scientific applications such as the population dynamics, spread of epidemics and semi-conductor devices. It was shown that Volterra integral equations can be derived from initial value problems. In this chapter we will apply the Adomian decomposition method, the modified decomposition method, the noise terms phenomenon, differential equations method and successive approximations method. The theorems of existence and uniqueness of the solutions are given in the last section of this chapter.

3.1 Generalized Volterra Integral Equations of the Second Kind

3.1.1 The Adomian Decomposition Method

The Adomian decomposition method (ADM) consists of decomposing the unknown function $\phi(x)$ of any generalized Volterra integral equation into a sum of infinite number of components by the series

$$\phi(x) = \sum_{n=0}^{\infty} \phi_n(x) \tag{3.1}$$

S.G. Georgiev, *Integral Equations on Time Scales*, Atlantis Studies in Dynamical Systems 5, DOI 10.2991/978-94-6239-228-1_3

or equivalently

$$\phi(x) = \phi_0(x) + \phi_1(x) + \phi_2(x) + \cdots,$$

where the components $\phi_l(x)$, $l \in \mathscr{N}_0$, are to be determined in a recursive manner.

To establish the recursive relation, we substitute (3.1) into the generalized Volterra integral equation

$$\phi(x) = u(x) + \lambda \int_a^x K(x, y)\phi(y)\Delta y. \tag{3.2}$$

We obtain

$$\sum_{n=0}^{\infty} \phi_n(x) = u(x) + \lambda \int_a^x K(x, y) \sum_{n=0}^{\infty} \phi_n(y)\Delta y,$$

or equivalently,

$$\begin{aligned}\phi_0(x) + \phi_1(x) + \phi_2(x) + \cdots = u(x) &+ \lambda \int_a^x K(x, y)\phi_0(y)\Delta y \\ &+ \lambda \int_a^x K(x, y)\phi_1(y)\Delta y \\ &+ \lambda \int_a^x K(x, y)\phi_2(y)\Delta y \\ &+ \cdots.\end{aligned}$$

We set

$$\begin{aligned}\phi_0(x) &= u(x), \\ \phi_1(x) &= \lambda \textstyle\int_a^x K(x, y)\phi_0(y)\Delta y, \\ \phi_2(x) &= \lambda \textstyle\int_a^x K(x, y)\phi_1(y)\Delta y,\end{aligned}$$

and so on for the other components, or equivalently,

$$\begin{aligned}\phi_0(x) &= u(x), \\ \phi_n(x) &= \lambda \textstyle\int_a^x K(x, y)\phi_{n-1}(y)\Delta y, \quad n \in \mathscr{N}.\end{aligned} \tag{3.3}$$

In view of (3.3), the components $\phi_0(x)$, $\phi_1(x)$, $\phi_2(x)$, ... are completely determined. The solution $\phi(x)$ of (3.2) in a series form is obtained by using the series (3.1).

In other words, the Adomian decomposition method converts the generalized Volterra integral equation into a determination of computable components. Note that if an exact solution exists for (3.2), then the obtained series converges to that solution. However, for some equations, when a closed form solution is not obtainable, a number of terms of (3.1) can be used for numerical purposes. The question for convergence of $\phi_n(x)$ will be discussed in the last section of this chapter.

Example 1 Let $\mathscr{T} = \mathscr{Z}$. Consider the generalized Volterra integral equation

$$\phi(x) = x + \int_0^x \phi(y)\Delta y.$$

Here

$$\sigma(x) = x + 1, \quad x \in \mathscr{T}, \quad u(x) = x, \quad K(x, y) = 1, \quad \lambda = 1.$$

Then

$$\begin{aligned}
\phi_0(x) &= x,\\
\phi_1(x) &= \int_0^x \phi_1(y)\Delta y\\
&= \int_0^x y\Delta y\\
&= \int_0^x \left(\frac{1}{2}(y^2)^\Delta - \frac{1}{2}\right)\Delta y\\
&= \frac{1}{2}\int_0^x (y^2)^\Delta \Delta y - \frac{1}{2}\int_0^x \Delta y\\
&= \frac{1}{2}x^2 - \frac{1}{2}x,
\end{aligned}$$

$$\begin{aligned}
\phi_2(x) &= \int_0^x \phi_1(y)\Delta y\\
&= \int_0^x \left(\frac{1}{2}y^2 - \frac{1}{2}y\right)\Delta y\\
&= \frac{1}{2}\int_0^x y^2\Delta y - \frac{1}{2}\int_0^x y\Delta y\\
&= \frac{1}{2}\int_0^x \left(\frac{1}{3}(y^3)^\Delta - \frac{1}{2}(y^2)^\Delta + \frac{1}{6}\right)\Delta y\\
&\quad - \frac{1}{2}\int_0^x \left(\frac{1}{2}(y^2)^\Delta - \frac{1}{2}\right)\Delta y\\
&= \frac{1}{6}\int_0^x (y^3)^\Delta \Delta y - \frac{1}{4}\int_0^x (y^2)^\Delta \Delta y + \frac{1}{12}\int_0^x \Delta y\\
&\quad - \frac{1}{4}\int_0^x (y^2)^\Delta \Delta y + \frac{1}{4}\int_0^x \Delta y\\
&= \frac{1}{6}\int_0^x (y^3)^\Delta \Delta y - \frac{1}{2}\int_0^x (y^2)^\Delta \Delta y + \frac{1}{3}\int_0^x \Delta y\\
&= \frac{1}{6}x^3 - \frac{1}{2}x^2 + \frac{1}{3}x,
\end{aligned}$$

$$
\begin{aligned}
\phi_3(x) &= \int_0^x \phi_2(y)\Delta y \\
&= \int_0^x \left(\frac{1}{6}y^3 - \frac{1}{2}y^2 + \frac{1}{3}y\right)\Delta y \\
&= \frac{1}{6}\int_0^x y^3\Delta y - \frac{1}{2}\int_0^x y^2\Delta y + \frac{1}{3}\int_0^x y\Delta y \\
&= \frac{1}{6}\int_0^x \left(\frac{1}{4}(y^4)^\Delta - \frac{1}{2}(y^3)^\Delta + \frac{1}{4}(y^2)^\Delta\right)\Delta y \\
&\quad - \frac{1}{2}\int_0^x \left(\frac{1}{3}(y^3)^\Delta - \frac{1}{2}(y^2)^\Delta + \frac{1}{6}\right)\Delta y \\
&\quad + \frac{1}{3}\int_0^x \left(\frac{1}{2}(y^2)^\Delta - \frac{1}{2}\right)\Delta y \\
&= \frac{1}{24}x^4 - \frac{1}{12}x^3 + \frac{1}{24}x^2 - \frac{1}{6}x^3 + \frac{1}{4}x^2 \\
&\quad - \frac{1}{12}x + \frac{1}{6}x^2 - \frac{1}{6}x \\
&= \frac{1}{24}x^4 - \frac{1}{4}x^3 + \frac{11}{24}x^2 - \frac{1}{4}x.
\end{aligned}
$$

Example 2 Let $\mathscr{T} = 2^{\mathscr{N}_0} \cup \{0\}$. Consider the equation

$$\phi(x) = 1 + x\int_0^x \phi(y)\Delta y.$$

Here

$$K(x, y) = x, \quad u(x) = 1, \quad \sigma(x) = 2x, \quad x \in \mathscr{T}.$$

Then

$$
\begin{aligned}
\phi_0(x) &= 1, \\
\phi_1(x) &= x\int_0^x \phi_0(y)\Delta y \\
&= x\int_0^x \Delta y \\
&= x^2, \\
\phi_2(x) &= x\int_0^x \phi_1(y)\Delta y \\
&= x\int_0^x y^2\Delta y \\
&= \frac{1}{7}x\int_0^x (y^3)^\Delta\Delta y \\
&= \frac{1}{7}x^4,
\end{aligned}
$$

$$\begin{aligned}\phi_3(x) &= x\int_0^x \phi_2(y)\Delta y\\ &= \frac{1}{7}x\int_0^x y^4\Delta y\\ &= \frac{1}{217}x\int_0^x (y^5)^\Delta \Delta y\\ &= \frac{1}{217}x^6.\end{aligned}$$

Exercise 1 Let $\mathscr{T} = 3^{\mathscr{N}_0} \cup \{0\}$. Consider the equation

$$\phi(x) = x + 2\int_0^x \phi(y)\Delta y.$$

Find $\phi_i(x)$, $i \in \{0, 1, 2, 3\}$.

Answer

$$\phi_0(x) = x, \quad \phi_1(x) = \frac{1}{2}x^2, \quad \phi_2(x) = \frac{1}{13}x^3, \quad \phi_3(x) = \frac{1}{260}x^4.$$

Example 3 Let $\mathscr{T} = 3\mathscr{Z}$. Consider the equation

$$\phi(x) = 1 + 2\int_0^x \phi(y)\Delta y.$$

We will find its solution in a series form.
Here

$$u(x) = 1, \quad K(x, y) = 1, \quad \lambda = 2, \quad \sigma(x) = x + 3, \quad x \in \mathscr{T}.$$

Using the ADM we set the recurrence formula

$$\begin{cases}\phi_0(x) = 1,\\ \phi_n(x) = 2\int_0^x \phi_{n-1}(y)\Delta y, \quad n \in \mathscr{N}.\end{cases}$$

This in turn gives

$$\begin{aligned}\phi_1(x) &= 2\int_0^x \phi_0(y)\Delta y\\ &= 2\int_0^x \Delta y\\ &= 2x,\end{aligned}$$

$$\begin{aligned}\phi_2(x) &= 2\int_0^x \phi_1(y)\Delta y\\ &= 4\int_0^x y\Delta y\\ &= 4\int_0^x \left(\frac{1}{2}(y^2)^\Delta - \frac{3}{2}\right)\Delta y\\ &= 2\int_0^x (y^2)^\Delta \Delta y - 6\int_0^x \Delta y\\ &= 2x^2 - 6x,\end{aligned}$$

$$\begin{aligned}\phi_3(x) &= 2\int_0^x \phi_2(y)\Delta y\\ &= 4\int_0^x (y^2 - 3y)\Delta y\\ &= 4\int_0^x \left(\frac{1}{3}(y^3)^\Delta - \frac{3}{2}(y^2)^\Delta + \frac{3}{2} - 3\left(\frac{1}{2}(y^2)^\Delta - \frac{3}{2}\right)\right)\Delta y\\ &= 4\int_0^x \left(\frac{1}{3}(y^3)^\Delta - \frac{3}{2}(y^2)^\Delta + \frac{3}{2} - \frac{3}{2}(y^2)^\Delta + \frac{9}{2}\right)\Delta y\\ &= 4\int_0^x \left(\frac{1}{3}(y^3)^\Delta - 3(y^2)^\Delta + 6\right)\Delta y\\ &= \frac{4}{3}\int_0^x (y^3)^\Delta \Delta y - 12\int_0^x (y^2)^\Delta \Delta y + 24\int_0^x \Delta y\\ &= \frac{4}{3}x^3 - 12x^2 + 24x.\end{aligned}$$

Then the solution in a series form is given by

$$\phi(x) = 1 + 2x + (2x^2 - 6x) + \left(\frac{4}{3}x^3 - 12x^2 + 24x\right) + \cdots.$$

Example 4 Let $\mathscr{T} = 2^{\mathscr{N}_0} \cup \{0\}$. Consider the equation

$$\phi(x) = x + 1 - x\int_0^x \phi(y)\Delta y.$$

We will find its solution in a series form.
Here

$$u(x) = x + 1, \quad \lambda = -1, \quad K(x, y) = x, \quad \sigma(x) = 2x, \quad x \in \mathscr{T}.$$

Using the ADM we set the recurrence formula

$$\begin{cases} \phi_0(x) = x+1, \\ \phi_n(x) = -x\int_0^x \phi_{n-1}(y)\Delta y, \quad n \in \mathcal{N}. \end{cases}$$

This in turn gives

$$\begin{aligned}
\phi_1(x) &= -x\int_0^x \phi_0(y)\Delta y \\
&= -x\int_0^x (y+1)\Delta y \\
&= -x\int_0^x \left(\frac{1}{3}(y^2)^\Delta + 1\right)\Delta y \\
&= -\frac{1}{3}x\int_0^x (y^2)^\Delta \Delta y - x\int_0^x \Delta y \\
&= -\frac{1}{3}x^3 - x^2, \\
\phi_2(x) &= -x\int_0^x \phi_1(y)\Delta y \\
&= x\int_0^x \left(\frac{1}{3}y^3 + y^2\right)\Delta y \\
&= x\int_0^x \left(\frac{1}{45}(y^4)^\Delta + \frac{1}{7}(y^3)^\Delta\right)\Delta y \\
&= \frac{1}{45}x\int_0^x (y^4)^\Delta \Delta y + \frac{1}{7}x\int_0^x (y^3)^\Delta \Delta y \\
&= \frac{1}{45}x^5 + \frac{1}{7}x^4, \\
\phi_3(x) &= -x\int_0^x \phi_2(y)\Delta y \\
&= -x\int_0^x \left(\frac{1}{45}y^5 + \frac{1}{7}y^4\right)\Delta y \\
&= -x\int_0^x \left(\frac{1}{2835}(y^6)^\Delta + \frac{1}{217}(y^5)^\Delta\right)\Delta y \\
&= -\frac{1}{2835}x\int_0^x (y^6)^\Delta \Delta y - \frac{1}{217}x\int_0^x (y^5)^\Delta \Delta y \\
&= -\frac{1}{2835}x^7 - \frac{1}{217}x^6,
\end{aligned}$$

Then the solution in a series form is given by

$$\phi(x) = x+1+\left(-\frac{1}{3}x^3 - x^2\right) + \left(\frac{1}{45}x^5 + \frac{1}{7}x^4\right) + \left(-\frac{1}{2835}x^7 - \frac{1}{217}x^6\right) + \cdots.$$

Exercise 2 Let $\mathscr{T} = 4^{\mathscr{N}_0} \cup \{0\}$. Find a solution in a series form of the following equation

$$\phi(x) = 1 + \int_0^x \phi(t)\Delta t.$$

Answer

$$\phi(x) = 1 + x + \frac{1}{5}x^2 + \frac{1}{105}x^3 + \cdots .$$

3.1.2 *The Modified Decomposition Method*

For many cases, the function $u(x)$ can be represented as the sum of two partial functions, namely, $u_1(x)$ and $u_2(x)$, i.e.,

$$u(x) = u_1(x) + u_2(x).$$

The modified decomposition method(MDM) introduces the modified recurrence relation.

$$\begin{aligned} \phi_0(x) &= u_1(x), \\ \phi_1(x) &= u_2(x) + \lambda \int_a^x K(x, y)\phi_0(y)\Delta y, \\ \phi_n(x) &= \lambda \int_a^x K(x, y)\phi_{n-1}(y)\Delta y, \quad n \in \mathscr{N}\backslash\{1\}. \end{aligned} \tag{3.4}$$

The difference between the standard recurrence (3.3) and the modified recurrence relation (3.4) is in the information of the first two components $\phi_0(x)$ and $\phi_1(x)$. The other components $\phi_n(x)$, $n \geq 2$, remain the same in the two recurrence formulas. The success of this modification depends only on the proper choice of $u_1(x)$ and $u_2(x)$. A rule for such choice of $u_1(x)$ and $u_2(x)$ could not be found yet. If $u(x)$ consists of one term only, the standard decomposition method can be used in this case.

Example 5 Let $\mathscr{T} = \mathscr{Z}$. Consider the equation

$$\phi(x) = x^2 + x + \int_0^x \phi(y)\Delta y.$$

Here

$$u(x) = x^2 + x, \quad \lambda = 1, \quad K(x, y) = 1, \quad \sigma(x) = x + 1, \quad x \in \mathscr{T}.$$

Let

$$u_1(x) = x, \quad u_2(x) = x^2.$$

Then, using (3.4), we have

$$
\begin{aligned}
\phi_0(x) &= x,\\
\phi_1(x) &= x^2 + \int_0^x \phi_0(y)\Delta y\\
&= x^2 + \int_0^x y\Delta y\\
&= x^2 + \int_0^x \left(\frac{1}{2}(y^2)^\Delta - \frac{1}{2}\right)\Delta y\\
&= x^2 + \frac{1}{2}x^2 - \frac{1}{2}x\\
&= \frac{3}{2}x^2 - \frac{1}{2}x,\\
\phi_2(x) &= \int_0^x \phi_1(y)\Delta y\\
&= \int_0^x \left(\frac{3}{2}y^2 - \frac{1}{2}y\right)\Delta y\\
&= \int_0^x \left(\frac{3}{2}\left(\frac{1}{3}(y^3)^\Delta - \frac{1}{2}(y^2)^\Delta + \frac{1}{6}\right) - \frac{1}{2}\left(\frac{1}{2}(y^2)^\Delta - \frac{1}{2}\right)\right)\Delta y\\
&= \int_0^x \left(\frac{1}{2}(y^3)^\Delta - \frac{3}{4}(y^2)^\Delta + \frac{1}{4} - \frac{1}{4}(y^2)^\Delta + \frac{1}{4}\right)\Delta y\\
&= \int_0^x \left(\frac{1}{2}(y^3)^\Delta - (y^2)^\Delta + \frac{1}{2}\right)\Delta y\\
&= \frac{1}{2}\int_0^x (y^3)^\Delta\Delta y - \int_0^x (y^2)^\Delta \Delta y + \frac{1}{2}\int_0^x \Delta y\\
&= \frac{1}{2}x^3 - x^2 + \frac{1}{2}x,\\
\phi_3(x) &= \int_0^x \phi_2(y)\Delta y\\
&= \int_0^x \left(\frac{1}{2}y^3 - y^2 + \frac{1}{2}y\right)\Delta y\\
&= \int_0^x \left(\frac{1}{2}\left(\frac{1}{4}(y^4)^\Delta - \frac{1}{2}(y^3)^\Delta + \frac{1}{4}(y^2)^\Delta\right)\right.\\
&\quad \left. - \frac{1}{3}(y^3)^\Delta + \frac{1}{2}(y^2)^\Delta - \frac{1}{6} + \frac{1}{2}\left(\frac{1}{2}(y^2)^\Delta - \frac{1}{2}\right)\right)\Delta y\\
&= \int_0^x \left(\frac{1}{8}(y^4)^\Delta - \frac{1}{4}(y^3)^\Delta + \frac{1}{8}(y^2)^\Delta - \frac{1}{3}(y^3)^\Delta\right.\\
&\quad \left. + \frac{1}{2}(y^2)^\Delta - \frac{1}{6} + \frac{1}{4}(y^2)^\Delta - \frac{1}{4}\right)\Delta y
\end{aligned}
$$

$$
\begin{aligned}
&= \int_0^x \left(\frac{1}{8}(y^4)^\Delta - \frac{7}{12}(y^3)^\Delta + \frac{7}{8}(y^2)^\Delta - \frac{5}{12}\right)\Delta y \\
&= \frac{1}{8}\int_0^x (y^4)^\Delta \Delta y - \frac{7}{12}\int_0^x (y^3)^\Delta \Delta y + \frac{7}{8}\int_0^x (y^2)^\Delta \Delta y - \frac{5}{12}\int_0^x \Delta y \\
&= \frac{1}{8}x^4 - \frac{7}{12}x^3 + \frac{7}{8}x^2 - \frac{5}{12}x.
\end{aligned}
$$

Example 6 Let $\mathscr{T} = 2^{\mathscr{N}_0} \cup \{0\}$. Consider the equation

$$\phi(x) = x - 2 + 3x\int_0^x \phi(y)\Delta y.$$

Here

$$u(x) = x - 2, \quad K(x, y) = 3x, \quad \sigma(x) = 2x, \quad x \in \mathscr{T}.$$

Let

$$u_1(x) = -2, \quad u_2(x) = x.$$

Then, using (3.4), we have

$$
\begin{aligned}
\phi_0(x) &= -2, \\
\phi_1(x) &= x + 3x\int_0^x \phi_0(y)\Delta y \\
&= x - 6x\int_0^x \Delta y \\
&= x - 6x^2, \\
\phi_2(x) &= 3x\int_0^x \phi_1(y)\Delta y \\
&= 3x\int_0^x (y - 6y^2)\Delta y \\
&= 3x\int_0^x \left(\frac{1}{3}(y^2)^\Delta - \frac{6}{7}(y^3)^\Delta\right)\Delta y \\
&= x\int_0^x (y^2)^\Delta \Delta y - \frac{18}{7}x\int_0^x (y^3)^\Delta \Delta y \\
&= x^3 - \frac{18}{7}x^4,
\end{aligned}
$$

$$\begin{aligned}
\phi_3(x) &= 3x \int_0^x \phi_2(y)\Delta y \\
&= 3x \int_0^x \left(y^3 - \frac{18}{7} y^4 \right) \Delta y \\
&= 3x \int_0^x \left(\frac{1}{15} (y^4)^{\Delta} - \frac{18}{217} (y^5)^{\Delta} \right) \Delta y \\
&= \frac{1}{5} x \int_0^x (y^4)^{\Delta} \Delta y - \frac{54}{217} x \int_0^x (y^5)^{\Delta} \Delta y \\
&= \frac{1}{5} x^5 - \frac{54}{217} x^6.
\end{aligned}$$

Exercise 3 Let $\mathscr{T} = 3\mathscr{Z}$. Consider the equation

$$\phi(x) = x + 1 + \int_0^x y\phi(y)\Delta y.$$

Using MDM, find $\phi_0(x)$, $\phi_1(x)$, $\phi_2(x)$ and $\phi_3(x)$.

Answer

$$\begin{aligned}
\phi_0(x) &= 1, \quad \phi_1(x) = -\frac{1}{2}x + \frac{1}{2}x^2, \\
\phi_2(x) &= \frac{1}{8}x^4 - \frac{11}{12}x^3 + \frac{15}{8}x^2 - \frac{3}{4}x, \\
\phi_3(x) &= \frac{1}{48}x^6 - \frac{89}{240}x^5 + \frac{37}{16}x^4 - \frac{93}{16}x^3 + \frac{9}{2}x^2 + \frac{27}{20}x.
\end{aligned}$$

Now we consider the following generalized Volterra integral equation of the second kind

$$\phi(x) = u(x) + \lambda \int_a^x K(x, y)\phi(\sigma(y))\Delta y, \quad a \in \mathscr{T}. \tag{3.5}$$

We define

$$\begin{aligned}
&\phi_0(x) = u(x), \\
&\phi_n(x) = \lambda \textstyle\int_a^x K(x, y)\phi_{n-1}(\sigma(y))\Delta y, \quad n \in \mathscr{N}.
\end{aligned} \tag{3.6}$$

Then a series solution of the Eq. (3.5) is given by

$$\phi(x) = \sum_{n=0}^{\infty} \phi_n(x),$$

where $\phi_n(x)$, $n \in \mathscr{N}_0$, are defined by (3.6).

Example 7 Let $\mathscr{T} = \mathscr{Z}$. Consider the equation

$$\phi(x) = 1 + 2\int_0^x y\phi(y+1)\Delta y.$$

Here

$$u(x) = 1, \quad \lambda = 2, \quad a = 0, \quad K(x, y) = y.$$

Then

$$\begin{aligned}
\phi_0(x) &= 1,\\
\phi_1(x) &= 2\int_0^x y\phi_0(y+1)\Delta y\\
&= 2\int_0^x y\Delta y\\
&= 2\int_0^x \left(\frac{1}{2}(y^2)^\Delta - \frac{1}{2}\right)\Delta y\\
&= \int_0^x \left((y^2)^\Delta - 1\right)\Delta y\\
&= \int_0^x (y^2)^\Delta \Delta y - \int_0^x \Delta y\\
&= x^2 - x,\\
\phi_2(x) &= 2\int_0^x y\phi_1(y+1)\Delta y\\
&= 2\int_0^x y\left((y+1)^2 - (y+1)\right)\Delta y\\
&= 2\int_0^x y(y^2 + 2y + 1 - y - 1)\Delta y\\
&= 2\int_0^x y(y^2 + y)\Delta y\\
&= 2\int_0^x (y^2 + y^3)\Delta y\\
&= 2\int_0^x \left(\frac{1}{3}(y^3)^\Delta - \frac{1}{2}(y^2)^\Delta + \frac{1}{6} + \frac{1}{4}(y^4)^\Delta - \frac{1}{2}(y^3)^\Delta + \frac{1}{4}(y^2)^\Delta\right)\Delta y\\
&= 2\int_0^x \left(-\frac{1}{4}(y^2)^\Delta - \frac{1}{6}(y^3)^\Delta + \frac{1}{4}(y^4)^\Delta + \frac{1}{6}\right)\Delta y\\
&= -\frac{1}{2}\int_0^x (y^2)^\Delta \Delta y - \frac{1}{3}\int_0^x (y^3)^\Delta \Delta y + \frac{1}{2}\int_0^x (y^4)^\Delta \Delta y + \frac{1}{3}\int_0^x \Delta y\\
&= \frac{1}{3}x - \frac{1}{2}x^2 - \frac{1}{3}x^3 + \frac{1}{2}x^4.
\end{aligned}$$

Example 8 Let $\mathscr{T} = 2^{\mathscr{N}_0} \cup \{0\}$. Consider the equation

$$\phi(x) = x + 3\int_0^x y\phi(2y)\Delta y.$$

Here

$$u(x) = x, \quad \lambda = 3, \quad K(x,y) = y, \quad a = 0.$$

$$\begin{aligned}
\phi_0(x) &= x,\\
\phi_1(x) &= 3\int_0^x y\phi_0(2y)\Delta y\\
&= 6\int_0^x y^2\Delta y\\
&= \frac{6}{7}\int_0^x (y^3)^\Delta \Delta y\\
&= \frac{6}{7}x^3,\\
\phi_2(x) &= 3\int_0^x y\phi_1(2y)\Delta y\\
&= \frac{18}{7}\int_0^x y(2y)^3\Delta y\\
&= \frac{144}{7}\int_0^x y^4\Delta y\\
&= \frac{144}{217}\int_0^x (y^5)^\Delta \Delta y\\
&= \frac{144}{217}x^5.
\end{aligned}$$

Exercise 4 Using ADM, find $\phi_0(x)$, $\phi_1(x)$ and $\phi_2(x)$ for the equation

$$\phi(x) = x^2 + \int_0^x y\phi(3y)\Delta y, \quad \mathscr{T} = 3^{\mathscr{N}_0} \cup \{0\}.$$

Answer

$$\phi_0(x) = x^2, \quad \phi_1(x) = \frac{9}{40}x^4, \quad \phi_2(x) = \frac{729}{14560}x^6.$$

3.1.3 The Noise Terms Phenomenon

The noise terms are defined as the identical terms with opposite signs which can appear in the components $\phi_0(x)$ and $\phi_1(x)$ and in the other components as well. Noise terms may appear if the exact solutions of the considered equation is part of the zeroth component $\phi_0(x)$. By cancelling the noise terms between $\phi_0(x)$ and $\phi_1(x)$, the remaining non-cancelled terms of $\phi_0(x)$ may give the exact solution of the given equation. Verification that the remaining non-cancelled terms satisfy the integral equation is necessary and essential. Note that the noise terms appear for specific cases of inhomogeneous integral equations. Homogeneous integral equations do not give rise to noise terms. The question for convergence of $\phi_n(x)$ will be discussed in the last section of this chapter.

Example 9 Let $\mathcal{T} = 2^{\mathcal{N}_0} \cup \{0\}$. Consider the equation

$$\phi(x) = x^2 - 3x + \int_0^x \phi(y)\Delta y.$$

Using ADM, we have

$$\begin{aligned}
\phi_0(x) &= x^2 - 3x,\\
\phi_1(x) &= \int_0^x \phi_0(y)\Delta y\\
&= \int_0^x (y^2 - 3y)\Delta y\\
&= \int_0^x \left(\frac{1}{7}(y^3)^{\Delta} - (y^2)^{\Delta}\right)\Delta y\\
&= \frac{1}{7}x^3 - x^2.
\end{aligned}$$

The noise term is x^2. Therefore we will check that $\phi(x) = -3x$ is a solution of the given equation. We have

$$\begin{aligned}
x^2 - 3x + \int_0^x (-3y)\Delta y &= x^2 - 3x - \int_0^x (y^2)^{\Delta}\Delta y\\
&= x^2 - 3x - x^2\\
&= -3x\\
&= \phi(x).
\end{aligned}$$

Example 10 Let $\mathcal{T} = 3\mathcal{Z}$. Consider the equation

$$\phi(x) = x + \frac{1}{2} + \int_0^x \phi(y)\Delta y.$$

Using ADM, we have

$$\begin{aligned}
\phi_0(x) &= x + \frac{1}{2},\\
\phi_1(x) &= \int_0^x \left(y + \frac{1}{2}\right) \Delta y\\
&= \int_0^x \left(\frac{1}{2}(y^2)^\Delta - \frac{3}{2} + \frac{1}{2}\right) \Delta y\\
&= \int_0^x \left(\frac{1}{2}(y^2)^\Delta - 1\right) \Delta y\\
&= \frac{1}{2}\int_0^x (y^2)^\Delta \Delta y - \int_0^x \Delta y\\
&= \frac{1}{2}x^2 - x.
\end{aligned}$$

The noise term is $-x$. We will check if $\phi(x) = \frac{1}{2}$ is a solution of the given equation. We have

$$\begin{aligned}
x + \frac{1}{2} + \int_0^x \frac{1}{2}\Delta y &= x + \frac{1}{2} + \frac{1}{2}x\\
&= \frac{3}{2}x + \frac{1}{2}\\
&\neq \phi(x).
\end{aligned}$$

Therefore $\phi(x) = \frac{1}{2}$ is not a solution of the given equation.

Exercise 5 Check if the noise terms phenomenon gives a solution to the equation

$$\phi(x) = x^2 - x + \int_0^x \phi(y)\Delta y, \quad \mathscr{T} = \mathscr{Z}.$$

Answer No.

Exercise 6 Use the noise terms phenomenon to solve the following equation

$$\phi(x) = x - 1 + \int_0^x \phi(y)\Delta y, \quad \mathscr{T} = 2^{\mathscr{N}_0} \cup \{0\}.$$

Answer $\phi(x) = -1$.

3.1.4 Differential Equations Method

In this section we will solve some generalized Volterra integral equations reducing them to IVPs for some differential equations. This method is known as Differential Equation Method (DEM).

Example 11 Let $\mathscr{T} = \mathscr{Z}$. Consider the equation

$$\phi(x) = 1 + \int_{-1}^{x} y^2 \phi(y) \Delta y. \tag{3.7}$$

We differentiate the Eq. (3.7) with respect to x and we get

$$\phi^{\Delta}(x) = x^2 \phi(x).$$

Substituting $x = -1$ in (3.7) we find $\phi(-1) = 1$. In this way we obtain the following IVP

$$\begin{cases} \phi^{\Delta}(x) = x^2 \phi(x) \\ \phi(-1) = 1. \end{cases} \tag{3.8}$$

Also, $\sigma(x) = x + 1, \quad x \in \mathscr{T}$. Hence $\mu(x) = \sigma(x) - x = 1$. Therefore

$$1 + \mu(x)x^2 = 1 + x^2 \neq 0 \quad for\, any \quad x \in \mathscr{T}.$$

Consequently the solution of (3.8) is given by

$$\phi(x) = e_{x^2}(x, -1),$$

which is a solution of (3.7).

Example 12 Let $\mathscr{T} = 3\mathscr{Z}$. Consider the equation

$$\phi(x) = -3 + 2 \int_{-1}^{x} \phi(y) \Delta y. \tag{3.9}$$

We differentiate (3.9) with respect to x and we find

$$\phi^{\Delta}(x) = 2\phi(x).$$

We substitute $x = -1$ in (3.9) and we obtain

$$\phi(-1) = -3.$$

Consequently we go to the following IVP

$$\begin{cases} \phi^{\Delta}(x) = 2\phi(x) \\ \phi(-1) = -3. \end{cases} \tag{3.10}$$

Here $\sigma(x) = x + 3$. Hence,

$$\mu(x) = \sigma(x) - x = x + 3 - x = 3 \quad and \quad 1 + 2\mu(x) = 7 \neq 0.$$

Therefore the solution of (3.10) is given by

$$\phi(x) = -3e_2(x, -1).$$

Example 13 Let $\mathcal{T} = 3^{\mathcal{N}_0}$. Consider the equation

$$\phi(x) = 10 + 2\int_1^x \phi(y)\Delta y. \tag{3.11}$$

We differentiate the Eq. (3.11) with respect to x and we get

$$\phi^{\Delta}(x) = 2\phi(x).$$

We substitute $x = -1$ in (3.11) and we find $\phi(1) = 10$.
In this way we go to the following IVP

$$\begin{cases} \phi^{\Delta}(x) = 2\phi(x) \\ \phi(1) = 10. \end{cases} \tag{3.12}$$

Here

$$\begin{aligned} &\sigma(x) = 3x, \quad \mu(x) = \sigma(x) - x = 3x - x = 2x, \\ &1 + 2\mu(x) = 1 + 4x \neq 0 \quad for\, any \quad x \in \mathcal{T}. \end{aligned}$$

Consequently the solution of (3.12) is given by

$$\begin{aligned} \phi(x) &= 10e_2(x, 1) \\ &= 10 \prod_{s\in[1,x)} (1 + (3-1)2s) \\ &= 10 \prod_{s\in[1,x)} (1 + 4s), \end{aligned}$$

which is a solution to the Eq. (3.11).

Example 14 Let $\mathscr{T} = \dfrac{1}{3}\mathscr{Z}$. Consider the equation

$$\phi(x) = -2 + \int_0^x \phi(y)\Delta y. \tag{3.13}$$

We differentiate the Eq. (3.13) with respect to x and we get

$$\phi^{\Delta}(x) = \phi(x).$$

We substitute $x = 0$ in (3.13) and we find $\phi(0) = -2$.
Therefore we obtain the following IVP

$$\begin{cases} \phi^{\Delta}(x) = \phi(x) \\ \phi(0) = -2. \end{cases} \tag{3.14}$$

Here

$$\sigma(x) = x + \tfrac{1}{3}, \quad \mu(x) = \sigma(x) - x = x + \tfrac{1}{3} - x = \tfrac{1}{3}.$$

Consequently the solution of (3.14) is given by

$$\begin{aligned} \phi(x) &= -2e_1(x, 0) \\ &= -2\left(1 + \frac{1}{3}\right)^{3x} \\ &= -2\left(\frac{4}{3}\right)^{3x}. \end{aligned}$$

Example 15 Let $\mathscr{T} = q^{\mathscr{N}_0}, \quad q > 1$. Consider the integral equation

$$\phi(x) = b + \int_1^x \frac{1-y}{(q-1)y^2}\phi(y)\Delta y, \tag{3.15}$$

where b is a real constant. We differentiate (3.15) with respect to x and we find

$$\phi^{\Delta}(x) = \frac{1-x}{(q-1)x^2}\phi(x).$$

We substitute $x = 1$ in (3.15) and we find $\phi(1) = b$.
Therefore we go to the following IVP

$$\begin{cases} \phi^{\Delta}(x) = \frac{1-x}{(q-1)x^2}\phi(x) \\ \phi(1) = b. \end{cases} \tag{3.16}$$

Here

$$\sigma(x) = qx, \quad \mu(x) = \sigma(x) - x = (q-1)x,$$
$$1 + \mu(x)\frac{1-x}{(q-1)x} = 1 + (q-1)x\frac{1-x}{(q-1)x^2} = 1 + \frac{1-x}{x} = \frac{1}{x} \neq 0$$

for any $x \in \mathscr{T}$.
Therefore the solution of (3.16) is given by

$$\phi(x) = be_{\frac{1-x}{(q-1)x^2}}(x, 1),$$

which is a solution of the Eq. (3.15).

Example 16 Let $\mathscr{T} = \mathscr{N}_0^2$. Consider the equation

$$\phi(x) = 1 + \int_0^x \phi(y)\Delta y. \tag{3.17}$$

We differentiate the Eq. (3.17) with respect to x and we find

$$\phi^{\Delta}(x) = \phi(x).$$

We substitute $x = 0$ in (3.17) and we obtain $\phi(0) = 1$. In this way we go to the following IVP

$$\begin{cases} \phi^{\Delta}(x) = \phi(x) \\ \phi(0) = 1. \end{cases} \tag{3.18}$$

Here

$$\sigma(x) = (\sqrt{x}+1)^2, \quad \mu(x) = \sigma(x) - x = 2\sqrt{x} + 1,$$
$$1 + \mu(x) = 2 + 2\sqrt{x} \neq 0 \quad \textit{for any} \quad x \in \mathscr{T}.$$

Therefore the solution of (3.18) is given by

$$\phi(x) = e_1(x, 0),$$

which is a solution of the Eq. (3.17).

Exercise 7 Find a solution of the following generalized Volterra integral equations.

1. $\phi(x) = -1 + \int_0^x y^2\phi(y)\Delta y, \quad \mathscr{T} = \mathscr{Z},$
2. $\phi(x) = 2 + \int_0^x (y^4+1)\phi(y)\Delta y, \quad \mathscr{T} = 2\mathscr{Z},$
3. $\phi(x) = 10 + 4\int_1^x (1+y)\phi(y)\Delta y, \quad \mathscr{T} = 2^{\mathscr{N}_0}.$

Answer

1. $-e_{x^2}(x, 0)$,
2. $2e_{x^4+1}(x, 0)$,
3. $10e_{4+4x}(x, 1)$.

Theorem 1 *Let $f, g : \mathscr{T} \longmapsto \mathscr{R}$ be rd-continuous functions and g be delta differentiable in $\mathscr{T}$. Then a solution of the generalized Volterra integral equation*

$$\phi(x) = f(x) + \int_a^x g(y)\phi(y)\Delta y, \quad a \in \mathscr{T}, \tag{3.19}$$

is given by the expression

$$\phi(x) = e_g(x, a)f(a) + \int_a^x \frac{e_g(x, \tau)}{1 + \mu(\tau)g(\tau)} f^{\Delta}(\tau)\Delta\tau. \tag{3.20}$$

Proof We differentiate the Eq. (3.19) with respect to x and we get

$$\phi^{\Delta}(x) = f^{\Delta}(x) + g(x)\phi(x).$$

We substitute $x = a$ in (3.19) and we find

$$\phi(a) = f(a).$$

Therefore we get to the following IVP

$$\begin{cases} \phi^{\Delta}(x) = g(x)\phi(x) + f^{\Delta}(x) \\ \phi(a) = f(a). \end{cases}$$

The solution of the last IVP is given by (3.20).

Example 17 Let $\mathscr{T} = \mathscr{Z}$. Consider the equation

$$\phi(x) = x^2 + x + \int_0^x y\phi(y)\Delta y.$$

Here

$$f(x) = x^2 + x, \quad g(x) = x, \quad a = 0,$$
$$\sigma(x) = x + 1, \quad \mu(x) = \sigma(x) - x = x + 1 - x = 1, \quad x \in \mathscr{T}.$$

Hence,

$$\begin{aligned} f^{\Delta}(x) &= \sigma(x) + x + 1 \\ &= x + 1 + x + 1 \\ &= 2x + 2, \\ f(0) &= 0. \end{aligned}$$

Then, using (3.20), we find

$$\begin{aligned}\phi(x) &= \int_0^x \frac{e_x(x,\tau)}{1+\tau}(2\tau+2)\Delta\tau\\ &= 2\int_0^x e_x(x,\tau)\Delta\tau.\end{aligned}$$

Example 18 Let $\mathscr{T} = \mathscr{N}_0^3$. Consider the equation

$$\phi(x) = 2x^2 - 3x + \int_1^x (y^2+y)\phi(y)\Delta y.$$

Here

$$\begin{aligned}f(x) &= 2x^2 - 3x, \quad g(x) = x^2 + x, \quad a = 1,\\ \sigma(x) &= (\sqrt[3]{x}+1)^3,\\ \mu(x) &= \sigma(x) - x\\ &= (\sqrt[3]{x}+1)^3 - x\\ &= x + 3\sqrt[3]{x^2} + 3\sqrt[3]{x} + 1 - x\\ &= 3\sqrt[3]{x^2} + 3\sqrt[3]{x} + 1,\\ \frac{1}{1+\mu(x)g(x)} &= \frac{1}{1+(3\sqrt[3]{x^2}+3\sqrt[3]{x}+1)(x^2+x)},\\ f^{\Delta}(x) &= 2(\sigma(x)+x) - 3\\ &= 2(\sqrt[3]{x}+1)^3 + 2x - 3\\ &= 2x + 6\sqrt[3]{x^2} + 6\sqrt[3]{x} + 2 + 2x - 3\\ &= 4x + 6\sqrt[3]{x^2} + 6\sqrt[3]{x} - 1,\\ f(1) &= -1.\end{aligned}$$

Then, using (3.20), we find

$$\phi(x) = -e_{x^2+x}(x,1) + \int_1^x \frac{e_{x^2+x}(x,\tau)}{1+(3\sqrt[3]{\tau^2}+3\sqrt[3]{\tau}+1)(\tau^2+\tau)}(4\tau + 6\sqrt[3]{\tau^2} + 6\sqrt[3]{\tau} - 1)\Delta\tau.$$

Exercise 8 Using DEM, find a solution of the following generalized Volterra integral equations.

1. $\phi(x) = x^3 - 4x + \int_1^x y\phi(y)\Delta y, \quad \mathscr{T} = 2^{\mathscr{N}_0},$
2. $\phi(x) = x^2 + 2x + 1 + \int_1^x (y^2+y)\phi(y)\Delta y, \quad \mathscr{T} = 3\mathscr{Z},$
3. $\phi(x) = 2x + 3 + \int_0^x (y^4+y^2+1)\phi(y)\Delta y, \quad \mathscr{T} = \mathscr{Z}.$

Answer

1. $\phi(x) = -3e_x(x, 1) + \int_1^x \frac{e_x(x, \tau)}{1+\tau^2}(7\tau^2 - 4)\Delta\tau,$
2. $\phi(x) = 4e_{x^2+x}(x, 1) + \int_1^x \frac{e_{x^2+x}(x, \tau)}{3\tau^2 + 3\tau + 1}(2\tau + 5)\Delta\tau,$
3. $\phi(x) = 3e_{x^4+x^2+1}(x, 0) + 2\int_0^x \frac{e_{x^4+x^2+1}(x, \tau)}{2 + \tau^2 + \tau^4}\Delta\tau.$

Theorem 2 *Let $f, g, h : \mathscr{T} \longmapsto \mathscr{R}$ be rd-continuous functions, and f and h be delta differentiable in $\mathscr{T}$, and $h(x) \neq 0$ for all $x \in \mathscr{T}$. Then the generalized Volterra integral equation*

$$\phi(x) = f(x) + h(x)\int_a^x g(y)\phi(y)\Delta y, \quad a \in \mathscr{T}, \tag{3.21}$$

has a solution given by the expression

$$\phi(x) = e_l(x, a)f(a) + \int_a^x \frac{e_l(x, \tau)}{1 + \mu(\tau)l(\tau)}\left(f^\Delta(\tau) - \frac{h^\Delta(\tau)f(\tau)}{h(\tau)}\right)\Delta\tau, \tag{3.22}$$

where

$$l(x) = \frac{h^\Delta(x)}{h(x)} + h(\sigma(x))g(x)$$

and

$$1 + l(x)\mu(x) \neq 0 \quad \textit{for any} \quad x \in \mathscr{T}.$$

Proof From (3.21) we get

$$\phi(x) - f(x) = h(x)\int_a^x g(y)\phi(y)\Delta y,$$

whereupon

$$\int_a^x g(y)\phi(y)\Delta y = \frac{\phi(x) - f(x)}{h(x)}. \tag{3.23}$$

We differentiate (3.21) with respect to x and we get

$$\phi^\Delta(x) = f^\Delta(x) + h^\Delta(x)\int_a^x g(y)\phi(y)\Delta y + h(\sigma(x))g(x)\phi(x).$$

Applying (3.23), we obtain

$$\begin{aligned}
\phi^{\Delta}(x) &= f^{\Delta}(x) + h^{\Delta}(x)\frac{\phi(x) - f(x)}{h(x)} + h(\sigma(x))g(x)\phi(x) \\
&= f^{\Delta}(x) - \frac{h^{\Delta}(x)f(x)}{h(x)} + \frac{h^{\Delta}(x)}{h(x)}\phi(x) + h(\sigma(x))g(x)\phi(x) \\
&= \left(\frac{h^{\Delta}(x)}{h(x)} + h(\sigma(x))g(x)\right)\phi(x) + f^{\Delta}(x) - \frac{h^{\Delta}(x)f(x)}{h(x)} \\
&= l(x)\phi(x) + f^{\Delta}(x) - \frac{h^{\Delta}(x)f(x)}{h(x)}.
\end{aligned}$$

We substitute $x = a$ in (3.21) and we find

$$\phi(a) = f(a).$$

In this way we go to the following IVP

$$\begin{cases} \phi^{\Delta}(x) = l(x)\phi(x) + f^{\Delta}(x) - \frac{h^{\Delta}(x)f(x)}{h(x)} \\ \phi(a) = f(a). \end{cases}$$

The solution of the last IVP is given by (3.22).

Example 19 Let $\mathscr{T} = 2\mathscr{Z}$. Consider the equation

$$\phi(x) = x + 2 + (x^2 + 1)\int_0^x y\phi(y)\Delta y.$$

Here

$$\begin{aligned}
&\sigma(x) = x + 2, \quad \mu(x) = \sigma(x) - x = x + 2 - x = 2, \quad a = 0, \\
&f(x) = x + 2, \quad g(x) = x, \quad h(x) = x^2 + 1.
\end{aligned}$$

Then

$$\begin{aligned}
h^{\Delta}(x) &= \sigma(x) + x \\
&= x + 2 + x \\
&= 2x + 2, \\
h(\sigma(x)) &= \sigma^2(x) + 1 \\
&= (x + 2)^2 + 1 \\
&= x^2 + 4x + 5,
\end{aligned}$$

$$\begin{aligned}
l(x) &= \frac{h^{\Delta}(x)}{h(x)} + h(\sigma(x))g(x)\\
&= \frac{2x+2}{x^2+1} + (x^2+4x+5)x\\
&= \frac{2x+2+x^5+4x^4+5x^3+x^3+4x^2+5x}{x^2+1}\\
&= \frac{x^5+4x^4+6x^3+4x^2+7x+2}{x^2+1},
\end{aligned}$$

$$\begin{aligned}
f^{\Delta}(x) - \frac{h^{\Delta}(x)}{h(x)}f(x) &= 1 - \frac{2x+2}{x^2+1}(x+2)\\
&= \frac{x^2+1-2x^2-4x-2x-4}{x^2+1}\\
&= -\frac{x^2+6x+3}{x^2+1},\\
f^{\Delta}(x) &= 1,\\
f(0) &= 2,\\
1 + l(\tau)\mu(\tau) &= 1 + 2\frac{\tau^5+4\tau^4+6\tau^3+4\tau^2+7\tau+2}{\tau^2+1}\\
&= \frac{\tau^2+1+2\tau^5+8\tau^4+12\tau^3+8\tau^2+14\tau+4}{\tau^2+1}\\
&= \frac{2\tau^5+8\tau^4+12\tau^3+9\tau^2+14\tau+5}{\tau^2+1}.
\end{aligned}$$

Then, using (3.22), we obtain

$$\begin{aligned}
\phi(x) &= 2e_{\frac{x^5+4x^4+6x^3+4x^2+7x+2}{x^2+1}}(x,0)\\
&\quad - \int_0^x e_{\frac{x^5+4x^4+6x^3+4x^2+7x+2}{x^2+1}}(x,\tau)\frac{\tau^2+6\tau+3}{2\tau^5+8\tau^4+12\tau^3+9\tau^2+14\tau+5}\Delta\tau.
\end{aligned}$$

Example 20 Let $\mathscr{T} = 3^{\mathscr{N}_0}$. Consider the equation

$$\phi(x) = x^2 + x\int_1^x y\phi(y)\Delta y.$$

Here

$$\begin{aligned}
&\sigma(x) = 3x, \quad \mu(x) = \sigma(x) - x = 3x - x = 2x, \quad a = 1,\\
&f(x) = x^2, \quad g(x) = h(x) = x.
\end{aligned}$$

Then

$$\begin{aligned} h^{\Delta}(x) &= 1, \\ h(\sigma(x)) &= \sigma(x) \\ &= 3x, \end{aligned}$$

$$\begin{aligned} l(x) &= \frac{h^{\Delta}(x)}{h(x)} + h(\sigma(x))g(x) \\ &= \frac{1}{x} + 3x^2 \\ &= \frac{1+3x^3}{x}, \\ 1+\mu(\tau)l(\tau) &= 1 + \frac{1+3\tau^3}{\tau}2\tau \\ &= 1 + 2(1+3\tau^3) \\ &= 3 + 6\tau^3, \\ f^{\Delta}(\tau) &= \sigma(\tau) + \tau \\ &= 3\tau + \tau \\ &= 4\tau, \\ f(1) &= 1, \\ f^{\Delta}(\tau) - \frac{h^{\Delta}(\tau)f(\tau)}{h(\tau)} &= 4\tau - \frac{\tau^2}{\tau} \\ &= 4\tau - \tau \\ &= 3\tau. \end{aligned}$$

Hence, using (3.22), we get

$$\phi(x) = e_{\frac{1+3x^3}{x}}(x, 1) + \int_1^x e_{\frac{1+3x^3}{x}}(x, \tau)\frac{\tau}{1+2\tau^3}\Delta\tau.$$

Exercise 9 Using DEM, find a solution of the following generalized Volterra integral equations.

1. $\phi(x) = x - 3 + 2x^2 \int_1^x y\phi(y)\Delta y, \quad \mathscr{T} = \mathscr{N},$
2. $\phi(x) = x + (x+1) \int_1^x (y+1)\phi(y)\Delta y, \quad \mathscr{T} = \mathscr{N}_0,$
3. $\phi(x) = x - 2x \int_1^x y\phi(y)\Delta y, \quad \mathscr{T} = 2^{\mathscr{N}_0}.$

Answer

1.

$$\begin{aligned}\phi(x) &= -2e_{\frac{2x^5+4x^4+2x^3+2x+1}{x^2}}(x,1)\\ &\quad -\int_1^x e_{\frac{2x^5+4x^4+2x^3+2x+1}{x^2}}(x,\tau)\frac{\tau^2-5\tau-3}{2\tau^5+4\tau^4+2\tau^3+\tau^2+2\tau+1}\Delta\tau,\end{aligned}$$

2.

$$\begin{aligned}\phi(x) &= e_{\frac{x^3+4x^2+5x+3}{x+1}}(x,0)\\ &\quad +\int_1^x e_{\frac{x^3+4x^2+5x+3}{x+1}}(x,\tau)\frac{1}{\tau^3+4\tau^2+6\tau+4}\Delta\tau,\end{aligned}$$

3. $\phi(x) = e_{\frac{1-4x^3}{x}}(x,1)$.

Theorem 3 *Let $f, g : \mathscr{T} \longmapsto \mathscr{R}$ be rd-continuous, and f be delta differentiable in $\mathscr{T}$, $1+\mu(x)g(x) \neq 0$ for all $x \in \mathscr{T}$. Then a solution of the equation*

$$\phi(x) = f(x) - \int_a^x g(y)\phi(\sigma(y))\Delta y, \quad a \in \mathscr{T}, \tag{3.24}$$

has a solution in the following form

$$\begin{aligned}\phi(x) &= e_{\ominus g}(x,a)f(a) + \textstyle\int_a^x e_{\ominus g}(x,\tau)f^{\Delta}(\tau)\Delta\tau\\ &= e_{\ominus g}(x,a)f(a) + \textstyle\int_a^x \frac{e_g(\tau,a)}{e_g(x,a)}f^{\Delta}(\tau)\Delta\tau.\end{aligned} \tag{3.25}$$

Proof We differentiate the Eq. (3.24) with respect to x and we get

$$\phi^{\Delta}(x) = f^{\Delta}(x) - g(x)\phi(\sigma(x)).$$

We substitute $x = a$ in (3.24) and we find

$$\phi(a) = f(a).$$

In this way we get the following IVP

$$\begin{cases}\phi^{\Delta}(x) = -g(x)\phi(\sigma(x)) + f^{\Delta}(x)\\ \phi(a) = f(a).\end{cases}$$

The solution of the last IVP is given by (3.25).

Example 21 Let $\mathscr{T} = \mathscr{Z}$. Consider the equation

$$\phi(x) = x^2 + 1 - \int_0^x y\phi(y+1)\Delta y.$$

Here

$$\begin{aligned}&\sigma(x) = x+1, \quad \mu(x) = \sigma(x) - x = x + 1 - x = 1, \quad a = 0,\\ &f(x) = x^2 + 1, \quad g(x) = x.\end{aligned}$$

Then

$$\begin{aligned}\ominus g &= -\frac{g(x)}{1+\mu(x)g(x)}\\ &= -\frac{x}{1+x},\\ f^{\Delta}(x) &= \sigma(x) + x\\ &= x + 1 + x\\ &= 2x + 1,\\ f(0) &= 1.\end{aligned}$$

Using (3.25), we get

$$\phi(x) = e_{-\frac{x}{x+1}}(x, 0) + \int_0^x e_{-\frac{x}{x+1}}(x, \tau)(2\tau + 1)\Delta\tau.$$

Example 22 Let $\mathscr{T} = 3^{\mathscr{N}_0}$. Consider the equation

$$\phi(x) = x^2 + 2\int_1^x y^3\phi(3y)\Delta y.$$

Here

$$\begin{aligned}&\sigma(x) = 3x, \quad \mu(x) = \sigma(x) - x = 3x - x = 2x, \quad a = 1,\\ &f(x) = x^2, \quad g(x) = -2x^3.\end{aligned}$$

Then

$$\begin{aligned}\ominus g(x) &= -\frac{g(x)}{1+\mu(x)g(x)}\\ &= -\frac{-2x^3}{1+2x(-2x^3)}\\ &= \frac{2x^3}{1-4x^4},\end{aligned}$$

$$\begin{aligned} f^{\Delta}(x) &= \sigma(x) + x \\ &= 3x + x \\ &= 4x, \\ f(1) &= 1. \end{aligned}$$

Hence, using (3.25), we obtain

$$\phi(x) = e_{\frac{2x^3}{1-4x^4}}(x, 1) + 4\int_1^x e_{\frac{2x^3}{1-4x^4}}(x, \tau)\tau\Delta\tau.$$

Exercise 10 Using DEM, find a solution to the following generalized Volterra integral equations.

1. $\phi(x) = x^2 - 3\int_0^x y\phi(y+1)\Delta y, \quad \mathscr{T} = \mathscr{Z},$
2. $\phi(x) = x^2 + 4x + 1 - \int_1^x (y^2 + 2y)\phi(2y)\Delta y, \quad \mathscr{T} = 2^{\mathscr{N}_0},$
3. $\phi(x) = x^3 + 7x^2 + 6x - \int_1^x y\phi(3y)\Delta y, \quad \mathscr{T} = 3^{\mathscr{N}_0}.$

Answer

1. $\phi(x) = \int_0^x e_{-\frac{3x}{1+3x}}(x, \tau)(2\tau + 1)\Delta\tau,$
2. $\phi(x) = 6e_{-\frac{x^2+2x}{x^3+2x^2+1}}(x, 1) + \int_1^x e_{-\frac{x^2+2x}{x^3+2x^2+1}}(x, \tau)(3\tau + 4)\Delta\tau,$
3. $\phi(x) = 14e_{-\frac{x}{1+2x^2}}(x, 1) + \int_1^x e_{-\frac{x}{1+2x^2}}(x, \tau)(13\tau^2 + 28\tau + 6)\Delta\tau.$

3.1.5 *The Successive Approximations Method*

The successive approximations method (SAM), also called the Picard iteration method, provides a scheme that can be used for solving initial value problems or generalized integral equations.

Given the linear generalized Volterra integral equation of the second kind

$$\phi(x) = u(x) + \lambda\int_a^x K(x, y)\phi(y)\Delta y, \quad a \in \mathscr{T},$$

where $\phi(x)$ is unknown function to be determined, $K(x, y)$ is the kernel, and λ is a parameter. The successive approximations method introduces the recurrence relation

$$\phi_n(x) = u(x) + \lambda \int_a^x K(x, y)\phi_{n-1}(y)\Delta y, \quad n \in \mathscr{N}, \tag{3.26}$$

when the zeroth approximation $\phi_0(x)$ can be any selective real-valued function. We select $0, 1, x$ for $\phi_0(x)$, and using (3.26), several successive approximations $\phi_k(x)$, $k \geq 1$, will be determined as follows.

$$\begin{aligned}
\phi_1(x) &= u(x) + \lambda \int_a^x K(x, y)\phi_0(y)\Delta y \\
\phi_2(x) &= u(x) + \lambda \int_a^x K(x, y)\phi_1(y)\Delta y \\
\phi_3(x) &= u(x) + \lambda \int_a^x K(x, y)\phi_2(y)\Delta y \\
&\vdots \\
\phi_n(x) &= u(x) + \lambda \int_a^x K(x, y)\phi_{n-1}(y)\Delta y.
\end{aligned}$$

We point out that the successive approximations method for the equation

$$\phi(x) = u(x) + \lambda \int_a^x K(x, y)\phi(\sigma(y))\Delta y$$

introduces the recurrence formula

$$\phi_n(x) = u(x) + \lambda \int_a^x K(x, y)\phi_{n-1}(\sigma(y))\Delta y, \quad n \in \mathscr{N}. \tag{3.27}$$

For the zeroth approximation $\phi_0(x)$ we select $0, 1, x$.
The question of convergence $\phi_n(x)$, defined by (3.26) or (3.27), will be discussed in the last section of this chapter.

Example 23 Let $\mathscr{T} = \mathscr{Z}$. Consider the equation

$$\phi(x) = x + \int_0^x y\phi(y)\Delta y.$$

We select $\phi_0(x) = 1$. Then

$$\begin{aligned}
\phi_1(x) &= x + \int_0^x y\phi_0(y)\Delta y \\
&= x + \int_0^x y\Delta y \\
&= x + \int_0^x \left(\frac{1}{2}(y^2)^\Delta - \frac{1}{2}\right)\Delta y \\
&= x + \frac{1}{2}\int_0^x (y^2)^\Delta \Delta y - \frac{1}{2}\int_0^x \Delta y \\
&= x - \frac{1}{2}x + \frac{1}{2}x^2 \\
&= \frac{1}{2}x + \frac{1}{2}x^2, \\
\phi_2(x) &= x + \int_0^x y\phi_1(y)\Delta y \\
&= x + \frac{1}{2}\int_0^x y(y + y^2)\Delta y \\
&= x + \frac{1}{2}\int_0^x (y^2 + y^3)\Delta y \\
&= x + \frac{1}{2}\int_0^x \left(\frac{1}{3}(y^3)^\Delta - \frac{1}{2}(y^2)^\Delta + \frac{1}{6} + \frac{1}{4}(y^4)^\Delta - \frac{1}{2}(y^3)^\Delta + \frac{1}{4}(y^2)^\Delta\right)\Delta y \\
&= x + \frac{1}{2}\int_0^x \left(-\frac{1}{4}(y^2)^\Delta - \frac{1}{6}(y^3)^\Delta + \frac{1}{4}(y^4)^\Delta + \frac{1}{6}\right)\Delta y \\
&= x - \frac{1}{8}\int_0^x (y^2)^\Delta \Delta y - \frac{1}{12}\int_0^x (y^3)^\Delta \Delta y + \frac{1}{8}\int_0^x (y^4)^\Delta \Delta y + \frac{1}{12}\int_0^x \Delta y \\
&= x - \frac{1}{8}x^2 - \frac{1}{12}x^3 + \frac{1}{8}x^4 + \frac{1}{12}x \\
&= \frac{13}{12}x - \frac{1}{8}x^2 - \frac{1}{12}x^3 + \frac{1}{8}x^4.
\end{aligned}$$

Example 24 Let $\mathscr{T} = 2^{\mathscr{N}_0} \cup \{0\}$. Consider the equation

$$\phi(x) = x^2 + x\int_0^x y^2\phi(y)\Delta y.$$

We select $\phi_0(x) = 1$. Then

$$\begin{aligned}
\phi_1(x) &= x^2 + x\int_0^x y^2\phi_0(y)\Delta y \\
&= x^2 + x\int_0^x y^2\Delta y \\
&= x^2 + \frac{1}{7}x\int_0^x (y^3)^\Delta \Delta y \\
&= x^2 + \frac{1}{7}x^4, \\
\phi_2(x) &= x^2 + x\int_0^x y^2\phi_1(y)\Delta y \\
&= x^2 + x\int_0^x y^2\left(y^2 + \frac{1}{7}y^4\right)\Delta y \\
&= x^2 + x\int_0^x \left(y^4 + \frac{1}{7}y^6\right)\Delta y \\
&= x^2 + x\int_0^x \left(\frac{1}{31}(y^5)^\Delta + \frac{1}{889}(y^7)^\Delta\right)\Delta y \\
&= x^2 + \frac{1}{31}x\int_0^x (y^5)^\Delta\Delta y + \frac{1}{889}x\int_0^x (y^7)^\Delta\Delta y \\
&= x^2 + \frac{1}{31}x^6 + \frac{1}{889}x^8.
\end{aligned}$$

Exercise 11 Let $\mathscr{T} = 3^{\mathscr{N}_0} \cup \{0\}$. Consider the equation

$$\phi(x) = 2x + 2\int_0^x y^2\phi(y)\Delta y.$$

Using SAM and $\phi_0(x) = 1$, find $\phi_1(x)$ and $\phi_2(x)$.

Answer

$$\phi_1(x) = 2x + \frac{2}{13}x^3, \quad \phi_2(x) = 2x + \frac{1}{10}x^4 + \frac{1}{1183}x^6.$$

Example 25 Let $\mathscr{T} = \mathscr{Z}$. Consider the equation

$$\phi(x) = x^2 + 2\int_0^x y\phi(y+1)\Delta y.$$

Let $\phi_0(x) = 0$. Then

$$
\begin{aligned}
\phi_1(x) &= x^2, \\
\phi_2(x) &= x^2 + 2\int_0^x y\phi_1(y+1)\Delta y \\
&= x^2 + 2\int_0^x y(y+1)^2\Delta y \\
&= x^2 + 2\int_0^x y(y^2+2y+1)\Delta y \\
&= x^2 + 2\int_0^x (y^3+2y^2+y)\Delta y \\
&= x^2 + 2\int_0^x \Big(\frac{1}{4}(y^4)^\Delta - \frac{1}{2}(y^3)^\Delta + \frac{1}{4}(y^2)^\Delta + 2\left(\frac{1}{3}(y^3)^\Delta - \frac{1}{2}(y^2)^\Delta + \frac{1}{6}\right) \\
&\quad + \frac{1}{2}(y^2)^\Delta - \frac{1}{2}\Big)\Delta y \\
&= x^2 + 2\int_0^x \Big(\frac{1}{4}(y^4)^\Delta - \frac{1}{2}(y^3)^\Delta + \frac{1}{4}(y^2)^\Delta + \frac{2}{3}(y^3)^\Delta - (y^2)^\Delta \\
&\quad + \frac{1}{3} + \frac{1}{2}(y^2)^\Delta - \frac{1}{2}\Big)\Delta y \\
&= x^2 + 2\int_0^x \left(\frac{1}{4}(y^4)^\Delta + \frac{1}{6}(y^3)^\Delta - \frac{1}{4}(y^2)^\Delta - \frac{1}{6}\right)\Delta y \\
&= x^2 + \frac{1}{2}\int_0^x (y^4)^\Delta\Delta y + \frac{1}{3}\int_0^x (y^3)^\Delta\Delta y - \frac{1}{2}\int_0^x (y^2)^\Delta\Delta y - \frac{1}{3}\int_0^x \Delta y \\
&= x^2 + \frac{1}{2}x^4 + \frac{1}{3}x^3 - \frac{1}{2}x^2 - \frac{1}{3}x \\
&= \frac{1}{2}x^4 + \frac{1}{3}x^3 + \frac{1}{2}x^2 - \frac{1}{3}x.
\end{aligned}
$$

Example 26 Let $\mathscr{T} = 2^{\mathscr{N}_0} \cup \{0\}$. Consider the equation

$$\phi(x) = x + (2x+1)\int_0^x y\phi(2y)\Delta y.$$

Let $\phi_0(x) = 0$. Then

$$
\begin{aligned}
\phi_1(x) &= x, \\
\phi_2(x) &= x + (2x+1)\int_0^x y\phi_1(2y)\Delta y \\
&= x + 2(2x+1)\int_0^x y^2\Delta y \\
&= x + \frac{2}{7}(2x+1)\int_0^x (y^3)^\Delta\Delta y \\
&= x + \frac{2}{7}x^3(2x+1).
\end{aligned}
$$

Exercise 12 Let $\mathscr{T} = \mathscr{Z}$. Consider the equation

$$\phi(x) = 1 + \int_0^x y\phi(y+1)\Delta y.$$

Using SAM and $\phi_0(x) = 1$, find $\phi_1(x)$ and $\phi_2(x)$.

Answer

$$\phi_1(x) = 1 - \frac{1}{2}x + \frac{1}{2}x^2, \quad \phi_2(x) = 1 - \frac{5}{12}x + \frac{3}{8}x^2 - \frac{1}{12}x^3 + \frac{1}{8}x^4.$$

3.2 Conversion of a Generalized Volterra Integral Equation of the First Kind to a Generalized Volterra Integral Equation of the Second Kind

In this section we will represent a method that will convert Volterra integral equations of the first kind to Volterra integral equations of the second kind. Having converted the Volterra integral equation of the first kind to the Volterra integral equation of the second kind, we then can use any method that was presented before.
Consider the generalized Volterra integral equation of the first kind

$$u(x) = \lambda \int_a^x K(x,y)\phi(y)\Delta y, \quad a \in \mathscr{T}, \tag{3.28}$$

where $\phi : \mathscr{T} \longmapsto \mathscr{R}$ is unknown function to be determined, $K : \mathscr{T} \times \mathscr{T} \longmapsto \mathscr{R}$ is the kernel, $\lambda \neq 0$ is a parameter, $u : \mathscr{T} \longmapsto \mathscr{R}$ is a given function.
Suppose that $K_x^{\Delta}(x,y)$ exists for any $(x,y) \in \mathscr{T}^{\kappa} \times \mathscr{T}$, $K_x^{\Delta}(\sigma(x),x) \neq 0$ for any $x \in \mathscr{T}^{\kappa}$. Also, assume that $u^{\Delta}(x)$ exists for any $x \in \mathscr{T}^{\kappa}$.
Differentiating the Eq. (3.28) with respect to x we obtain

$$u^{\Delta}(x) = \lambda \int_a^x K_x^{\Delta}(x,y)\phi(y)\Delta y + \lambda K(\sigma(x),x)\phi(x)$$

or

$$\lambda K(\sigma(x),x)\phi(x) = u^{\Delta}(x) - \lambda \int_a^x K_x^{\Delta}(x,y)\phi(y)\Delta y,$$

whereupon

$$\phi(x) = \frac{1}{\lambda K(\sigma(x),x)}u^{\Delta}(x) - \int_a^x \frac{K_x^{\Delta}(x,y)}{K(\sigma(x),x)}\phi(y)\Delta y, \tag{3.29}$$

which is a generalized Volterra integral equation of the second kind.

Example 27 Let $\mathscr{T} = \mathscr{Z}$. Consider the equation

$$x^3 = 2\int_0^x (x+y)^2\phi(y)\Delta y,$$

which is a generalized Volterra integral equation of the first kind.
Here

$$\sigma(x) = x+1, \quad x \in \mathscr{T}, \quad \lambda = 2, \quad a = 0,$$
$$u(x) = x^3, \quad K(x,y) = (x+y)^2.$$

Then

$$\begin{aligned}
u^{\Delta}(x) &= \sigma^2(x) + x\sigma(x) + x^2\\
&= (x+1)^2 + x(x+1) + x^2\\
&= x^2 + 2x + 1 + x^2 + x + x^2\\
&= 3x^2 + 3x + 1,\\
K(\sigma(x), x) &= (\sigma(x) + x)^2\\
&= (x+1+x)^2\\
&= (2x+1)^2\\
&= 4x^2 + 4x + 1,\\
K_x^{\Delta}(x,y) &= \sigma(x) + y + x + y\\
&= x + 1 + x + 2y\\
&= 2x + 2y + 1.
\end{aligned}$$

Hence, using (3.29), we go to the following generalized Volterra integral equation of the second kind

$$\phi(x) = \frac{3x^2+3x+1}{8x^2+8x+2} - \int_0^x \frac{2x+2y+1}{4x^2+4x+1}\phi(y)\Delta y.$$

Example 28 Let $\mathscr{T} = 2^{\mathscr{N}_0}$. Consider the equation

$$x^2 + x = \int_1^x (x^2 + 2xy)\phi(y)\Delta y.$$

Here

$$\sigma(x) = 2x, \quad x \in \mathscr{T}, \quad a = 1, \quad \lambda = 1,$$
$$u(x) = x^2 + x, \quad K(x,y) = x^2 + 2xy.$$

Then

$$
\begin{aligned}
u^{\Delta}(x) &= \sigma(x) + x + 1 \\
&= 2x + x + 1 \\
&= 1 + 3x, \\
K_x^{\Delta}(x, y) &= \sigma(x) + x + 2y \\
&= 2x + x + 2y \\
&= 3x + 2y, \\
K(\sigma(x), x) &= \sigma^2(x) + 2\sigma(x)x \\
&= 4x^2 + 4x^2 \\
&= 8x^2.
\end{aligned}
$$

Then, using (3.29), we get the following generalized Volterra integral equation of the second kind.

$$
\phi(x) = \frac{1 + 3x}{8x^2} - \int_1^x \frac{3x + 2y}{8x^2} \phi(y) \Delta y.
$$

Example 29 Let $\mathscr{T} = 3^{\mathscr{N}_0}$. Consider the generalized Volterra integral equation of the first kind

$$
e_x(x, 1) + x^4 + x = 2 \int_1^x (\sinh_x(x, 2) + x^2 + xy) \phi(y) \Delta y.
$$

Here

$$
\begin{aligned}
&\sigma(x) = 3x, \quad x \in \mathscr{T}, \quad \lambda = 2, \quad a = 1, \\
&u(x) = e_x(x, 1) + x^4 + x, \quad K(x, y) = \sinh_x(x, 2) + x^2 + xy.
\end{aligned}
$$

Then

$$
\begin{aligned}
u^{\Delta}(x) &= x e_x(x, 1) + \sigma^3(x) + x\sigma^2(x) + x^2\sigma(x) + x^3 + 1 \\
&= x e_x(x, 1) + 27x^3 + 9x^3 + 3x^3 + x^3 + 1 \\
&= x e_x(x, 1) + 40x^3 + 1, \\
K_x^{\Delta}(x, y) &= x \cosh_x(x, 2) + \sigma(x) + x + y \\
&= x \cosh_x(x, 2) + 4x + y, \\
K(\sigma(x), x) &= \sinh_{3x}(3x, 2) + 9x^2 + 3x^2 \\
&= \sinh_{3x}(3x, 2) + 12x^2.
\end{aligned}
$$

Then, using (3.29), we get the following generalized Volterra integral equation of the second kind.

$$
\phi(x) = \frac{x e_x(x, 1) + 40x^3 + 1}{2(\sinh_{3x}(3x, 2) + 12x^2)} - \int_1^x \frac{x \cosh_x(x, 2) + 4x + y}{\sinh_{3x}(3x, 2) + 12x^2} \phi(y) \Delta y.
$$

Exercise 13 Convert the following generalized Volterra integral equations of the first kind to generalized Volterra integral equations of the second kind.

1. $x^2 + 2x + 3 = 2\int_0^x (x+y)\phi(y)\Delta y, \quad \mathscr{T} = \mathscr{Z},$
2. $x^2 + 2x = \int_1^x (2x+y)^2\phi(y)\Delta y, \quad \mathscr{T} = 2^{\mathscr{N}_0}.$

Answer

1. $\phi(x) = \dfrac{2x+3}{4x+2} - \displaystyle\int_0^x \frac{1}{2x+1}\phi(y)\Delta y,$
2. $\phi(x) = \dfrac{3x+2}{25x^2} - 2\displaystyle\int_1^x \frac{6x+2y}{25x^2}\phi(y)\Delta y.$

3.3 Existence and Uniqueness of Solutions

In this section we will state and prove the existence and uniqueness of the solutions of generalized Volterra integral equations of the first and second kind. For this purpose we need of some preliminary results.

3.3.1 *Preliminary Results*

Let $s, t \in \mathscr{T}$. Define the polynomials.

$$g_0(t,s) = h_0(t,s) = 1,$$
$$g_{k+1}(t,s) = \int_s^t g_k(\sigma(\tau), s)\Delta\tau, \quad h_{k+1}(t,s) = \int_s^t h_k(\tau, s)\Delta\tau, \quad k = 0, 1, 2, \ldots.$$

We have

$$\begin{aligned}
g_1(t,s) &= \int_s^t g_0(\sigma(\tau), s)\Delta\tau \\
&= \int_s^t \Delta\tau \\
&= t - s, \\
g_2(t,s) &= \int_s^t g_1(\sigma(\tau), s)\Delta\tau \\
&= \int_s^t (\sigma(\tau) - s)\Delta\tau, \\
h_1(t,s) &= \int_s^t h_0(\tau, s)\Delta\tau \\
&= \int_s^t \Delta\tau \\
&= t - s,
\end{aligned}$$

$$\begin{aligned} h_2(t,s) &= \int_s^t h_1(\tau,s)\Delta\tau \\ &= \int_s^t (\tau - s)\Delta\tau, \end{aligned}$$

and so on.
Also,

$$g_k^{\Delta}(t,s) = g_{k-1}(\sigma(t),s), \quad h_k^{\Delta}(t,s) = h_{k-1}(t,s), \quad k \in \mathscr{N}.$$

Lemma 1 *Let $n \in \mathscr{N}$. If f is n times differentiable and p_k, $0 \le k \le n-1$, are differentiable at some $t \in \mathscr{T}$ with*

$$p_{k+1}^{\Delta}(t) = p_k^{\sigma}(t) \quad \textit{for all} \quad 0 \le k \le n-2, \quad n \ge 2,$$

then we have

$$\left(\sum_{k=0}^{n-1}(-1)^k f^{\Delta^k}(t)p_k(t)\right)^{\Delta} = (-1)^{n-1} f^{\Delta^n}(t)p_{n-1}^{\sigma}(t) + f(t)p_0^{\Delta}(t).$$

Proof We have

$$\begin{aligned} \left(\sum_{k=0}^{n-1}(-1)^k f^{\Delta^k}(t)p_k(t)\right)^{\Delta} &= \sum_{k=0}^{n-1}(-1)^k \left(f^{\Delta^k}(t)p_k(t)\right)^{\Delta} \\ &= \sum_{k=0}^{n-1}(-1)^k \left(f^{\Delta^{k+1}}(t)p_k^{\sigma}(t) + f^{\Delta^k}(t)p_k^{\Delta}(t)\right) \\ &= \sum_{k=0}^{n-1}(-1)^k f^{\Delta^{k+1}}(t)p_k^{\sigma}(t) + \sum_{k=0}^{n-1}(-1)^k f^{\Delta^k}(t)p_k^{\Delta}(t) \\ &= \sum_{k=0}^{n-2}(-1)^k f^{\Delta^{k+1}}(t)p_k^{\sigma}(t) + (-1)^{n-1} f^{\Delta^n}(t)p_{n-1}^{\sigma}(t) \\ &\quad + \sum_{k=1}^{n-1}(-1)^k f^{\Delta^k}(t)p_k^{\Delta}(t) + f^{\Delta^0}(t)p_0^{\Delta}(t) \\ &= \sum_{k=0}^{n-2}(-1)^k f^{\Delta^{k+1}}(t)p_{k+1}^{\Delta}(t) + (-1)^{n-1} f^{\Delta^n}(t)p_{n-1}^{\sigma}(t) \\ &\quad + \sum_{k=0}^{n-2}(-1)^{k+1} f^{\Delta^{k+1}}(t)p_{k+1}^{\Delta}(t) + f(t)p_0^{\Delta}(t) \\ &= (-1)^{n-1} f^{\Delta^n}(t)p_{n-1}^{\sigma}(t) + f(t)p_0^{\Delta}(t). \end{aligned}$$

Lemma 2 *The functions $g_n(t,s)$ satisfy for all $t \in \mathscr{T}$ the relationship*

$$g_n(\rho^k(t), t) = 0 \quad \textit{for all} \quad n \in \mathscr{N} \quad \textit{and all} \quad 0 \le k \le n-1.$$

Here $\rho^k(t) = \rho\left(\rho^{k-1}(t)\right)$.

Proof Let $n \in \mathscr{N}$ be arbitrarily chosen. Then

$$\begin{aligned} g_n(\rho^0(t), t) &= g_n(t,t) \\ &= \int_t^t g_{n-1}(\sigma(\tau), t)\Delta\tau \\ &= 0. \end{aligned}$$

Assume that

$$g_{n-1}(\rho^k(t), t) = 0 \quad \textit{and} \quad g_n(\rho^k(t), t) = 0$$

for some $0 \le k < n-1$.
We will prove that

$$g_n(\rho^{k+1}(t), t) = 0.$$

1. case. $\rho^k(t)$ is left-dense. Then

$$\rho^{k+1}(t) = \rho(\rho^k(t)) = \rho^k(t).$$

Consequently, using the induction assumption, we have

$$g_n(\rho^{k+1}(t), t) = g_n(\rho^k(t), t) = 0.$$

2. case. $\rho^k(t)$ is left-scattered. Then

$$\rho(\rho^k(t)) < \rho^k(t)$$

and there is no $s \in \mathscr{T}$ such that $\rho^{k+1}(t) < s < \rho^k(t)$. Hence,

$$\sigma(\rho^{k+1}(t)) = \rho^k(t).$$

Therefore

$$g_n(\sigma(\rho^{k+1}(t)), t) = g_n(\rho^{k+1}(t), t) + \mu(\rho^{k+1}(t))g_n^{\Delta}(\rho^{k+1}(t), t)$$

or

$$g_n(\rho^k(t), t) = g_n(\rho^{k+1}(t), t) + \mu(\rho^{k+1}(t))g_n^{\Delta}(\rho^{k+1}(t), t),$$

whereupon

$$\begin{aligned} g_n(\rho^{k+1}(t), t) &= g_n(\rho^k(t), t) - \mu(\rho^{k+1}(t)) g_n^{\Delta}(\rho^{k+1}(t), t) \\ &= g_n(\rho^k(t), t) - \mu(\rho^{k+1}(t)) g_{n-1}(\sigma(\rho^{k+1}(t)), t) \\ &= g_n(\rho^k(t), t) - \mu(\rho^{k+1}(t)) g_{n-1}(\rho^k(t), t) \\ &= 0. \end{aligned}$$

Lemma 3 *Let $n \in \mathcal{N}$, and suppose that f is $(n-1)$-times differentiable at $\rho^{n-1}(t)$. Then*

$$f(t) = \sum_{k=0}^{n-1} (-1)^k f^{\Delta^k}(\rho^{n-1}(t)) g_k(\rho^{n-1}(t), t).$$

Proof 1. Let $n = 1$. Then

$$\begin{aligned} \sum_{k=0}^{0} (-1)^k f^{\Delta^k}(\rho^0(t)) g_k(\rho^0(t), t) &= (-1)^0 f^{\Delta^0}(t) g_0(t, t) \\ &= f(t). \end{aligned}$$

2. Assume that

$$f(t) = \sum_{k=0}^{m-1} (-1)^k f^{\Delta^k}(\rho^{m-1}(t)) g_k(\rho^{m-1}(t), t)$$

for some $m \in \mathcal{N}$.

3. We will prove that

$$f(t) = \sum_{k=0}^{m} (-1)^k f^{\Delta^k}(\rho^m(t)) g_k(\rho^m(t), t).$$

1. case. $\rho^{m-1}(t)$ is left-dense. Then

$$\rho^m(t) = \rho(\rho^{m-1}(t)) = \rho^{m-1}(t).$$

Hence and the induction assumption, we obtain

$$\begin{aligned} &\textstyle\sum_{k=0}^{m} (-1)^k f^{\Delta^k}(\rho^m(t)) g_k(\rho^m(t), t) \\ &= \textstyle\sum_{k=0}^{m-1} (-1)^k f^{\Delta^k}(\rho^m(t)) g_k(\rho^m(t), t) + (-1)^m f^{\Delta^m}(\rho^m(t)) g_m(\rho^m(t), t) \\ &= \textstyle\sum_{k=0}^{m-1} (-1)^k f^{\Delta^k}(\rho^{m-1}(t)) g_k(\rho^{m-1}(t), t) \\ &+ (-1)^m f^{\Delta^m}(\rho^{m-1}(t)) g_m(\rho^{m-1}(t), t) \\ &\textit{now we apply Lemma 2 } (g_m(\rho^{m-1}(t), t) = 0) \\ &= \textstyle\sum_{k=0}^{m-1} (-1)^k f^{\Delta^k}(\rho^{m-1}(t)) g_k(\rho^{m-1}(t), t) \\ &\textit{now we apply the induction assumption} \\ &= f(t). \end{aligned}$$

2. case. $\rho^{m-1}(t)$ is left-scattered. Then

$$\rho^m(t) = \rho(\rho^{m-1}(t)) < \rho^{m-1}(t)$$

and there is no $s \in \mathscr{T}$ such that

$$\rho^m(t) < s < \rho^{m-1}(t).$$

Also,

$$\sigma(\rho^m(t)) = \rho^{m-1}(t).$$

Hence,

$$g_k(\sigma(\rho^m(t)), t) = g_k(\rho^{m-1}(t), t).$$

Therefore

$$\begin{aligned} g_k(\rho^{m-1}(t), t) &= g_k(\sigma(\rho^m(t)), t) \\ &= g_k(\rho^m(t), t) + \mu(\rho^m(t)) g_k^{\Delta}(\rho^m(t), t) \\ &= g_k(\rho^m(t), t) + \mu(\rho^m(t)) g_{k-1}(\sigma(\rho^m(t)), t) \\ &= g_k(\rho^m(t), t) + \mu(\rho^m(t)) g_{k-1}(\rho^{m-1}(t), t), \end{aligned}$$

whereupon

$$g_k(\rho^m(t), t) = g_k(\rho^{m-1}(t), t) - \mu(\rho^m(t)) g_{k-1}(\rho^{m-1}(t), t).$$

Consequently

$$\begin{aligned} &\sum_{k=0}^{m}(-1)^k f^{\Delta^k}(\rho^m(t)) g_k(\rho^m(t), t) \\ &= f(\rho^m(t)) + \sum_{k=1}^{m}(-1)^k f^{\Delta^k}(\rho^m(t)) g_k(\rho^m(t), t) \\ &= f(\rho^m(t)) + \sum_{k=1}^{m}(-1)^k f^{\Delta^k}(\rho^m(t)) g_k(\rho^{m-1}(t), t) \\ &+ \sum_{k=1}^{m}(-1)^{k-1} f^{\Delta^k}(\rho^m(t)) \mu(\rho^m(t)) g_{k-1}(\rho^{m-1}(t), t) \\ &= f(\rho^m(t)) + \sum_{k=1}^{m-1}(-1)^k f^{\Delta^k}(\rho^m(t)) g_k(\rho^{m-1}(t), t) \\ &+ (-1)^m f^{\Delta^m}(\rho^m(t)) g_m(\rho^{m-1}(t), t) \\ &+ \sum_{k=0}^{m-1}(-1)^k f^{\Delta^{k-1}}(\rho^m(t)) \mu(\rho^m(t)) g_k(\rho^{m-1}(t), t) \\ &= \sum_{k=0}^{m-1}(-1)^k f^{\Delta^k}(\rho^m(t)) g_k(\rho^{m-1}(t), t) \\ &+ \sum_{k=0}^{m-1}(-1)^k \mu(\rho^m(t)) f^{\Delta^{k+1}}(\rho^m(t)) g_k(\rho^{m-1}(t), t) \\ &= \sum_{k=0}^{m-1}(-1)^k \left(f^{\Delta^k}(\rho^m(t)) + \mu(\rho^m(t))(f^{\Delta^k})^{\Delta}(\rho^m(t))\right) g_k(\rho^{m-1}(t), t) \\ &= \sum_{k=0}^{m-1}(-1)^k f^{\Delta^k}(\sigma(\rho^m(t))) g_k(\rho^{m-1}(t), t) \\ &= \sum_{k=0}^{m-1}(-1)^k f^{\Delta^k}(\rho^{m-1}(t)) g_k(\rho^{m-1}(t), t) \\ &= f(t). \end{aligned}$$

Theorem 4 (Taylor's Formula) *Let $n \in \mathcal{N}$. Suppose f is n-times differentiable on $\mathcal{T}^{\kappa^n}$. Let $\alpha \in \mathcal{T}^{\kappa^{n-1}}$, $t \in \mathcal{T}$. Then*

$$f(t) = \sum_{k=0}^{n-1}(-1)^k g_k(\alpha, t) f^{\Delta^k}(\alpha)$$
$$+ \int_\alpha^{\rho^{n-1}(t)} (-1)^{n-1} g_{n-1}(\sigma(\tau), t) f^{\Delta^n}(\tau)\Delta\tau.$$

Proof We note that, applying Lemma 1 for $p_k = g_k$, we have

$$\begin{aligned}&\left(\sum_{k=0}^{n-1}(-1)^k g_k(\tau, t) f^{\Delta^k}(\tau)\right)_\tau^\Delta \\ &= (-1)^{n-1} f^{\Delta^n}(\tau) g_{n-1}(\sigma(\tau), t) + f(\tau) g_0^\Delta(\tau, t) \\ &= (-1)^{n-1} f^{\Delta^n}(\tau) g_{n-1}(\sigma(\tau), t) \quad \textit{for all} \quad \tau \in \mathcal{T}^{\kappa^n}.\end{aligned}$$

The last relation we integrate from α to $\rho^{n-1}(t)$ and we get

$$\int_\alpha^{\rho^{n-1}(t)} \left(\sum_{k=0}^{n-1}(-1)^k g_k(\tau, t) f^{\Delta^k}(\tau)\right)_\tau^\Delta \Delta\tau = \int_\alpha^{\rho^{n-1}(t)} (-1)^{n-1} f^{\Delta^n}(\tau) g_{n-1}(\sigma(\tau), t)\Delta\tau$$

or

$$\begin{aligned}&\sum_{k=0}^{n-1}(-1)^k g_k(\rho^{n-1}(t), t) f^{\Delta^k}(\rho^{n-1}(t)) - \sum_{k=0}^{n-1}(-1)^k g_k(\alpha, t) f^{\Delta^k}(\alpha) \\ &= \int_\alpha^{\rho^{n-1}(t)} (-1)^{n-1} f^{\Delta^n}(\tau) g_{n-1}(\sigma(\tau), t)\Delta\tau.\end{aligned}$$

Hence, applying Lemma 3,

$$f(t) - \sum_{k=0}^{n-1}(-1)^k g_k(\alpha, t) f^{\Delta^k}(\alpha) = \int_\alpha^{\rho^{n-1}(t)} (-1)^{n-1} f^{\Delta^n}(\tau) g_{n-1}(\sigma(\tau), t)\Delta\tau.$$

Theorem 5 *The functions g_n and h_n satisfy the relationship*

$$h_n(t, s) = (-1)^n g_n(s, t)$$

for all $t \in \mathcal{T}$ and all $s \in \mathcal{T}^{\kappa^n}$.

Proof Let $t \in \mathcal{T}$ and $s \in \mathcal{T}^{\kappa^n}$ be arbitrarily chosen. We apply Theorem 4 for $\alpha = s$ and $f(\tau) = h_n(\tau, s)$. We observe that

$$f^{\Delta^k}(\tau) = h_{n-k}(\tau, s), \quad 0 \le k \le n.$$

Hence,

$$\begin{aligned}&f^{\Delta^k}(s) = h_{n-k}(s, s) = 0, \quad 0 \le k \le n-1, \\ &f^{\Delta^n}(s) = h_0(s, s) = 1, \quad f^{\Delta^{n+1}}(\tau) = 0.\end{aligned}$$

From here, using Taylor's formula, we get

$$\begin{aligned}
f(t) &= h_n(t,s)\\
&= \sum_{k=0}^{n}(-1)^k g_k(\alpha,t)f^{\Delta^k}(\alpha) + \int_\alpha^{\rho^n(t)} (-1)^n g_n(\sigma(\tau),t)f^{\Delta^{n+1}}(\tau)\Delta\tau\\
&= \sum_{k=0}^{n}(-1)^k g_k(s,t)f^{\Delta^k}(s) + \int_s^{\rho^n(t)} (-1)^n g_n(\sigma(\tau),t)f^{\Delta^{n+1}}(\tau)\Delta\tau\\
&= \sum_{k=0}^{n-1}(-1)^k g_k(s,t)f^{\Delta^k}(s) + (-1)^n g_n(s,t)f^{\Delta^n}(s)\\
&= (-1)^n g_n(s,t)f^{\Delta^n}(s)\\
&= (-1)^n g_n(s,t),
\end{aligned}$$

i.e.,

$$h_n(t,s) = (-1)^n g_n(s,t).$$

From Theorems 4 and 5, it follows the following theorem.

Theorem 6 (Taylor's Formula) *Let $n \in \mathcal{N}$. Suppose f is n-times differentiable on $\mathcal{T}^{\kappa^n}$. Let also, $\alpha \in \mathcal{T}^{\kappa^{n-1}}$, $t \in \mathcal{T}$. Then*

$$f(t) = \sum_{k=0}^{n-1} h_k(t,\alpha)f^{\Delta^k}(\alpha) + \int_\alpha^{\rho^{n-1}(t)} h_{n-1}(t,\sigma(\tau))f^{\Delta^n}(\tau)\Delta\tau.$$

Corollary 1 *Let $\alpha \in [a,b]$. Then for $|\lambda| < \infty$*

$$\sum_{k=0}^{\infty} \lambda^k h_k(t,\alpha)$$

is absolutely and uniformly convergent on the interval $a \le t \le b$.

3.3.2 Existence of Solutions of Generalized Volterra Integral Equations of the Second Kind

Theorem 7 *Let $K(x,y)$ be a real-valued rd-continuous function defined on $a \le x, y \le b$, $u(x)$ be a real-valued rd-continuous function defined on $a \le x \le b$. Let also,*

$$|u(x)| \le M \quad \text{for all} \quad x \in [a,b],$$
$$|K(x,y)| \le N \quad \text{for all} \quad x,y \in [a,b].$$

Then for $|\lambda| < \infty$ the generalized Volterra integral equation of the second kind

$$\phi(x) = u(x) + \lambda \int_a^x K(x, y)\phi(y)\Delta y, \quad a \le x \le b, \tag{3.30}$$

has a rd-continuous solution $\phi(x)$ defined on $[a, b]$.

Proof Suppose that a formal power series

$$\phi(x) = \sum_{n=0}^{\infty} \lambda^n \phi_n(x), \quad a \le x \le b, \tag{3.31}$$

satisfies (3.30). Then substituting (3.31) in (3.30) we get

$$\sum_{n=0}^{\infty} \lambda^n \phi_n(x) = u(x) + \lambda \int_a^x K(x, y) \sum_{n=0}^{\infty} \lambda^n \phi_n(y)\Delta y$$

or

$$\sum_{n=0}^{\infty} \lambda^n \phi_n(x) = u(x) + \sum_{n=0}^{\infty} \lambda^{n+1} \int_a^x K(x, y)\phi_n(y)\Delta y,$$

or

$$\begin{aligned}
&\phi_0(x) + \lambda\phi_1(x) + \lambda^2\phi_2(x) + \cdots + \lambda^n\phi_n(x) + \cdots \\
&= u(x) + \lambda \int_a^x K(x, y)\phi_0(y)\Delta y + \lambda^2 \int_a^x K(x, y)\phi_1(y)\Delta y \\
&\quad + \cdots + \lambda^n \int_a^x K(x, y)\phi_{n-1}(y)\Delta y + \cdots .
\end{aligned}$$

Hence, by a comparison of coefficients, we get the following relations

$$\begin{aligned}
\phi_0(x) &= u(x), \\
\phi_1(x) &= \int_a^x K(x, y)\phi_0(y)\Delta y, \\
\phi_2(x) &= \int_a^x K(x, y)\phi_1(y)\Delta y, \\
&\vdots \\
\phi_n(x) &= \int_a^x K(x, y)\phi_{n-1}(y)\Delta y, \\
&\vdots
\end{aligned}$$

or

$$\begin{aligned} \phi_0(x) &= u(x), \\ \phi_n(x) &= \int_a^x K(x, y)\phi_{n-1}(y)\Delta y, \quad n \in \mathscr{N}. \end{aligned} \tag{3.32}$$

For all $x \in [a, b]$, we have

$$\begin{aligned} |\phi_0(x)| &= |u(x)| \\ &\le M, \\ |\phi_1(x)| &= \left|\int_a^x K(x, y)\phi_0(y)\Delta y\right| \\ &\le \int_a^x |K(x, y)||\phi_0(y)|\Delta y \\ &\le MN \int_a^x \Delta y \\ &= MN(x - a) \\ &= MNh_1(x, a), \\ |\phi_2(x)| &= \left|\int_a^x K(x, y)\phi_1(y)\Delta y\right| \\ &\le \int_a^x |K(x, y)||\phi_1(y)|\Delta y \\ &\le MN^2 \int_a^x h_1(y, a)\Delta y \\ &= MN^2 h_2(x, a). \end{aligned}$$

Assume for some $m \in \mathscr{N}$

$$|\phi_m(x)| \le MN^m h_m(x, a) \quad \textit{for all} \quad x \in [a, b].$$

We will prove that

$$|\phi_{m+1}(x)| \le MN^{m+1} h_{m+1}(x, a) \quad \textit{for all} \quad x \in [a, b].$$

Really, we have

$$\begin{aligned} |\phi_{m+1}(x)| &= \left|\int_a^x K(x, y)\phi_m(y)\Delta y\right| \\ &\le \int_a^x |K(x, y)||\phi_m(y)|\Delta y \\ &\le MN^{m+1} \int_a^x h_m(y, a)\Delta y \\ &= MN^{m+1} h_{m+1}(x, a), \quad x \in [a, b]. \end{aligned}$$

Consequently

$$|\phi_m(x)| \leq MN^m h_m(x, a)$$

for all $m \in \mathcal{N}$ and all $x \in [a, b]$.
Therefore for $|\lambda| < \infty$ the series (3.31) is absolutely and uniformly convergent on the interval $[a, b]$. Hence, the formal solution (3.31) is a genuine solution of (3.30).

3.3.3 Uniqueness of Solutions of Generalized Volterra Integral Equations of Second Kind

Theorem 8 *Let $x_0 \in \mathcal{T}$ and $\phi, f, g \in \mathcal{C}_{rd}(\mathcal{T})$, and $1 + \mu(x)f(x) > 0$ for all $x \in \mathcal{T}$. Then*

$$\phi^{\Delta}(x) \leq f(x)\phi(x) + g(x) \quad \text{for all} \quad x \in \mathcal{T} \tag{3.33}$$

implies

$$\phi(x) \leq \phi(x_0)e_f(x, x_0) + \int_{x_0}^{x} e_f(x, \sigma(y))g(y)\Delta y \quad \text{for all} \quad x \in \mathcal{T}, \quad x \geq x_0. \tag{3.34}$$

Proof We calculate

$$\begin{aligned}
\left(\phi(x)e_{\ominus f}(x, x_0)\right)^{\Delta} &= \phi^{\Delta}(x)e_{\ominus f}(\sigma(x), x_0) + \phi(x)(\ominus f(x))e_{\ominus f}(x, x_0) \\
&= \phi^{\Delta}(x)e_{\ominus f}(\sigma(x), x_0) + \phi(x)\frac{\ominus f(x)}{1 + \mu(x)(\ominus f(x))}e_{\ominus f}(\sigma(x), x_0) \\
&= \left(\phi^{\Delta}(x) + \phi(x)\frac{\ominus f(x)}{1 + \mu(x)(\ominus f(x))}\right)e_{\ominus f}(\sigma(x), x_0) \\
&= \left(\phi^{\Delta}(x) + \phi(x)\frac{-\frac{f(x)}{1+\mu(x)f(x)}}{1 - \frac{\mu(x)f(x)}{1+\mu(x)f(x)}}\right)e_{\ominus f}(\sigma(x), x_0) \\
&= \left(\phi^{\Delta}(x) - f(x)\phi(x)\right)e_{\ominus f}(\sigma(x), x_0).
\end{aligned}$$

The last relation we integrate from x_0 to x and we get

$$\int_{x_0}^{x} \left(\phi(y)e_{\ominus f}(y, x_0)\right)_y^{\Delta} \Delta y = \int_{x_0}^{x} \left(\phi^{\Delta}(y) - f(y)\phi(y)\right)e_{\ominus f}(\sigma(y), x_0)\Delta y$$

or

$$\phi(x)e_{\ominus f}(x,x_0)-\phi(x_0)e_{\ominus f}(x_0,x_0)=\int_{x_0}^{x}\left(\phi^{\Delta}(y)-f(y)\phi(y)\right)e_{\ominus f}(\sigma(y),x_0)\Delta y,$$

or

$$\phi(x)e_{\ominus f}(x,x_0)=\phi(x_0)+\int_{x_0}^{x}\left(\phi^{\Delta}(y)-f(y)\phi(y)\right)e_{\ominus f}(\sigma(y),x_0)\Delta y.$$

Note that $e_{\ominus f}(\sigma(y),x_0)>0$ for all $y\in\mathscr{T}$ because $1+\mu(y)f(y)>0$ for all $y\in\mathscr{T}$. Hence, using (3.33), we get

$$\begin{aligned}\phi(x)e_{\ominus f}(x,x_0)&\leq\phi(x_0)+\int_{x_0}^{x}e_{\ominus f}(\sigma(y),x_0)g(y)\Delta y\\&=\phi(x_0)+\int_{x_0}^{x}e_f(x_0,\sigma(y))g(y)\Delta y.\end{aligned}$$

From the last inequality, using that $e_{\ominus f}(x,x_0)>0$ and

$$\frac{1}{e_{\ominus f}(x,x_0)}=e_f(x,x_0),\quad e_f(x,x_0)e_f(x_0,\sigma(y))=e_f(x,\sigma(y))$$

for all $x,y\in\mathscr{T}$, we obtain the inequality (3.34).

Theorem 9 (Gronwall's Inequality) *Let* $x_0\in\mathscr{T}$ *and* $\phi,f,g\in\mathscr{C}_{rd}(\mathscr{T})$, *and*

$$1+\mu(x)g(x)>0,\quad g(x)\geq 0\ \textit{for all}\ \ x\in\mathscr{T}.$$

Then

$$\phi(x)\leq f(x)+\int_{x_0}^{x}\phi(y)g(y)\Delta y\ \ \textit{for all}\ \ x\in\mathscr{T},\quad x\geq x_0, \tag{3.35}$$

implies

$$\phi(x)\leq f(x)+\int_{x_0}^{x}e_g(x,\sigma(y))f(y)g(y)\Delta y\ \ \textit{for all}\ \ x\in\mathscr{T},\quad x\geq x_0. \tag{3.36}$$

Proof Define

$$\psi(x)=\int_{x_0}^{x}\phi(y)g(y)\Delta y\ \ \textit{for}\ \ x\in\mathscr{T},\quad x\geq x_0.$$

Then $\psi(x_0)=0$. Also, using (3.35), we get

$$\phi(x)\leq f(x)+\psi(x)\ \ \textit{for}\ \ x\in\mathscr{T},\quad x\geq x_0. \tag{3.37}$$

Hence, we get

$$\begin{aligned}\psi^{\Delta}(x) &= \phi(x)g(x)\\ &\le f(x)g(x) + g(x)\psi(x), \quad x \in \mathscr{T}.\end{aligned}$$

From here and Theorem 8, we obtain

$$\begin{aligned}\psi(x) &\le \psi(x_0)e_g(x, x_0) + \int_{x_0}^{x} e_g(x, \sigma(y))f(y)g(y)\Delta y\\ &= \int_{x_0}^{x} e_g(x, \sigma(y))f(y)g(y)\Delta y.\end{aligned}$$

From the last inequality and from (3.37), we get the inequality (3.36).

Theorem 10 *Suppose that all conditions of Theorem 7 are fulfilled. Then the Eq. (3.30) has unique rd-continuous solution $\phi(x)$ defined on $[a, b]$.*

Proof The existence of a rd-continuous solution $\phi(x)$ of the Eq. (3.30) is ensured by Theorem 7.
Assume that $\phi_1(x)$ and $\phi_2(x)$ are rd-continuous solutions of (3.30) defined on $[a, b]$. Then $\psi(x) = \phi_1(x) - \phi_2(x)$ is a rd-continuous solution of the equation

$$\psi(x) = \lambda \int_a^x K(x, y)\psi(y)\Delta y, \quad x \in [a, b].$$

Hence,

$$\begin{aligned}|\psi(x)| &= \left|\lambda \int_a^x K(x, y)\psi(y)\Delta y\right|\\ &\le |\lambda| \int_a^x |K(x, y)||\psi(y)|\Delta y\\ &\le N|\lambda| \int_a^x |\psi(y)|\Delta y, \quad x \in [a, b].\end{aligned}$$

From the last inequality and from Theorem 9, we get that

$$\psi(x) = \phi_1(x) - \phi_2(x) = 0 \quad \textit{for}\, x \in [a, b].$$

3.3.4 Existence and Uniqueness of Solutions of Generalized Volterra Integral Equations of the First Kind

Theorem 11 *Let $u(x)$ be differentiable in $[a, b]^{\kappa}$, $K(x, y)$ be rd-continuous in x and y, $x, y \in [a, b]$, and $K_x^{\Delta}(x, y)$ exists for all $(x, y) \in [a, b]^{\kappa} \times [a, b]$. Let also,*

$K(\sigma(x), x) \neq 0$ for all $x \in [a, b]$, and

$$|u^\Delta(x)| \leq M_1, \quad \left|\frac{1}{K(\sigma(x),x)}\right| \geq M_2 \quad \textit{for} \quad x \in [a, b],$$
$$|K_x^\Delta(x, y)| \leq M_3 \quad \textit{for}\, x, y \in [a, b].$$

Then the generalized Volterra integral equation of the first kind (3.28) has unique rd-continuous solution $\phi(x)$ defined on $[a, b]$.

Proof We reduce the Eq. (3.28) to the Eq. (3.29). Now the result follows from Theorems 7 and 10.

3.4 Resolvent Kernels

We shall give another expression of the solution $\phi(x)$ of the Eq. (3.30) as follows. We will start with the following useful lemma.

Lemma 4 *Let $f : \mathscr{T} \times \mathscr{T} \longmapsto \mathscr{R}$ be rd-continuous function and $a \in \mathscr{T}$ be fixed. Then*

$$\int_a^x \int_a^{x_1} f(x_1, y)\Delta y \Delta x_1 = \int_a^x \int_{\sigma(y)}^x f(x_1, y)\Delta x_1 \Delta y. \tag{3.38}$$

Proof Let

$$G(x) = \int_a^x \int_a^{x_1} f(x_1, y)\Delta y \Delta x_1 - \int_a^x \int_{\sigma(y)}^x f(x_1, y)\Delta x_1 \Delta y.$$

Then

$$G^\Delta(x) = \int_a^x f(x, y)\Delta y - \int_a^x f(x, y)\Delta y - \int_{\sigma(x)}^{\sigma(x)} f(x_1, x)\Delta x_1 = 0.$$

Since $G(a) = 0$, we get (3.38).

From (3.32) it follows that

$$\phi_1(x) = \int_a^x K(x, y)\phi_0(y)\Delta y$$
$$= \int_a^x K(x, y)u(y)\Delta y,$$

$$\begin{aligned}
\phi_2(x) &= \int_a^x K(x,y)\phi_1(y)\Delta y \\
&= \int_a^x K(x,y)\int_a^y K(y,z)u(z)\Delta z\Delta y \\
&= \int_a^x \int_a^y K(x,y)K(y,z)u(z)\Delta z\Delta y \\
&= \int_a^x \int_{\sigma(y)}^x K(x,z)K(z,y)u(y)\Delta z\Delta y \\
&= \int_a^x u(y)\int_{\sigma(y)}^x K(x,z)K(z,y)\Delta z\Delta y.
\end{aligned}$$

Let

$$K^{(2)}(x,y) = \int_{\sigma(y)}^x K(x,z)K(z,y)\Delta z.$$

Therefore

$$\phi_2(x) = \int_a^x K^{(2)}(x,y)u(y)\Delta y.$$

Hence,

$$\begin{aligned}
\phi_3(x) &= \int_a^x K(x,y)\phi_2(y)\Delta y \\
&= \int_a^x K(x,y)\int_a^y K^{(2)}(y,z)u(z)\Delta z\Delta y \\
&= \int_a^x \int_a^y K(x,y)K^{(2)}(y,z)u(z)\Delta z\Delta y \\
&= \int_a^x \int_{\sigma(y)}^x K(x,z)K^{(2)}(z,y)u(y)\Delta z\Delta y \\
&= \int_a^x u(y)\int_{\sigma(y)}^x K(x,z)K^{(2)}(z,y)\Delta z\Delta y.
\end{aligned}$$

Let

$$K^{(3)}(x,y) = \int_{\sigma(y)}^x K(x,z)K^{(2)}(z,y)\Delta z.$$

Then

$$\phi_3(x) = \int_a^x K^{(3)}(x,y)u(y)\Delta y.$$

Repeating the same argument, we obtain the iterated kernels.

$$\begin{aligned}
K^{(1)}(x,y) &= K(x,y), \\
K^{(n)}(x,y) &= \textstyle\int_{\sigma(y)}^x K(x,z)K^{(n-1)}(z,y)\Delta z.
\end{aligned} \tag{3.39}$$

We set

$$\Gamma(x, y; \lambda) = K(x, y) + \lambda K^{(2)}(x, y) + \lambda^2 K^{(3)}(x, y) + \cdots + \lambda^{n-1} K^{(n)}(x, y) + \cdots .$$

Definition 1 The function $\Gamma(x, y; \lambda)$ is called the resolvent kernel of the kernel $K(x, y)$.

Theorem 12 *Let $K(x, y)$ be defined and rd-continuous on $[a, b] \times [a, b]$. Let also, $|K(x, y)| \leq M$ for all $(x, y) \in [a, b] \times [a, b]$, for some positive constant M. Then $\Gamma(x, y; \lambda)$ converges uniformly with respect to (x, y) for $|\lambda| < \infty$.*

Proof Let $(x, y) \in [a, b] \times [a.b]$ be arbitrarily chosen. Without loss of generality we suppose that $y \leq x$. Then

$$\begin{aligned}
|K^{(1)}(x, y)| &\leq M, \\
|K^{(2)}(x, y)| &= \left| \int_{\sigma(y)}^{x} K(x, z) K^{(1)}(z, y) \Delta z \right| \\
&\leq \int_{\sigma(y)}^{x} |K(x, z)| |K^{(1)}(z, y)| \Delta z \\
&\leq M^2 (x - \sigma(y)) \\
&= M^2 h_1(x, \sigma(y)), \\
|K^{(3)}(x, y)| &= \left| \int_{\sigma(y)}^{x} K(x, z) K^{(2)}(z, y) \Delta z \right| \\
&\leq \int_{\sigma(y)}^{x} |K(x, z)| |K^{(2)}(z, y)| \Delta z \\
&\leq M^3 \int_{\sigma(y)}^{x} h_1(z, \sigma(y)) \Delta z \\
&= M^3 h_2(x, \sigma(y)).
\end{aligned}$$

Assume that

$$|K^{(m)}(x, y)| \leq M^m h_{m-1}(x, \sigma(y))$$

for some $m \in \mathscr{N}$.
We will prove that

$$|K^{(m+1)}(x, y)| \leq M^{m+1} h_m(x, \sigma(y)).$$

Really, we have

$$\begin{aligned} |K^{(m+1)}(x, y)| &= \left| \int_{\sigma(y)}^{x} K(x, z) K^{(m)}(z, y) \Delta z \right| \\ &\leq \int_{\sigma(y)}^{x} |K(x, z)||K^{(m)}(z, y)| \Delta z \\ &\leq M^{m+1} \int_{\sigma(y)}^{x} h_{m-1}(z, \sigma(y)) \Delta z \\ &= M^{m+1} h_m(x, \sigma(y)). \end{aligned}$$

Consequently

$$|K^{(m)}(x, y)| \leq M^m h_{m-1}(x, \sigma(y)) \quad \textit{for all} \quad m \in \mathcal{N},$$

which completes the proof.

Theorem 13 *The resolvent kernel $\Gamma(x, y; \lambda)$ of the kernel $K(x, y)$ satisfies the generalized Volterra integral equation of the second kind*

$$\Gamma(x, y; \lambda) = K(x, y) + \lambda \int_{\sigma(y)}^{x} K(x, z) \Gamma(z, y; \lambda) \Delta z. \tag{3.40}$$

Proof We have

$$\begin{aligned} \Gamma(x, y; \lambda) &= K(x, y) + \lambda K^{(2)}(x, y) + \lambda^2 K^{(3)}(x, y) + \cdots \\ &= K(x, y) + \lambda \left(K^{(2)}(x, y) + \lambda K^{(3)}(x, y) + \cdots \right) \\ &= K(x, y) + \lambda \left(\int_{\sigma(y)}^{x} K(x, z) K^{(1)}(z, y) \Delta z + \int_{\sigma(y)}^{x} K(x, z) \left(\lambda K^{(2)}(z, y) \right) \Delta z + \cdots \right) \\ &= K(x, y) + \lambda \int_{\sigma(y)}^{x} K(x, z) \left(K^{(1)}(z, y) + \lambda K^{(2)}(z, y) + \cdots \right) \Delta z \\ &= K(x, y) + \lambda \int_{\sigma(y)}^{x} K(x, z) \Gamma(z, y; \lambda) \Delta z. \end{aligned}$$

Exercise 14 Prove that the resolvent kernel $\Gamma(x, y; \lambda)$ of the kernel $K(x, y)$ satisfies the generalized Volterra integral equation of the second kind

$$\Gamma(x, y; \lambda) = K(x, y) + \lambda \int_{\sigma(y)}^{x} \Gamma(x, z; \lambda) K(z, y) \Delta z. \tag{3.41}$$

Remark 1 Assume that the kernel $K(x, y)$ satisfies all conditions of Theorem 7. Hence, from Theorems 7 and 10, it follows that there exists unique rd-continuous function $\Gamma(x, y; \lambda)$ with respect to $(x, y) \in [a, b] \times [a, b]$ which satisfies the Eqs. (3.40) and (3.41) for $|\lambda| < \infty$.

Remark 2 Assume that the resolvent kernel $\Gamma(x, y; \lambda)$ satisfies all conditions of Theorem 7 for $|\lambda| < \infty$. Hence, from Theorems 7 and 10, it follows that there exists unique rd-continuous function $K(x, y)$ in $[a, b] \times [a, b]$ which satisfies the Eqs. (3.40) and (3.41).

From Remarks 1 and 2 we get the following theorem.

Theorem 14 (Reciprocity Theorem) *If $\Gamma(x, y; \lambda)$ is the resolvent kernel of $K(x, y)$, then the resolvent kernel of $\Gamma(x, y; \lambda)$ is the kernel $K(x, y)$ itself.*

Example 30 Let $\mathscr{T} = \mathscr{Z}$, $K(x, y) = 1$ for $(x, y) \in [0, 4] \times [0, 4]$. Then

$$\begin{aligned}
K^{(1)}(x, y) &= K(x, y) = 1,\\
K^{(2)}(x, y) &= \int_{\sigma(y)}^{x} K(x, z)K(z, y)\Delta z\\
&= \int_{\sigma(y)}^{x} \Delta z\\
&= x - y - 1,\\
K^{(3)}(x, y) &= \int_{\sigma(y)}^{x} K(x, z)K^{(2)}(z, y)\Delta z\\
&= \int_{\sigma(y)}^{x} (z - y - 1)\Delta z\\
&= \int_{\sigma(y)}^{x} \left(\frac{1}{2}(z^2)^{\Delta} - \frac{1}{2} - y - 1\right)\Delta z\\
&= \frac{1}{2}\int_{\sigma(y)}^{x} (z^2)^{\Delta}\Delta z - \left(\frac{3}{2} + y\right)\int_{\sigma(y)}^{x} \Delta z\\
&= \frac{1}{2}(x^2 - (y + 1)^2) - \left(\frac{3}{2} + y\right)(x - y - 1)\\
&= (x - y - 1)\left(\frac{1}{2}x + \frac{1}{2}y + \frac{1}{2} - \frac{3}{2} - y\right)\\
&= \frac{1}{2}(x - y - 1)(x - y - 2).
\end{aligned}$$

Example 31 Let $\mathscr{T} = 2^{\mathscr{N}_0}$ and $K(x, y) = x$, $(x, y) \in [1, 8] \times [1, 8]$. Then

$$\begin{aligned}
K^{(1)}(x, y) &= K(x, y) \\
&= x, \\
K^{(2)}(x, y) &= \int_{2y}^{x} K(x, z)K^{(1)}(z, y)\Delta z \\
&= \int_{2y}^{x} xz\Delta z \\
&= x\int_{2y}^{x} z\Delta z \\
&= \frac{1}{3}x\int_{2y}^{x} (z^2)^{\Delta}\Delta z \\
&= \frac{1}{3}x(x^2 - 4y^2), \\
K^{(3)}(x, y) &= \int_{2y}^{x} K(x, z)K^{(2)}(z, y)\Delta z \\
&= \frac{1}{3}x\int_{2y}^{x} z(z^2 - 4y^2)\Delta z \\
&= \frac{1}{3}x\int_{2y}^{x} (z^3 - 4zy^2)\Delta z \\
&= \frac{1}{3}x\int_{2y}^{x} \left(\frac{1}{15}(z^4)^{\Delta} - \frac{4}{3}y^2(z^2)^{\Delta}\right)\Delta z \\
&= \frac{1}{45}x\int_{2y}^{x} (z^4)^{\Delta}\Delta z - \frac{4}{9}xy^2\int_{2y}^{x} (z^2)^{\Delta}\Delta z \\
&= \frac{1}{45}x(x^4 - (2y)^4) - \frac{4}{9}xy^2(x^2 - (2y)^2) \\
&= \frac{1}{45}x(x^2 - (2y)^2)(x^2 + (2y)^2 - 20y^2) \\
&= \frac{1}{45}x(x^2 - 4y^2)(x^2 - 16y^2) \\
&= \frac{1}{45}x(x - 2y)(x - 4y)(x + 4y)(x + 2y).
\end{aligned}$$

Exercise 15 Let $\mathscr{T} = 3^{\mathscr{N}_0}$ and $K(x, y) = x + y$, $(x, y) \in [1, 9] \times [1, 9]$. Find $K^{(2)}(x, y)$.

Answer

$$(x - 3y)\frac{17x^2 + 116xy + 75y^2}{52}.$$

3.5 Application to Linear Dynamic Equations

It was shown in Sect. 2.2 that initial value problems for second-order and first-order dynamic equations can be reduced to generalized Volterra integral equations of the second kind. In this section we will show that initial value problems for nth order dynamic equations can be reduced to Volterra integral equations of the second kind. We will start with a formula that reduces multiple integrals to single integrals.

Lemma 5 *Let $a \in \mathscr{T}$ and $f : \mathscr{T} \longmapsto \mathscr{R}$ be integrable. Then*

$$\begin{aligned}&\int_a^x \int_a^{x_1} \int_a^{x_2} \dots \int_a^{x_n} f(y)\Delta y \Delta x_n \dots \Delta x_2 \Delta x_1 \\ &= \int_a^x (h_n(x,y) - \mu(y)h_{n-1}(x,y))f(y)\Delta y, \quad x \in \mathscr{T}.\end{aligned} \tag{3.42}$$

Proof 1. $n = 1$. Then, using Theorem 1 in Chap. 2, we have

$$\begin{aligned}\int_a^x \int_a^{x_1} f(y)\Delta y \Delta x_1 &= \int_a^x (x - \sigma(y))f(y)\Delta y \\ &= \int_a^x (x - y + y - \sigma(y))f(y)\Delta y \\ &= \int_a^x (h_1(x,y) - \mu(y)h_0(x,y))f(y)\Delta y.\end{aligned}$$

2. Assume that (3.42) holds for some $n \in \mathscr{N}$.
3. We will prove that

$$\begin{aligned}&\int_a^x \int_a^{x_1} \int_a^{x_2} \dots \int_a^{x_{n+1}} f(y)\Delta y \Delta x_{n+1} \dots \Delta x_2 \Delta x_1 \\ &= \int_a^x (h_{n+1}(x,y) - \mu(y)h_n(x,y))f(y)\Delta y, \quad x \in \mathscr{T}.\end{aligned}$$

Really, for $x \in \mathscr{T}$, using (3.42), we get

$$\int_a^{x_1} \int_a^{x_2} \dots \int_a^{x_n} f(y)\Delta \dots \Delta x_2 = \int_a^{x_1} (h_{n-1}(x_1,y) - \mu(y)h_{n-2}(x_1,y))f(y)\Delta y.$$

Hence,

$$\begin{aligned}&\int_a^x \int_a^{x_1} \int_a^{x_2} \dots \int_a^{x_n} f(y)\Delta y \dots \Delta x_2 \Delta x_1 \\ &= \int_a^x \int_a^{x_1} (h_{n-1}(x_1,y) - \mu(y)h_{n-2}(x_1,y))f(y)\Delta y \Delta x_1 \\ &= \int_a^x \int_y^x (h_{n-1}(x_1,y) - \mu(y)h_{n-2}(x_1,y))f(y)\Delta x_1 \Delta y \\ &= \int_a^x \left(\int_y^x (h_{n-1}(x_1,y) - \mu(y)h_{n-2}(x_1,y))\Delta x_1\right) f(y)\Delta y \\ &= \int_a^x \left(\int_y^x h_{n-1}(x_1,y)\Delta x_1 - \mu(y)\int_y^x h_{n-2}(x_1,y)\Delta x_1\right) f(y)\Delta y \\ &= \int_a^x (h_n(x,y) - \mu(y)h_{n-1}(x,y))f(y)\Delta y.\end{aligned}$$

Consequently (3.42) holds for all $n \in \mathscr{N}$.

We consider a linear dynamic equation of nth order, in unknown z,

$$z^{\Delta^n}(x) + p_1(x)z^{\Delta^{n-1}}(x) + \cdots + p_n(x)z(x) = f(x) \tag{3.43}$$

where $f(x)$ and $p_l(x)$, $l = 1, 2, \ldots, n$, are rd-continuous in a neighbourhood of the point $x = a$.
Setting

$$z^{\Delta^n}(x) = \phi(x),$$

we obtain that

$$\begin{aligned}
z^{\Delta^{n-1}}(x) &= \int_a^x \phi(y)\Delta y + c_1, \\
z^{\Delta^{n-2}}(x) &= \int_a^x z^{\Delta^{n-1}}(y)\Delta y + c_2 \\
&= \int_a^x \left(\int_a^{x_1} \phi(y)\Delta y + c_1 \right) \Delta x_1 + c_2 \\
&= \int_a^x \int_a^{x_1} \phi(y)\Delta y \Delta x_1 + c_1(x-a) + c_2 \\
&= \int_a^x (h_1(x,y) - \mu(y)h_0(x,y))\Delta y + c_1 h_1(x,a) + c_2, \\
z^{\Delta^{n-3}}(x) &= \int_a^x z^{\Delta^{n-2}}(y)\Delta y + c_3 \\
&= \int_a^x \left(\int_a^{x_1} (h_1(x_1,y) - \mu(y)h_0(x_1,y))\phi(y)\Delta y \right. \\
&\quad \left. + c_1 h_1(x_1,a) + c_2 \right) \Delta x_1 + c_3 \\
&= \int_a^x \int_a^{x_1} (h_1(x_1,y) - \mu(y)h_0(x_1,y))\phi(y)\Delta y \Delta x_1 \\
&\quad + c_1 \int_a^x h_1(x_1,a)\Delta x_1 + c_2 \int_a^x \Delta x_1 + c_3 \\
&= \int_a^x \int_y^x (h_1(x_1,y) - \mu(y)h_0(x_1,y))\phi(y)\Delta x_1 \Delta y \\
&\quad + c_1 h_2(x,a) + c_2(x-a) + c_3 \\
&= \int_a^x \left(\int_y^x (h_1(x_1,y) - \mu(y)h_0(x_1,y))\Delta x_1 \right) \phi(y)\Delta y \\
&\quad + c_1 h_2(x,a) + c_2 h_1(x,a) + c_3 \\
&= \int_a^x \left(\int_y^x h_1(x_1,y)\Delta x_1 - \mu(y) \int_y^x h_0(x_1,y)\Delta x_1 \right) \phi(y)\Delta y \\
&\quad + c_1 h_2(x,a) + c_2 h_1(x,a) + c_3
\end{aligned}$$

$$= \int_a^x (h_2(x,y) - \mu(y)h_1(x,y))\phi(y)\Delta y + c_1 h_2(x,a) + c_2 h_1(x,a) + c_3,$$

and so on. We find

$$z^{\Delta^{n-m}}(x) = \int_a^x (h_{m-1}(x,y) - \mu(y)h_{m-2}(x,y))\phi(y)\Delta y + c_1 h_{m-1}(x,a) + c_2 h_{m-2}(x,a) + \cdots + c_{m-1}h_1(x,a) + c_m, \quad m = 2, \ldots, n.$$

Accordingly the Eq. (3.43) is reduced to an equation, in unknown ϕ, of the form

$$\begin{aligned}\phi(x) & + p_1(x)\left(\int_a^x \phi(y)\Delta y + c_1\right) \\ & + p_2(x)\left(\int_a^x (h_1(x,y) - \mu(y)h_0(x,y)\phi(y)\Delta y + c_1 h_1(x,a) + c_2\right) \\ & + p_3(x)\left(\int_a^x (h_2(x,y) - \mu(y)h_1(x,y))\phi(y)\Delta y + c_1 h_2(x,a) + c_2 h_1(x,a) + c_3\right) \\ & + \cdots \\ & + p_n(x)\left(\int_a^x (h_{n-1}(x,y) - \mu(y)h_{n-2}(x,y))\phi(y)\Delta y + c_1 h_{n-1}(x,a) \right. \\ & \left. + c_2 h_{n-2}(x,a) + \cdots + c_{n-1}h_1(x,a) + c_n\right) = f(x),\end{aligned}$$

or

$$\phi(x) + \int_a^x \left(\sum_{l=1}^n p_l(x)h_{l-1}(x,y) - \mu(y)\sum_{l=2}^n p_l(x)h_{l-2}(x,y)\right)\phi(y)\Delta y + c_1 \sum_{l=1}^n p_l(x)h_{l-1}(x,a) + c_2 \sum_{l=2}^n p_l(x)h_{l-2}(x,a) + \cdots + c_n p_n(x) = f(x),$$

or

$$\phi(x) + \int_a^x \left(\sum_{l=1}^n p_l(x)h_{l-1}(x,y) - \mu(y)\sum_{l=2}^n p_l(x)h_{l-2}(x,y)\right)\phi(y)\Delta y + \sum_{m=1}^n \sum_{l=m}^n c_m p_l(x)h_{l-m}(x,a) = f(x),$$

or

$$\begin{aligned}&\phi(x) + \textstyle\int_a^x \left(\sum_{l=1}^n p_l(x)h_{l-1}(x,y) - \mu(y)\sum_{l=2}^n p_l(x)h_{l-2}(x,y)\right)\phi(y)\Delta y \\ &= f(x) - \textstyle\sum_{m=1}^n \sum_{l=m}^n c_m p_l(x)h_{l-m}(x,a).\end{aligned} \tag{3.44}$$

We note that the Eqs. (3.43) and (3.44) are equivalent. The existence and uniqueness of the solution $z(x)$ of the initial value problem for the nth order linear dynamic equation (3.43) corresponds exactly to the existence and uniqueness of the solution $\phi(x)$ of the generalized Volterra integral equation of the second kind (3.44). The constants $c_1, c_2, \ldots, c_n$ in (3.44) are determined by the initial conditions for (3.43).

Remark 3 The above arguments suggest that we can define a linear dynamic equation of infinite order by

$$\begin{aligned}&\phi(x) + \int_a^x \left(\sum_{l=1}^{\infty} p_l(x)h_{l-1}(x,y) - \mu(y)\sum_{l=2}^{\infty} p_l(x)h_{l-2}(x,y)\right)\phi(y)\Delta y\\ &= f(x) - \sum_{m=1}^{\infty}\sum_{l=m}^{\infty} c_m p_l(x)h_{l-m}(x,a).\end{aligned} \tag{3.45}$$

If we suppose

$$\sup_{x\in[a,b]} |p_l(x)| \le A, \quad l = 1, 2, \ldots,$$

for some positive constant A, and $\sum_{m=1}^{\infty} |c_m| < \infty$, using that the series $\sum_{l=0}^{\infty} h_l(x,y)$ is absolutely and uniformly convergent on the interval $[a, b]$, we have that the series in (3.45) are absolutely and uniformly convergent on the interval $[a, b]$. Thus we can find a unique solution $\phi(x)$ of the Eq. (3.45).

Example 32 Let $\mathscr{T} = \mathscr{Z}$. Consider the initial value problem

$$\begin{aligned}&z^{\Delta^3}(x) + xz^{\Delta^2}(x) + z^{\Delta}(x) + z(x) = x^2\\ &z(0) = z^{\Delta}(0) = z^{\Delta^2}(0) = 1.\end{aligned} \tag{3.46}$$

Here $\sigma(x) = x + 1$, $x \in \mathscr{T}$.
We set

$$z^{\Delta^3}(x) = \phi(x). \tag{3.47}$$

Then

$$z^{\Delta^2}(x) - z^{\Delta^2}(0) = \int_0^x \phi(y)\Delta y$$

or

$$z^{\Delta^2}(x) = 1 + \int_0^x \phi(y)\Delta y. \tag{3.48}$$

Hence,

$$\begin{aligned}z^{\Delta}(x) - z^{\Delta}(0) &= \int_0^x \left(1 + \int_0^{x_1} \phi(y)\Delta y\right)\Delta x_1\\ &= \int_0^x \Delta x_1 + \int_0^x \int_0^{x_1} \phi(y)\Delta y\Delta x_1\end{aligned}$$

$$= x + \int_0^x (x - \sigma(y))\phi(y)\Delta y$$
$$= x + \int_0^x (x - y - 1)\phi(y)\Delta y,$$

or

$$z^\Delta(x) = x + 1 + \int_0^x (x - y - 1)\phi(y)\Delta y. \tag{3.49}$$

From here,

$$\begin{aligned}
z(x) - z(0) &= \int_0^x \left(x_1 + 1 + \int_0^{x_1} (x_1 - y - 1)\phi(y)\Delta y \right) \Delta x_1 \\
&= \int_0^x (x_1 + 1)\Delta x_1 + \int_0^x \int_0^{x_1} (x_1 - y - 1)\phi(y)\Delta y \Delta x_1 \\
&= \int_0^x \left(\frac{1}{2}(x_1^2)^\Delta - \frac{1}{2} + 1 \right) \Delta x_1 + \int_0^x \int_0^{x_1} (x_1 - y - 1)\phi(y)\Delta y \Delta x_1 \\
&= \frac{1}{2}\int_0^x (x_1^2)^\Delta \Delta x_1 + \frac{1}{2}\int_0^x \Delta x_1 + \int_0^x \int_0^{x_1} (x_1 - y - 1)\phi(y)\Delta y \Delta x_1 \\
&= \frac{1}{2}x^2 + \frac{1}{2}x + \int_0^x \int_0^{x_1} (x_1 - y - 1)\phi(y)\Delta y \Delta x_1 \\
&= \frac{1}{2}x^2 + \frac{1}{2}x + \int_0^x \int_y^x (x_1 - y - 1)\phi(y)\Delta x_1 \Delta y + \int_0^x \phi(y)\Delta y \\
&= \frac{1}{2}x^2 + \frac{1}{2}x + \int_0^x \left(\int_y^x (x_1 - y - 1)\Delta x_1 \right) \phi(y)\Delta y + \int_0^x \phi(y)\Delta y \\
&= \frac{1}{2}x^2 + \frac{1}{2}x + \int_0^x \left(\int_y^x \left(\frac{1}{2}(x_1^2)^\Delta - 1 - y - \frac{1}{2} \right) \Delta x_1 \right) \phi(y)\Delta y + \int_0^x \phi(y)\Delta y \\
&= \frac{1}{2}x^2 + \frac{1}{2}x + \int_0^x \left(\int_y^x \left(\frac{1}{2}(x_1^2)^\Delta - y - \frac{3}{2} \right) \Delta x_1 \right) \phi(y)\Delta y + \int_0^x \phi(y)\Delta y \\
&= \frac{1}{2}x^2 + \frac{1}{2}x + \int_0^x \left(\frac{1}{2}(x^2 - y^2) - \left(y + \frac{3}{2} \right)(x - y) + 1 \right) \phi(y)\Delta y \\
&= \frac{1}{2}x^2 + \frac{1}{2}x + \int_0^x \left(\frac{1}{2}x^2 - \frac{1}{2}y^2 - xy + y^2 - \frac{3}{2}x + \frac{3}{2}y + 1 \right) \phi(y)\Delta y \\
&= \frac{1}{2}x^2 + \frac{1}{2}x + \int_0^x \left(\frac{1}{2}x^2 + \frac{1}{2}y^2 - xy - \frac{3}{2}x + \frac{3}{2}y + 1 \right) \phi(y)\Delta y,
\end{aligned}$$

i.e.,

$$z(x) = \frac{1}{2}x^2 + \frac{1}{2}x + 1 + \int_0^x \left(\frac{1}{2}x^2 + \frac{1}{2}y^2 - xy - \frac{3}{2}x + \frac{3}{2}y + 1 \right) \phi(y)\Delta y. \tag{3.50}$$

We substitute (3.47)–(3.50) in (3.46), and we get

$$\begin{aligned}\phi(x) \; &+x\left(1+\int_0^x \phi(y)\Delta y\right)+(x+1)+\int_0^x (x-y-1)\phi(y)\Delta y\\ &+\frac{1}{2}x^2+\frac{1}{2}x+1+\int_0^x\left(\frac{1}{2}x^2+\frac{1}{2}y^2-xy-\frac{3}{2}x+\frac{3}{2}y+1\right)\phi(y)\Delta y=x^2,\end{aligned}$$

or

$$\begin{aligned}\phi(x) \; &+x+\int_0^x x\phi(y)\Delta y+x+1+\int_0^x (x-y-1)\phi(y)\Delta y\\ &+\frac{1}{2}x^2+\frac{1}{2}x+1+\int_0^x\left(\frac{1}{2}x^2+\frac{1}{2}y^2-xy-\frac{3}{2}x+\frac{3}{2}y+1\right)\phi(y)\Delta y=x^2,\end{aligned}$$

or

$$\begin{aligned}\phi(x) \; &+\int_0^x x\phi(y)\Delta y+\int_0^x (x-y-1)\phi(y)\Delta y\\ &+\int_0^x\left(\frac{1}{2}x^2+\frac{1}{2}y^2-xy-\frac{3}{2}x+\frac{3}{2}y+1\right)\phi(y)\Delta y+\;\frac{1}{2}x^2+\frac{5}{2}x+2=x^2,\end{aligned}$$

or

$$\phi(x)+\int_0^x\left(\frac{1}{2}x^2+\frac{1}{2}y^2-xy-\frac{3}{2}x+\frac{3}{2}y+x+x-y-1+1\right)\phi(y)\Delta y=\frac{1}{2}x^2-\frac{5}{2}x-2,$$

or

$$\phi(x)+\int_0^x\left(\frac{1}{2}x^2+\frac{1}{2}y^2-xy+\frac{1}{2}x+\frac{1}{2}y\right)\phi(y)\Delta y=\frac{1}{2}x^2-\frac{5}{2}x-2.$$

Example 33 Let $\mathscr{T}=2^{\mathscr{N}_0}\cup\{0\}$. Consider the initial value problem

$$\begin{aligned}&z^{\Delta^4}(x)+xz(x)=1\\ &z(0)=z^{\Delta}(0)=z^{\Delta^2}(0)=1,\quad z^{\Delta^3}(0)=0.\end{aligned} \tag{3.51}$$

Here $\sigma(x)=2x,\quad x\in\mathscr{T}$.
We set

$$z^{\Delta^4}(x)=\phi(x). \tag{3.52}$$

Then

$$z^{\Delta^3}(x)-z^{\Delta^3}(0)=\int_0^x \phi(y)\Delta y$$

or

$$z^{\Delta^3}(x) = \int_0^x \phi(y)\Delta y.$$

Hence,

$$\begin{aligned} z^{\Delta^2}(x) - z^{\Delta^2}(0) &= \int_0^x \int_0^{x_1} \phi(y)\Delta y \Delta x_1 \\ &= \int_0^x (x - \sigma(y))\phi(y)\Delta y \\ &= \int_0^x (x - 2y)\phi(y)\Delta y, \end{aligned}$$

or

$$z^{\Delta^2}(x) = 1 + \int_0^x (x - 2y)\phi(y)\Delta y.$$

From here,

$$\begin{aligned} z^{\Delta}(x) - z^{\Delta}(0) &= \int_0^x \left(1 + \int_0^{x_1} (x_1 - 2y)\phi(y)\Delta y\right) \Delta x_1 \\ &= \int_0^x \Delta x_1 + \int_0^x \int_0^{x_1} (x_1 - 2y)\phi(y)\Delta y \Delta x_1 \\ &= x + \int_0^x \int_0^{x_1} (x_1 - 2y)\phi(y)\Delta y \Delta x_1 \\ &= x + \int_0^x \int_y^x (x_1 - 2y)\phi(y)\Delta x_1 \Delta y + \int_0^x y^2\phi(y)\Delta y \\ &= x + \int_0^x \left(\int_y^x (x_1 - 2y)\Delta x_1\right) \phi(y)\Delta y + \int_0^x y^2\phi(y)\Delta y \\ &= x + \int_0^x \left(\int_y^x \left(\frac{1}{3}(x_1^2)^{\Delta} - 2y\right) \Delta x_1\right) \phi(y)\Delta y + \int_0^x y^2\phi(y)\Delta y \\ &= x + \int_0^x \left(\frac{1}{3}x^2 - \frac{1}{3}y^2 - 2y(x-y) + y^2\right) \phi(y)\Delta y \\ &= x + \int_0^x \left(\frac{1}{3}x^2 + \frac{2}{3}y^2 - 2xy + 2y^2\right) \phi(y)\Delta y \\ &= x + \int_0^x \left(\frac{1}{3}x^2 + \frac{8}{3}y^2 - 2xy\right) \phi(y)\Delta y, \end{aligned}$$

or

$$z^{\Delta}(x) = x + 1 + \int_0^x \left(\frac{1}{3}x^2 + \frac{8}{3}y^2 - 2xy\right) \phi(y)\Delta y.$$

Hence,

$$\begin{aligned}
z(x)-z(0) &= \int_0^x \left(x_1+1+\int_0^{x_1}\left(\frac{1}{3}x_1^2+\frac{8}{3}y^2-2x_1y\right)\phi(y)\Delta y\right)\Delta x_1\\
&= \int_0^x (x_1+1)\Delta x_1+\int_0^x\int_0^{x_1}\left(\frac{1}{3}x_1^2+\frac{8}{3}y^2-2x_1y\right)\phi(y)\Delta y\Delta x_1\\
&= \int_0^x \left(\frac{1}{3}(x_1^2)^\Delta+1\right)\Delta x_1+\int_0^x\int_0^{x_1}\left(\frac{1}{3}x_1^2+\frac{8}{3}y^2-2x_1y\right)\phi(y)\Delta y\Delta x_1\\
&= \frac{1}{3}\int_0^x (x_1^2)^\Delta\Delta x_1+\int_0^x\Delta x_1+\int_0^x\int_0^{x_1}\left(\frac{1}{3}x_1^2+\frac{8}{3}y^2-2x_1y\right)\phi(y)\Delta y\Delta x_1\\
&= \frac{1}{3}x^2+x+\int_0^x\int_0^{x_1}\left(\frac{1}{3}x_1^2+\frac{8}{3}y^2-2x_1y\right)\phi(y)\Delta y\Delta x_1\\
&= \frac{1}{3}x^2+x+\int_0^x\int_y^{x}\left(\frac{1}{3}x_1^2+\frac{8}{3}y^2-2x_1y\right)\phi(y)\Delta x_1\Delta y-\int_0^x y^3\phi(y)\Delta y\\
&= \frac{1}{3}x^2+x+\int_0^x\left(\int_y^{x}\left(\frac{1}{21}(x_1^3)^\Delta-\frac{2}{3}y(x_1^2)^\Delta+\frac{8}{3}y^2\right)\Delta x_1\right)\phi(y)\Delta y-\int_0^x y^3\phi(y)\Delta y\\
&= \frac{1}{3}x^2+x+\int_0^x\left(\frac{1}{21}x^3-\frac{1}{21}y^3-\frac{2}{3}x^2y+\frac{2}{3}y^3+\frac{8}{3}y^2x-\frac{8}{3}y^3-y^3\right)\phi(y)\Delta y\\
&= \frac{1}{3}x^2+x+\int_0^x\left(\frac{1}{21}x^3-\frac{64}{21}y^3-\frac{2}{3}x^2y+\frac{8}{3}y^2x\right)\phi(y)\Delta y,
\end{aligned}$$

or

$$z(x)=\frac{1}{3}x^2+x+1+\int_0^x\left(\frac{1}{21}x^3-\frac{64}{21}y^3-\frac{2}{3}x^2y+\frac{8}{3}y^2x\right)\phi(y)\Delta y. \tag{3.53}$$

We substitute (3.52) and (3.53) and we find

$$\phi(x)+x\left(\frac{1}{3}x^2+x+1+\int_0^x\left(\frac{1}{21}x^3-\frac{64}{21}y^3-\frac{2}{3}x^2y+\frac{8}{3}y^2x\right)\phi(y)\Delta y\right)=1,$$

or

$$\phi(x)+\int_0^x\left(\frac{1}{21}x^4-\frac{64}{21}xy^3-\frac{2}{3}x^3y+\frac{8}{3}y^2x^2\right)\phi(y)\Delta y=-\frac{1}{3}x^3-x^2-x+1.$$

Exercise 16 Let $\mathscr{T}=3^{\mathscr{N}_0}\cup\{0\}$. Convert the following IVP

$$\begin{cases} z^{\Delta^3}(x)+z^\Delta(x)=1\\ z(0)=z^\Delta(0)=z^{\Delta^2}(0)=0.\end{cases}$$

Answer

$$\phi(x)+\int_0^x (x-3y)\phi(y)\Delta y=1.$$

3.6 Advanced Practical Exercises

Problem 1 Let $\mathscr{T} = 2\mathscr{Z}$. Consider the equation

$$\phi(x) = x + x\int_0^x \phi(y)\Delta y.$$

Using ADM, find $\phi_0(x)$, $\phi_1(x)$ and $\phi_2(x)$.

Answer

$$\phi_0(x) = x, \quad \phi_1(x) = \frac{1}{2}x^3 - x^2, \quad \phi_2(x) = \frac{1}{8}x^5 - \frac{5}{6}x^4 + \frac{3}{2}x^3 - \frac{2}{3}x^2.$$

Problem 2 Let $\mathscr{T} = 4\mathscr{Z}$. Using ADM, find a solution in a series form of the equation

$$\phi(x) = x - 3\int_0^x \phi(t)\Delta t.$$

Answer

$$\phi(x) = x + \left(-\frac{3}{2}x^2 + 6x\right) + \left(\frac{3}{2}x^3 - 18x^2 + 48x\right) + \cdots.$$

Problem 3 Let $\mathscr{T} = 3^{\mathscr{N}_0} \cup \{0\}$. Consider the equation

$$\phi(x) = 1 + x^2 + \int_0^x \phi(y)\Delta y.$$

Using MDM, find $\phi_0(x)$, $\phi_1(x)$, $\phi_2(x)$ and $\phi_3(x)$.

Answer

$$\phi_0(x) = 1, \quad \phi_1(x) = x^2 + x,$$
$$\phi_2(x) = \frac{1}{13}x^3 + \frac{1}{4}x^2, \quad \phi_3(x) = \frac{1}{520}x^4 + \frac{1}{52}x^3.$$

Problem 4 Use the noise terms phenomenon to solve the following equation

$$\phi(x) = 2x - 2 + \int_0^x \phi(y)\Delta y, \quad \mathscr{T} = 4^{\mathscr{N}_0} \cup \{0\}.$$

Answer $\phi(x) = -2.$

Problem 5 Using DEM, find a solution of the following equations.

1. $\phi(x) = 3 + \int_1^x (y^3 + 1)\phi(y)\Delta y, \quad \mathscr{T} = \mathscr{N}_0^2,$
2. $\phi(x) = -8 + \int_0^x (y^4 + y^2 + 1)\phi(y)\Delta y, \quad \mathscr{T} = \mathscr{Z},$
3. $\phi(x) = 1 + \int_1^x y^3\phi(y)\Delta y, \quad \mathscr{T} = 2^{\mathscr{N}_0}.$

Answer

1. $\phi(x) = 3e_{x^3+1}(x, 1),$
2. $\phi(x) = -8e_{x^4+x^2+1}(x, 0),$
3. $\phi(x) = e_{x^3}(x, 1).$

Problem 6 Using DEM, find a solution of the following generalized Volterra integral equations,

1. $\phi(x) = x - 3 + 2\int_0^x y^2\phi(y)\Delta y, \quad \mathscr{T} = \mathscr{Z},$
2. $\phi(x) = x^2 + \int_1^x y^2\phi(y)\Delta y, \quad \mathscr{T} = 3^{\mathscr{N}_0},$
3. $\phi(x) = x^3 + \int_0^x y\phi(y)\Delta y, \quad \mathscr{T} = 4\mathscr{Z}.$

Answer

1. $\phi(x) = -3e_{2x^2}(x, 0) + \int_0^x \frac{e_{2x^2}(x, \tau)}{1 + 2\tau^2}\Delta\tau,$
2. $\phi(x) = e_{x^2}(x, 1) + 4\int_1^x \frac{e_{x^2}(x, \tau)}{1 + 2\tau^3}\tau\Delta\tau,$
3. $\phi(x) = \int_0^x \frac{e_x(x, 0)}{1 + 4\tau}(3\tau^2 + 12\tau + 16)\Delta\tau.$

Problem 7 Using DEM, find a solution of the following generalized Volterra integral equations.

1. $\phi(x) = 2x + 2\int_0^x y\phi(y)\Delta y, \quad \mathscr{T} = \mathscr{Z},$
2. $\phi(x) = 3 - 2x + 4x\int_1^x y\phi(y)\Delta y, \quad \mathscr{T} = 2^{\mathscr{N}_0},$
3. $\phi(x) = x^2 + x\int_1^x y\phi(y)\Delta y, \quad \mathscr{T} = 4^{\mathscr{N}_0}.$

Answer

1. $\phi(x) = 2\int_0^x e_{2x}(x,\tau)\frac{1}{1+2\tau}\Delta\tau,$
2. $\phi(x) = e_{\frac{1+8x^3}{x}}(x,1) - \frac{3}{2}\int_1^x e_{\frac{1+8x^3}{x}}(x,\tau)\frac{1}{\tau+4\tau^4}\Delta\tau,$
3. $\phi(x) = e_{\frac{1+4x^3}{x}}(x,1) + \int_1^x e_{\frac{1+4x^3}{x}}(x,\tau)\frac{\tau}{1+3\tau^3}\Delta\tau.$

Problem 8 Using DEM, find a solution of the following generalized Volterra integral equations.

1. $\phi(x) = x^2 - \int_0^x y^2\phi(y+1)\Delta y, \quad \mathscr{T} = \mathscr{Z},$
2. $\phi(x) = x^3 - \int_1^x y^3\phi(2y)\Delta y, \quad \mathscr{T} = 2^{\mathscr{N}_0},$
3. $\phi(x) = x^4 - \int_1^x y^4\phi(4y)\Delta y, \quad \mathscr{T} = 4^{\mathscr{N}_0}.$

Answer

1. $\phi(x) = \int_0^x e_{-\frac{x^2}{1+x^2}}(x,\tau)(2\tau+1)\Delta\tau,$
2. $\phi(x) = e_{-\frac{x^3}{1+x^4}}(x,1) + 7\int_1^x e_{-\frac{x^3}{1+x^4}}(x,\tau)\tau^2\Delta\tau,$
3. $\phi(x) = e_{-\frac{x^4}{1+3x^5}}(x,1) + 85\int_1^x e_{-\frac{x^4}{1+3x^5}}(x,\tau)\tau^3\Delta\tau.$

Problem 9 Let $\mathscr{T} = \mathscr{Z}$. Consider the equation

$$\phi(x) = 2x + 4 + \int_0^x y\phi(y)\Delta y.$$

Using SAM and $\phi_0(x) = 1$, find $\phi_1(x)$ and $\phi_2(x)$.

Answer

$$\phi_1(x) = \frac{1}{2}x^2 + \frac{3}{2}x + 4, \quad \phi_2(x) = 4 + \frac{1}{4}x + \frac{11}{8}x^2 + \frac{1}{4}x^3 + \frac{1}{8}x^4.$$

Problem 10 Let $\mathscr{T} = 2^{\mathscr{N}_0} \cup \{0\}$. Consider the equation

$$\phi(x) = 1 + \int_0^x \phi(y+1)\Delta y.$$

Using SAM and $\phi_0(x) = 1$, find $\phi_1(x)$ and $\phi_2(x)$.

Answer

$$\phi_1(x) = 1 + x, \quad \phi_2(x) = 1 + 2x + \frac{1}{3}x^2.$$

Problem 11 Convert the following generalized Volterra integral equation of the first kind to generalized Volterra integral equation of the second kind.

$$x^2 + x + 10 = -x + \int_1^x (2x + y)^2 \phi(y) \Delta y, \quad \mathscr{T} = 2^{\mathscr{N}_0}.$$

Answer

$$\phi(x) = \frac{3x + 2}{25x^2} - \int_1^x \frac{12x + 4y}{25x^2} \phi(y) \Delta y.$$

Problem 12 Let $\mathscr{T} = 2^{\mathscr{N}_0}$ and $K(x, y) = y$, $(x, y) \in [1, 16] \times [1, 16]$. Find $K^{(2)}(x, y)$.

Answer

$$\frac{1}{3} y(x - 2y)(x + 2y).$$

Problem 13 Let $\mathscr{T} = \mathscr{Z}$. Convert the following IVP

$$\begin{cases} z^{\Delta^5}(x) + z^{\Delta^3}(x) = x \\ z(0) = z^{\Delta}(0) = z^{\Delta^2}(0) = z^{\Delta^3}(0) = z^{\Delta^4}(0) = 0. \end{cases}$$

Answer

$$\phi(x) + \int_0^x (x - y - 1)\phi(y) \Delta y = x.$$

Chapter 4
Generalized Volterra Integro-Differential Equations

In this chapter we describe the Adomian decomposition method for generalized Volterra integro-differential equations of the second kind. They are given procedures for conversion of generalized Volterra integro-differential equations of the second kind to generalized Volterra integro-differential equations of the first kind and generalized Volterra integral equations.

Volterra integro-differential equations appear in many physical applications such as glassforming process, nanohydrodynamics, heat transfer, diffusion process in general, neutron diffusion and biological species coexisting together with increasing and decreasing rates of generating, and wind ripple in the desert. To find a solution for the integro-differential equation, the initial conditions should be given.

4.1 Generalized Volterra Integro-Differential Equations of the Second Kind

4.1.1 The Adomian Decomposition Method

The Adomian decomposition method (ADM) gives the solution in an infinite series that can be recurrently determined. The obtained series may give the exact solution if such a solution exists. Otherwise, the series gives an approximation for the solution that gives high accuracy level.

We consider a generalized Volterra integro-differential equation of the second kind given by

$$\begin{aligned} &\phi^{\Delta^n}(x) = u(x) + \int_a^x K(x, y)\phi(y)\Delta y, \\ &\phi(a) = a_0, \quad \phi^{\Delta}(a) = a_1, \quad \ldots, \quad \phi^{\Delta^{n-1}}(a) = a_{n-1}. \end{aligned} \tag{4.1}$$

S.G. Georgiev, *Integral Equations on Time Scales*, Atlantis Studies in Dynamical Systems 5, DOI 10.2991/978-94-6239-228-1_4

We integrate the Eq. (4.1) from a to x and we get

$$\int_a^x \phi^{\Delta^n}(t)\Delta t = \int_a^x u(t)\Delta t + \int_a^x \int_a^{x_1} K(x_1, y)\phi(y)\Delta y \Delta x_1$$

or

$$\phi^{\Delta^{n-1}}(x) = \phi^{\Delta^{n-1}}(a) + \int_a^x u(t)\Delta t + \int_a^x \int_a^{x_1} K(x_1, y)\phi(y)\Delta y \Delta x_1.$$

Again we integrate from a to x and we obtain

$$\int_a^x \phi^{\Delta^{n-1}}(y)\Delta y = \int_a^x \phi^{\Delta^{n-1}}(a)\Delta y + \int_a^x \int_a^{x_1} u(t)\Delta t \Delta x_1 + \int_a^x \int_a^{x_1} \int_a^{x_2} K(x_2, y)\phi(y)\Delta y \Delta x_2 \Delta x_1$$

or

$$\begin{aligned}\phi^{\Delta^{n-2}}(x) &= \phi^{\Delta^{n-2}}(a) + \phi^{\Delta^{n-1}}(a)(x-a) + \int_a^x \int_a^{x_1} u(t)\Delta t \Delta x_1 \\ &\quad + \int_a^x \int_a^{x_1} \int_a^{x_2} K(x_2, y)\phi(y)\Delta y \Delta x_2 \Delta x_1 \\ &= \phi^{\Delta^{n-2}}(a)h_0(x, a) + \phi^{\Delta^{n-1}}(a)h_1(x, a) + \int_a^x \int_a^{x_1} u(t)\Delta t \Delta x_1 \\ &\quad + \int_a^x \int_a^{x_1} \int_a^{x_2} K(x_2, y)\phi(y)\Delta y \Delta x_2 \Delta x_1,\end{aligned}$$

and so on,

$$\begin{aligned}\phi(x) &= \sum_{l=0}^{n-1} \phi^{\Delta^l}(a)h_l(x, a) + \int_a^x \int_a^{x_1} \cdots \int_a^{x_{n-1}} u(t)\Delta t \Delta x_{n-1} \ldots \Delta x_1 \\ &\quad + \int_a^x \int_a^{x_1} \cdots \int_a^{x_n} K(x_n, y)\phi(y)\Delta y \Delta x_n \ldots \Delta x_1,\end{aligned}$$

or

$$\begin{aligned}\phi(x) &= \textstyle\sum_{l=0}^{n-1} \phi^{\Delta^l}(a)h_l(x, a) + \int_a^x (h_{n-1}(x, y) - \mu(y)h_{n-2}(x, y))u(y)\Delta y \\ &\quad + \textstyle\int_a^x \int_a^{x_1} \cdots \int_a^{x_n} K(x_n, y)\phi(y)\Delta y \Delta x_n \ldots \Delta x_1\end{aligned} \tag{4.2}$$

We will search $\phi(x)$ in the form

$$\phi(x) = \sum_{m=0}^{\infty} \phi_m(x). \tag{4.3}$$

We substitute (4.3) in (4.2) and we find

$$\begin{aligned}\sum_{m=0}^{\infty} \phi_m(x) &= \sum_{l=0}^{n-1} \phi^{\Delta^l}(a)h_l(x,a) + \int_a^x (h_{n-1}(x,y) - \mu(y)h_{n-2}(x,y))u(y)\Delta y \\ &\quad + \int_a^x \int_a^{x_1} \cdots \int_a^{x_n} K(x_n,y) \sum_{m=0}^{\infty} \phi_m(y)\Delta y \Delta x_n \ldots \Delta x_1 \\ &= \sum_{l=0}^{n-1} \phi^{\Delta^l}(a)h_l(x,a) + \int_a^x (h_{n-1}(x,y) - \mu(y)h_{n-2}(x,y))u(y)\Delta y \\ &\quad + \sum_{m=0}^{\infty} \int_a^x \int_a^{x_1} \cdots \int_a^{x_n} K(x_n,y)\phi_m(y)\Delta y \Delta x_n \ldots \Delta x_1.\end{aligned}$$

To determine the components $\phi_0(x), \phi_2(x), \phi_3(x), \ldots$, we set the recurrence relation

$$\phi_0(x) = \sum_{l=0}^{n-1} \phi^{\Delta^l}(a)h_l(x,a) + \int_a^x (h_{n-1}(x,y) - \mu(y)h_{n-2}(x,y))u(y)\Delta y,$$

$$\phi_1(x) = \int_a^x \int_a^{x_1} \cdots \int_a^{x_n} K(x_n,y)\phi_0(y)\Delta y \Delta x_n \ldots \Delta x_1,$$

$$\ldots$$

$$\phi_k(x) = \int_a^x \int_a^{x_1} \cdots \int_a^{x_n} K(x_n,y)\phi_{k-1}(y)\Delta y \Delta x_n \ldots \Delta x_1, \quad k \geq 2.$$

Having determined the components $\phi_k(x)$, $k \geq 0$, the solution $\phi(x)$ of the equation (4.1) is then obtained in a series form. Using (4.3), the obtained series converges to the exact solution if such a solution exists.

Example 1 Let $\mathscr{T} = 2^{\mathscr{N}_0} \cup \{0\}$. Consider the equation

$$\phi^{\Delta}(x) = 1 + \int_0^x \phi(y)\Delta y, \quad \phi(0) = 0.$$

Here $\sigma(x) = 2x$, $\mu(x) = x$, $x \in\in \mathscr{T}$.
We integrate both sides of the given equation and we obtain

$$\int_0^x \phi^{\Delta}(y)\Delta y = \int_0^x \Delta y + \int_0^x \int_0^{x_1} \phi(y)\Delta y \Delta x_1$$

or

$$\phi(y)\Big|_{y=0}^{y=x} = y\Big|_{y=0}^{y=x} + \int_0^x \int_0^{x_1} \phi(y)\Delta y \Delta x_1,$$

or

$$\begin{aligned}\phi(x) &= x + \int_0^x (x - \sigma(y))\phi(y)\Delta y \\ &= x + \int_0^x (x - 2y)\phi(y)\Delta y.\end{aligned}$$

We set

$$\phi_0(x) = x,$$

$$\phi_k(x) = \int_0^x (x - 2y)\phi_{k-1}(y)\Delta y, \quad k \in \mathcal{N}.$$

Then

$$\begin{aligned}\phi_1(x) &= \int_0^x (x - 2y)\phi_0(y)\Delta y \\ &= \int_0^x (x - 2y)y\Delta y \\ &= \int_0^x (xy - 2y^2)\Delta y \\ &= \int_0^x \left(\frac{1}{3}x(y^2)^\Delta - \frac{2}{7}(y^3)^\Delta\right)\Delta y \\ &= \frac{1}{3}xy^2\Big|_{y=0}^{y=x} - \frac{2}{7}y^3\Big|_{y=0}^{y=x} \\ &= \frac{1}{3}x^3 - \frac{2}{7}x^3 \\ &= \frac{1}{21}x^3,\end{aligned}$$

$$\begin{aligned}\phi_2(x) &= \int_0^x (x - 2y)\phi_1(y)\Delta y \\ &= \frac{1}{21}\int_0^x (x - 2y)y^3\Delta y \\ &= \frac{1}{21}\int_0^x (xy^3 - 2y^4)\Delta y\end{aligned}$$

$$= \frac{1}{21}\int_0^x \left(x\frac{1}{15}(y^4)^\Delta - \frac{2}{31}(y^5)^\Delta\right)\Delta y$$

$$= \frac{1}{315}x\int_0^x (y^4)^\Delta \Delta y - \frac{2}{651}\int_0^x (y^5)^\Delta \Delta y$$

$$= \frac{1}{315}xy^4\Big|_{y=0}^{y=x} - \frac{2}{651}y^5\Big|_{y=0}^{y=x}$$

$$= \frac{x^5}{315} - \frac{2}{651}x^5$$

$$= \frac{1}{9765}x^5.$$

Example 2 Let $\mathcal{T} = \mathcal{Z}$. Consider the equation

$$\phi^\Delta(x) = 1 + x + \int_0^x y\phi(y)\Delta y, \quad \phi(0) = 0.$$

Here

$$\sigma(x) = x + 1, \quad \mu(x) = 1, \quad x \in \mathcal{T}.$$

We integrate both sides of the given equation from 0 to x and we get

$$\int_0^x \phi^\Delta(y)\Delta y = \int_0^x (1+y)\Delta y + \int_0^x \int_0^{x_1} y\phi(y)\Delta y \Delta x_1$$

or

$$\phi(y)\Big|_{y=0}^{y=x} = \int_0^x \left(1 + \frac{1}{2}(y^2)^\Delta - \frac{1}{2}\right)\Delta y + \int_0^x (x - \sigma(y))y\phi(y)\Delta y,$$

or

$$\phi(x) - \phi(0) = \frac{1}{2}\int_0^x \left(1 + (y^2)^\Delta\right)\Delta y + \int_0^x (x - y - 1)y\phi(y)\Delta y,$$

or

$$\phi(x) = \frac{1}{2}\int_0^x \Delta y + \frac{1}{2}\int_0^x (y^2)^\Delta \Delta y + \int_0^x (x - y - 1)y\phi(y)\Delta y$$

$$= \frac{1}{2}y\Big|_{y=0}^{y=x} + \frac{1}{2}y^2\Big|_{y=0}^{y=x} + \int_0^x (x - y - 1)y\phi(y)\Delta y$$

$$= \frac{1}{2}x + \frac{1}{2}x^2 + \int_0^x (x - y - 1)y\phi(y)\Delta y.$$

We set

$$\phi_0(x) = \tfrac{1}{2}x + \tfrac{1}{2}x^2,$$

$$\phi_k(x) = \textstyle\int_0^x (x - y - 1)y\phi_{k-1}(y)\Delta y, \quad k \in \mathscr{N}.$$

Example 3 Let $\mathscr{T} = \mathscr{Z}$. Consider the equation

$$\phi^{\Delta^2}(x) = 1 + \int_0^x \phi(y)\Delta y, \quad \phi(0) = \phi^{\Delta}(0) = 0.$$

Here $\sigma(x) = x + 1, \quad \mu(x) = 1, \quad x \in \mathscr{T}$.
We integrate both sides of the given equation and we get

$$\int_0^x \phi^{\Delta^2}(y)\Delta y = \int_0^x \Delta y + \int_0^x \int_0^{x_1} \phi(y)\Delta y \Delta x_1$$

or

$$\phi^{\Delta}(y)\Big|_{y=0}^{y=x} = x + \int_0^x \int_0^{x_1} \phi(y)\Delta y \Delta x_1,$$

or

$$\phi^{\Delta}(x) = x + \int_0^x \int_0^{x_1} \phi(y)\Delta y \Delta x_1.$$

We integrate the last equation from 0 to x and we obtain

$$\int_0^x \phi^{\Delta}(y)\Delta y = \int_0^x y\Delta y + \int_0^x \int_0^{x_1} \int_0^{x_2} \phi(y)\Delta y \Delta x_2 \Delta x_1,$$

or

$$\phi(x) - \phi(0) = \int_0^x \left(\frac{1}{2}(y^2)^{\Delta} - \frac{1}{2}\right)\Delta y + \int_0^x \int_0^{x_1} \int_0^{x_2} \phi(y)\Delta y \Delta x_2 \Delta x_1$$

$$= \frac{1}{2}\int_0^x (y^2)^{\Delta}\Delta y - \frac{1}{2}\int_0^x \Delta y + \int_0^x \int_0^{x_1} \int_0^{x_2} \phi(y)\Delta y \Delta x_2 \Delta x_1$$

$$= \frac{1}{2}y^2\Big|_{y=0}^{y=x} - \frac{1}{2}y\Big|_{y=0}^{y=x} + \int_0^x \int_0^{x_1} \int_0^{x_2} \phi(y)\Delta y \Delta x_2 \Delta x_1$$

$$= \frac{1}{2}x^2 - \frac{1}{2}x + \int_0^x \int_0^{x_1} \int_0^{x_2} \phi(y)\Delta y \Delta x_2 \Delta x_1$$

$$= \frac{1}{2}x^2 - \frac{1}{2}x + \int_0^x (h_2(x, y) - h_1(x, y))\phi(y)\Delta y,$$

i.e.,

$$\phi(x) = \frac{1}{2}x^2 - \frac{1}{2}x + \int_0^x (h_2(x, y) - h_1(x, y))\phi(y)\Delta y.$$

Note that

$$\begin{aligned} h_2(x, y) &= \int_y^x h_1(\tau, y)\Delta\tau \\ &= \int_y^x (\tau - y)\Delta\tau \\ &= \int_y^x \left(\frac{1}{2}(\tau^2)^\Delta - \frac{1}{2} - y\right)\Delta\tau \\ &= \frac{1}{2}\int_y^x (\tau^2)^\Delta \Delta\tau - \left(\frac{1}{2} + y\right)\int_y^x \Delta\tau \\ &= \frac{1}{2}\tau^2\Big|_{\tau=y}^{\tau=x} - \left(\frac{1}{2} + y\right)\tau\Big|_{\tau=y}^{\tau=x} \\ &= \frac{1}{2}(x^2 - y^2) - \left(\frac{1}{2} + y\right)(x - y) \\ &= \frac{1}{2}x^2 + \frac{1}{2}y^2 - xy - \frac{1}{2}x + \frac{1}{2}y, \\ h_2(x, y) - \mu(y)h_1(x, y) &= \frac{1}{2}x^2 + \frac{1}{2}y^2 - xy - \frac{1}{2}x + \frac{1}{2}y - x + y \\ &= \frac{1}{2}x^2 + \frac{1}{2}y^2 - xy - \frac{3}{2}x + \frac{3}{2}y. \end{aligned}$$

Then

$$\phi(x) = \frac{1}{2}x^2 - \frac{1}{2}x + \int_0^x \left(\frac{1}{2}x^2 + \frac{1}{2}y^2 - xy - \frac{3}{2}x + \frac{3}{2}y\right)\phi(y)\Delta y.$$

We set

$$\phi_0(x) = \tfrac{1}{2}x^2 - \tfrac{1}{2}x,$$

$$\phi_k(x) = \textstyle\int_0^x \left(\tfrac{1}{2}x^2 + \tfrac{1}{2}y^2 - xy - \tfrac{3}{2}x + \tfrac{3}{2}y\right)\phi_{k-1}(y)\Delta y, \quad k \in \mathcal{N}.$$

Example 4 Let $\mathscr{T} = 2^{\mathscr{N}_0} \cup \{0\}$. Consider the equation

$$\phi^{\Delta^2}(x) = 1 + \int_0^x \phi(y)\Delta y, \quad \phi(0) = \phi^{\Delta}(0) = 0.$$

Here $\sigma(x) = 2x, \quad \mu(x) = x, \quad x \in \mathscr{T}$.
We integrate both sides of the given equation from 0 to x and we get

$$\int_0^x \phi^{\Delta^2}(y)\Delta y = \int_0^x \Delta y + \int_0^x \int_0^{x_1} \phi(y)\Delta y \Delta x_1$$

or

$$\phi^{\Delta}(y)\Big|_{y=0}^{y=x} = y\Big|_{y=0}^{y=x} + \int_0^x \int_0^{x_1} \phi(y)\Delta y \Delta x_1,$$

or

$$\phi^{\Delta}(x) = x + \int_0^x \int_0^{x_1} \phi(y)\Delta y \Delta x_1.$$

Now we integrate the last equation from 0 to x and we obtain

$$\int_0^x \phi^{\Delta}(y)\Delta y = \int_0^x y\Delta y + \int_0^x \int_0^{x_1} \int_0^{x_2} \phi(y)\Delta y \Delta x_2 \Delta x_1$$

or

$$\phi(y)\Big|_{y=0}^{y=x} = \frac{1}{3}\int_0^x (y^2)^{\Delta}\Delta y + \int_0^x \int_0^{x_1} \int_0^{x_2} \phi(y)\Delta y \Delta x_2 \Delta x_1,$$

or

$$\phi(x) = \frac{1}{3}y^2\Big|_{y=0}^{y=x} + \int_0^x \int_0^{x_1} \int_0^{x_2} \phi(y)\Delta y \Delta x_2 \Delta x_1,$$

or

$$\phi(x) = \frac{1}{3}x^2 + \int_0^x (h_2(x, y) - \mu(y)h_1(x, y))\phi(y)\Delta y.$$

Note that

$$h_2(x, y) = \int_y^x h_1(\tau, y)\Delta\tau$$

$$= \int_y^x (\tau - y)\Delta\tau$$

$$= \int_y^x \left(\frac{1}{3}(\tau^2)^{\Delta} - y\right)\Delta\tau$$

$$= \frac{1}{3}\int_y^x (\tau^2)^{\Delta}\Delta\tau - y\int_y^x \Delta\tau$$

$$= \frac{1}{3}\tau^2\Big|_{\tau=y}^{\tau=x} - y\tau\Big|_{\tau=y}^{\tau=x}$$

$$= \frac{1}{3}(x^2 - y^2) - y(x - y)$$

$$= \frac{1}{3}x^2 - \frac{1}{3}y^2 - xy + y^2$$

$$= \frac{1}{3}x^2 + \frac{2}{3}y^2 - xy,$$

$$h_2(x, y) - \mu(y)h_1(x, y) = \frac{1}{3}x^2 + \frac{2}{3}y^2 - xy - y(x - y)$$

$$= \frac{1}{3}x^2 + \frac{2}{3}y^2 - xy - xy + y^2$$

$$= \frac{1}{3}x^2 + \frac{5}{3}y^2 - 2xy.$$

Consequently

$$\phi(x) = \frac{1}{3}x^2 + \int_0^x \left(\frac{1}{3}x^2 + \frac{5}{3}y^2 - 2xy\right)\phi(y)\Delta y.$$

We set

$$\phi_0(x) = \tfrac{1}{3}x^2,$$

$$\phi_k(x) = \int_0^x \left(\tfrac{1}{3}x^2 + \tfrac{5}{3}y^2 - 2xy\right)\phi_{k-1}(y)\Delta y, \quad k \in \mathcal{N}.$$

Remark 1 The modified decomposition method that we used before can be used for handling generalized Volterra integro-differential equations in any order.

Remark 2 The phenomenon of the noise terms that was applied before can be used here if noise terms appear.

Exercise 1 Using ADM, find the recurrence relation for $\{\phi_k(x)\}_{k=0}^{\infty}$ for the following equations.

1.
$$\begin{cases} \phi^{\Delta^3}(x) - 2\phi^{\Delta^2}(x) + \phi^{\Delta}(x) + \phi(x) = x^2 + x + \int_0^x y\phi(y)\Delta y, \\ \phi(0) = \phi^{\Delta}(0) = \phi^{\Delta^2}(0) = 0, \quad \mathscr{T} = \mathscr{Z}, \end{cases}$$

2.
$$\begin{cases} \phi^{\Delta^3}(x) + \phi^{\Delta^2}(x) = x^2 + 2x - 3 + x\int_0^x y^2\phi(y)\Delta y, \\ \phi(0) = \phi^{\Delta}(0) = 0, \quad \phi^{\Delta^2}(0) = 3, \quad \mathscr{T} = 2^{\mathscr{N}_0} \cup \{0\}, \end{cases}$$

3.
$$\begin{cases} \phi^{\Delta^2}(x) + x^2\phi^{\Delta}(x) = \phi(x) + 3 + \int_0^x y\phi(y)\Delta y, \\ \phi(0) = 1, \quad \phi^{\Delta}(0) = 2, \quad \mathscr{T} = 3^{\mathscr{N}_0} \cup \{0\}. \end{cases}$$

4.1.2 Converting Generalized Volterra Integro-Differential Equations of the Second Kind to Initial Value Problems

Consider the generalized Volterra integro-differential equation of the second kind

$$\phi^{\Delta^n}(x) = u(x) + g(x)\int_a^x f(y)\phi(y)\Delta y, \tag{4.4}$$

$$\phi^{\Delta^i}(a) = a_i, \quad 0 \le i \le n-1,$$

where $f, g : \mathscr{T} \longmapsto \mathscr{R}$, $g(x) \not\equiv 0$ for $x \in [a, b]$.
We substitute $x = a$ in (4.4) and we get

$$\begin{aligned} \phi^{\Delta^n}(a) &= u(a) + g(a)\int_a^a f(y)\phi(y)\Delta y \\ &= u(a). \end{aligned}$$

Also,

$$\phi^{\Delta^n}(x) - u(x) = g(x)\int_a^x f(y)\phi(y)\Delta y,$$

whereupon

$$\int_a^x f(y)\phi(y)\Delta y = \frac{\phi^{\Delta^n}(x) - u(x)}{g(x)}, \quad x \in [a, b]. \tag{4.5}$$

We differentiate (4.4) with respect to x and we get

$$\phi^{\Delta^{n+1}}(x) = u^{\Delta}(x) + g^{\Delta}(x)\int_a^x f(y)\phi(y)\Delta y + g(\sigma(x))f(x)\phi(x).$$

Now applying (4.5) we obtain

$$\phi^{\Delta^{n+1}}(x) = u^{\Delta}(x) + g^{\Delta}(x)\frac{\phi^{\Delta^n}(x) - u(x)}{g(x)} + g(\sigma(x))f(x)\phi(x), \quad x \in [a, b],$$

or

$$\phi^{\Delta^{n+1}}(x) - \frac{g^{\Delta}(x)}{g(x)}\phi^{\Delta^n}(x) - g(\sigma(x))f(x)\phi(x) = u^{\Delta}(x) - g^{\Delta}(x)\frac{u(x)}{g(x)},$$

$x \in [a, b]$.
Consequently we obtain the following initial value problem

$$\begin{cases} \phi^{\Delta^{n+1}}(x) - \frac{g^{\Delta}(x)}{g(x)}\phi^{\Delta^n}(x) - g(\sigma(x))f(x)\phi(x) = u^{\Delta}(x) - g^{\Delta}(x)\frac{u(x)}{g(x)} \\ \phi^{\Delta^i}(a) = a_i, \quad 0 \le i \le n-1, \quad \phi^{\Delta^n}(a) = u(a), \quad x \in [a, b]. \end{cases}$$

Example 5 Consider the equation

$$\phi^{\Delta^3}(x) = x^2 + 2x - 2x\textstyle\int_1^x y\phi(y)\Delta y, \quad x \in \mathscr{T} = \mathscr{N},$$

$$\phi(1) = 1, \quad \phi^{\Delta}(1) = -1, \quad \phi^{\Delta^2}(1) = 2.$$

Note that

$$\sigma(x) = x + 1, \quad \mu(x) = 1, \quad x \in \mathscr{T}.$$

Then

$$2x\int_1^x y\phi(y)\Delta y = x^2 + 2x - \phi^{\Delta^3}(x)$$

or

$$\int_1^x y\phi(y)\Delta y = \frac{x^2 + 2x - \phi^{\Delta^3}(x)}{2x}. \tag{4.6}$$

Now we integrate the given equation with respect to x and we find

$$\begin{aligned}
\phi^{\Delta^4}(x) &= \left(x^2 + 2x - 2x \int_1^x y\phi(y)\Delta y\right)^{\Delta} \\
&= (x^2+2x)^{\Delta} - 2\left(x\int_1^x y\phi(y)\Delta y\right)^{\Delta} \\
&= (x^2)^{\Delta} + 2x^{\Delta} - 2\left(x^{\Delta}\int_1^x y\phi(y)\Delta y + \sigma(x)\left(\int_1^x y\phi(y)\Delta y\right)^{\Delta}\right) \\
&= x + \sigma(x) + 2 - 2\left(\int_1^x y\phi(y)\Delta y + (x+1)x\phi(x)\right) \\
&= 2x + 3 - 2\int_1^x y\phi(y)\Delta y - 2x(x+1)\phi(x),
\end{aligned}$$

i.e.,

$$\phi^{\Delta^4}(x) = 2x + 3 - 2\int_1^x y\phi(y)\Delta y - 2x(x+1)\phi(x).$$

Now applying (4.6) we obtain

$$\phi^{\Delta^4}(x) = -2x(x+1)\phi(x) - 2\frac{x^2+2x-\phi^{\Delta^3}(x)}{2x} + 2x + 3$$

or

$$\phi^{\Delta^4}(x) = -2x(x+1)\phi(x) - x - 2 + \frac{1}{x}\phi^{\Delta^3}(x) + 2x + 3,$$

or

$$\phi^{\Delta^4}(x) - \frac{1}{x}\phi^{\Delta^3}(x) + 2x(x+1)\phi(x) = x + 1.$$

We substitute $x = 1$ in the given equation and we find

$$\phi^{\Delta^3}(1) = 3.$$

Therefore we get the following initial value problem

$$\begin{cases}
\phi^{\Delta^4}(x) - \frac{1}{x}\phi^{\Delta^3}(x) + 2x(x+1)\phi(x) = x + 1, \\
\phi(1) = 1, \quad \phi^{\Delta}(1) = -1, \quad \phi^{\Delta^2}(1) = 2, \quad \phi^{\Delta^3}(1) = 3.
\end{cases}$$

Example 6 Let $\mathscr{T} = 2^{\mathscr{N}_0}$. Consider the equation

$$\phi^{\Delta^2}(x) = e_{x^2}(x,1) + x\int_1^x y^2\phi(y)\Delta y, \quad x \in \mathscr{T},$$

$$\phi(1) = \phi^{\Delta}(1) = 1.$$

Here

$$\sigma(x) = 2x, \quad \mu(x) = x, \quad x \in \mathscr{T}.$$

We have

$$x\int_1^x y^2\phi(y)\Delta y = \phi^{\Delta^2}(x) - e_{x^2}(x,1),$$

whereupon

$$\int_1^x y^2\phi(y)\Delta y = \frac{\phi^{\Delta^2}(x) - e_{x^2}(x,1)}{x}. \tag{4.7}$$

We differentiate the given equation with respect to x and we get

$$\begin{aligned}
\phi^{\Delta^3}(x) &= \left(e_{x^2}(x,1) + x\int_1^x y^2\phi(y)\Delta y\right)^{\Delta} \\
&= e_{x^2}^{\Delta}(x,1) + \left(x\int_1^x y^2\phi(y)\Delta y\right)^{\Delta} \\
&= x^2 e_{x^2}(x,1) + x^{\Delta}\int_1^x y^2\phi(y)\Delta y + \sigma(x)\left(\int_1^x y^2\phi(y)\Delta y\right)^{\Delta} \\
&= x^2 e_{x^2}(x,1) + \int_1^x y^2\phi(y)\Delta y + 2x\left(x^2\phi(x)\right) \\
&= x^2 e_{x^2}(x,1) + \int_1^x y^2\phi(y)\Delta y + 2x^3\phi(x).
\end{aligned}$$

Applying (4.7) we obtain

$$\phi^{\Delta^3}(x) = x^2 e_{x^2}(x,1) + \frac{\phi^{\Delta^2}(x) - e_{x^2}(x,1)}{x} + 2x^3\phi(x),$$

whereupon

$$\phi^{\Delta^3}(x) - \frac{1}{x}\phi^{\Delta^2} - 2x^3\phi(x) = \left(x^2 - \frac{1}{x}\right)e_{x^2}(x,1).$$

We substitute $x = 1$ in the given equation and we find

$$\phi^{\Delta^2}(1) = e_1(1,1) = 1.$$

Consequently we obtain the following initial value problem

$$\begin{cases} \phi^{\Delta^3}(x) - \frac{1}{x}\phi^{\Delta^2} - 2x^3\phi(x) = \left(x^2 - \frac{1}{x}\right) e_{x^2}(x, 1) \\ \phi(1) = \phi^{\Delta}(1) = \phi^{\Delta^2}(1) = 1. \end{cases}$$

Example 7 Let $\mathscr{T} = 3^{\mathscr{N}_0}$. Consider the equation

$$\phi^{\Delta^2}(x) = x^3\phi^{\Delta}(x) + x^2 + 2x + 1 + x^2 \int_1^x y^3\phi(y)\Delta y$$

$$\phi(1) = \phi^{\Delta}(1) = 1.$$

Here

$$\sigma(x) = 3x, \quad \mu(x) = 2x, \quad x \in \mathscr{T}.$$

We have

$$x^2 \int_1^x y^3\phi(y)\Delta y = \phi^{\Delta^2}(x) - x^3\phi^{\Delta}(x) - x^2 - 2x - 1,$$

from where

$$\int_1^x y^3\phi(y)\Delta y = \frac{1}{x^2}\phi^{\Delta^2}(x) - x\phi^{\Delta}(x) - 1 - \frac{2}{x} - \frac{1}{x^2}. \tag{4.8}$$

We differentiate the given equation with respect to x and we find

$$\begin{aligned}
\phi^{\Delta^3}(x) &= \left(x^3\phi^{\Delta}(x) + x^2 + 2x + 1 + x^2 \int_1^x y^3\phi(y)\Delta y\right)^{\Delta} \\
&= \left(x^3\phi^{\Delta}(x)\right)^{\Delta} + (x^2)^{\Delta} + (2x)^{\Delta} + 1^{\Delta} + \left(x^2 \int_1^x y^3\phi(y)\Delta y\right)^{\Delta} \\
&= (x^3)^{\Delta}\phi^{\Delta}(x) + \sigma^3(x)\phi^{\Delta^2}(x) + \sigma(x) + x + 2 \\
&\quad +(x^2)^{\Delta} \int_1^x y^3\phi(y)\Delta y + \sigma^2(x)\left(\int_1^x y^3\phi(y)\Delta y\right)^{\Delta} \\
&= (\sigma^2(x) + x\sigma(x) + x^2)\phi^{\Delta}(x) + 27x^3\phi^{\Delta^2}(x) + 4x + 2 \\
&\quad + (\sigma(x) + x) \int_1^x y^3\phi(y)\Delta y + 9x^2\left(x^3\phi(x)\right)
\end{aligned}$$

$$= (9x^2 + 3x^2 + x^2)\phi^{\Delta}(x) + 27x^3\phi^{\Delta^2}(x) + 4x + 2$$

$$+ 4x \int_1^x ys\phi(y)\Delta y + 9x^5\phi(x)$$

$$= 13x^2\phi^{\Delta}(x) + 27x^3\phi^{\Delta^2}(x) + 4x + 2 + 4x \int_1^x y^3\phi(y)\Delta y + 9x^5\phi(x).$$

Applying (4.8) we obtain

$$\phi^{\Delta^3}(x) = 4x + 2 + 9x^5\phi(x) + 13x^2\phi^{\Delta}(x) + 27x^3\phi^{\Delta^2}(x)$$

$$+4x\left(\frac{1}{x^2}\phi^{\Delta^2}(x) - x\phi^{\Delta}(x) - 1 - \frac{2}{x} - \frac{1}{x^2}\right)$$

$$= 4x + 2 + 9x^5\phi(x) + 13x^2\phi^{\Delta}(x) + 27x^3\phi^{\Delta^2}(x)$$

$$+\frac{4}{x}\phi^{\Delta^2}(x) - 4x^2\phi^{\Delta}(x) - 4x - 8 - \frac{4}{x}$$

$$= -\frac{4}{x} - 6 + 9x^5\phi(x) + 9x^2\phi^{\Delta}(x)$$

$$+\left(27x^3 + \frac{4}{x}\right)\phi^{\Delta^2}(x),$$

i.e.,

$$\phi^{\Delta^3}(x) - \left(27x^3 + \frac{4}{x}\right)\phi^{\Delta^2}(x) - 9x^2\phi^{\Delta}(x) - 9x^5\phi(x) = -\frac{4}{x} - 6.$$

We substitute $x = 1$ in the given equation and we find

$$\phi^{\Delta^2}(1) = \phi^{\Delta}(1) + 3 = 4.$$

Consequently we obtain the following initial value problem

$$\begin{cases} \phi^{\Delta^3}(x) - \left(27x^3 + \frac{4}{x}\right)\phi^{\Delta^2}(x) - 9x^2\phi^{\Delta}(x) - 9x^5\phi(x) = -\frac{4}{x} - 6 \\ \phi(1) = \phi^{\Delta}(1) = 1, \quad \phi^{\Delta^2}(1) = 4. \end{cases}$$

Exercise 2 Convert the following equations into initial value problems.

1.

$$\begin{cases} \phi^{\Delta^2}(x) - 2\phi^{\Delta}(x) = x^4 + 2x^3 + 2x^2 + 2x + 1 + \int_0^x (x^2 - y^2)\phi(y)\Delta y, \\ \phi(0) = \phi^{\Delta}(0) = 1, \quad \mathscr{T} = \mathscr{Z}, \end{cases}$$

2.

$$\begin{cases} \phi^{\Delta^3}(x) + \phi^{\Delta^2}(x) + \phi^{\Delta}(x) - \phi(x) = 2x + 1 + x \int_0^x y^3\phi(y)\Delta y, \\ \phi(0) = \phi^{\Delta}(0) = \phi^{\Delta^2}(0) = -1, \quad \mathscr{T} = \mathscr{Z}, \end{cases}$$

3.

$$\begin{cases} \phi^{\Delta^4}(x) + \phi^{\Delta}(x) = \phi(x) - 2x^2 \int_0^x y\phi(y)\Delta y, \\ \phi(0) = \phi^{\Delta}(0) = \phi^{\Delta^2}(0) = \phi^{\Delta^3}(0) = 0, \quad \mathscr{T} = 2^{\mathscr{N}_0} \cup \{0\}. \end{cases}$$

4.1.3 Converting Generalized Volterra Integro-Differential Equations of the Second Kind to Generalized Volterra Integral Equations

Now we consider the equation

$$\begin{aligned} &\phi^{\Delta^n}(x) = u(x) + \lambda \int_a^x h_l(x, y)\phi(y)\Delta y \\ &\phi^{\Delta^i}(a) = a_i, \quad 0 \le i \le n - 1, \end{aligned} \tag{4.9}$$

where $u : \mathscr{T} \longmapsto \mathscr{R}$ is a given rd-continuous function, $\lambda \in \mathscr{C}$ is a given parameter, a_i, $i = 0, 1, \ldots, n - 1$, are given constants and $l \in \mathscr{N}$ is fixed.
We will reduce the Eq. (4.9) to a generalized Volterra integral equation of the second kind.
We integrate the Eq. (4.9) from a to x and we obtain

$$\int_a^x \phi^{\Delta^n}(\tau)\Delta\tau = \int_a^x u(y)\Delta y + \lambda \int_a^x \int_a^{x_1} h_l(x_1, y)\phi(y)\Delta y\Delta x_1.$$

Now we use Lemma 4 in Chap. 3 and we get

$$\phi^{\Delta^{n-1}}(\tau)\Big|_{\tau=a}^{\tau=x} = \int_a^x u(y)\Delta y + \lambda \int_a^x \int_{\sigma(y)}^x h_l(x_1, y)\phi(y)\Delta x_1\Delta y$$

$$= \int_a^x u(y)\Delta y + \lambda \int_a^x \int_y^x h_l(x_1, y)\phi(y)\Delta x_1 \Delta y$$

$$- \lambda \int_a^x \int_y^{\sigma(y)} h_l(x_1, y)\phi(y)\Delta x_1 \Delta y$$

$$= \int_a^x u(y)\Delta y + \lambda \int_a^x h_{l+1}(x, y)\phi(y)\Delta y - \lambda \int_a^x h_l(y, y)\mu(y)\phi(y)\Delta y$$

$$= \int_a^x u(y)\Delta y + \lambda \int_a^x h_{l+1}(x, y)\phi(y)\Delta y,$$

whereupon

$$\phi^{\Delta^{n-1}}(x) = a_{n-1} + \int_a^x u(y)\Delta y + \lambda \int_a^x h_{l+1}(x, y)\phi(y)\Delta y.$$

Then we integrate the last equation from a to x and so on while on the left side we obtain $\phi(x)$.

Remark 3 For every $l \in \mathscr{N}$ we have

$$\begin{aligned} \int_{\sigma(y)}^x h_l(x_1, y)\Delta x_1 &= \int_y^x h_l(x_1, y)\Delta x_1 - \int_y^{\sigma(y)} h_l(x_1, y)\Delta x_1 \\ &= h_{l+1}(x, y) - h_l(y, y)\mu(y) \\ &= h_{l+1}(x, y). \end{aligned}$$

Example 8 Let $\mathscr{T} = \mathscr{Z}$. Consider the equation

$$\phi^{\Delta^2}(x) = x + \int_0^x h_4(x, y)\phi(y)\Delta y$$

$$\phi(0) = \phi^{\Delta}(0) = 1.$$

We integrate the given equation from 0 to x and we find

$$\int_0^x \phi^{\Delta^2}(\tau)\Delta\tau = \int_0^x \tau\Delta\tau + \int_0^x \int_0^{x_1} h_4(x_1, y)\phi(y)\Delta y \Delta x_1$$

$$= \int_0^x \left(\frac{1}{2}(\tau^2)^{\Delta} - \frac{1}{2}\right)\Delta\tau + \int_0^x \left(\int_{\sigma(y)}^x h_4(x_1, y)\Delta x_1\right)\phi(y)\Delta y$$

$$= \frac{1}{2}\int_0^x (\tau^2)^\Delta \Delta\tau - \frac{1}{2}\int_0^x \Delta\tau + \int_0^x h_5(x, y)\phi(y)\Delta y$$

$$= \frac{1}{2}\tau^2\Big|_{\tau=0}^{\tau=x} - \frac{1}{2}\tau\Big|_{\tau=0}^{\tau=x} + \int_0^x h_5(x, y)\phi(y)\Delta y$$

$$= \frac{1}{2}x^2 - \frac{1}{2}x + \int_0^x h_5(x, y)\phi(y)\Delta y,$$

whereupon

$$\phi^\Delta(x) - \phi^\Delta(0) = \frac{1}{2}x^2 - \frac{1}{2}x + \int_0^x h_5(x, y)\phi(y)\Delta y$$

or

$$\phi^\Delta(x) = \frac{1}{2}x^2 - \frac{1}{2}x + 1 + \int_0^x h_5(x, y)\phi(y)\Delta y.$$

Now we integrate the last equation from 0 to x and we find

$$\int_0^x \phi^\Delta(\tau)\Delta\tau = \int_0^x \left(\frac{1}{2}\tau^2 - \frac{1}{2}\tau + 1\right)\Delta\tau + \int_0^x\int_0^{x_1} h_5(x_1, y)\phi(y)\Delta y\Delta x_1$$

$$= \int_0^x \left(\frac{1}{2}\left(\frac{1}{3}(\tau^3)^\Delta - \frac{1}{2}(\tau^2)^\Delta + \frac{1}{6}\right) - \frac{1}{2}\left(\frac{1}{2}(\tau^2)^\Delta - \frac{1}{2}\right) + 1\right)\Delta\tau$$

$$+ \int_0^x \left(\int_{\sigma(y)}^x h_5(x_1, y)\Delta x_1\right)\phi(y)\Delta y$$

$$= \int_0^x \left(\frac{1}{6}(\tau^3)^\Delta - \frac{1}{4}(\tau^2)^\Delta + \frac{1}{12} - \frac{1}{4}(\tau^2)^\Delta + \frac{1}{4} + 1\right)\Delta\tau$$

$$+ \int_0^x h_6(x, y)\phi(y)\Delta y$$

$$= \int_0^x \left(\frac{1}{6}(\tau^3)^\Delta - \frac{1}{2}(\tau^2)^\Delta + \frac{4}{3}\right)\Delta\tau + \int_0^x h_6(x, y)\phi(y)\Delta y$$

$$= \frac{1}{6}\int_0^x (\tau^3)^\Delta\Delta\tau - \frac{1}{2}\int_0^x (\tau^2)^\Delta\Delta\tau + \frac{4}{3}\int_0^x \Delta\tau + \int_0^x h_6(x, y)\phi(y)\Delta y$$

$$= \frac{1}{6}\tau^3\Big|_{\tau=0}^{\tau=x} - \frac{1}{2}\tau^2\Big|_{\tau=0}^{\tau=x} + \frac{4}{3}\tau\Big|_{\tau=0}^{\tau=x} + \int_0^x h_6(x, y)\phi(y)\Delta y$$

$$= \frac{1}{6}x^3 - \frac{1}{2}x^2 + \frac{4}{3}x + \int_0^x h_6(x, y)\phi(y)\Delta y,$$

from where

$$\phi(x) - \phi(0) = \frac{1}{6}x^3 - \frac{1}{2}x^2 + \frac{4}{3}x + \int_0^x h_6(x, y)\phi(y)\Delta y,$$

or

$$\phi(x) = \frac{1}{6}x^3 - \frac{1}{2}x^2 + \frac{4}{3}x + 1 + \int_0^x h_6(x, y)\phi(y)\Delta y.$$

Example 9 Let $\mathscr{T} = 2^{\mathscr{N}_0} \cup \{0\}$. Consider the equation

$$\phi^{\Delta^3}(x) = 1 + \int_0^x h_2(x, y)\phi(y)\Delta y,$$

$$\phi(0) = 1, \quad \phi^{\Delta}(0) = -1, \quad \phi^{\Delta^2}(0) = 2.$$

We integrate the given equation from 0 to x and we get

$$\begin{aligned}
\int_0^x \phi^{\Delta^3}(\tau)\Delta\tau &= \int_0^x \Delta\tau + \int_0^x \int_0^{x_1} h_2(x_1, y)\phi(y)\Delta y \Delta x_1 \\
&= \tau\Big|_{\tau=0}^{\tau=x} + \int_0^x \left(\int_{\sigma(y)}^x h_2(x_1, y)\Delta x_1\right)\phi(y)\Delta y \\
&= x + \int_0^x \left(\int_y^x h_2(x_1, y)\Delta x_1\right)\phi(y)\Delta y \\
&= x + \int_0^x h_3(x, y)\phi(y)\Delta y,
\end{aligned}$$

whereupon

$$\phi^{\Delta^2}(x) - \phi^{\Delta^2}(0) = x + \int_0^x h_3(x, y)\phi(y)\Delta y$$

or

$$\phi^{\Delta^2}(x) = x + 2 + \int_0^x h_3(x, y)\phi(y)\Delta y.$$

We integrate the last equation from 0 to x and we obtain

$$\begin{aligned}\int_0^x \phi^{\Delta^2}(\tau)\Delta\tau &= \int_0^x (\tau+2)\Delta\tau + \int_0^x \int_0^{x_1} h_3(x_1, y)\phi(y)\Delta y\Delta x_1 \\ &= \int_0^x \left(\frac{1}{3}(\tau^2)^\Delta + 2\right) + \int_0^x \left(\int_{\sigma(y)}^x h_3(x_1, y)\Delta x_1\right)\phi(y)\Delta y \\ &= \frac{1}{3}\int_0^x (\tau^2)^\Delta\Delta\tau + 2\int_0^x \Delta\tau + \int_0^x \left(\int_y^x h_3(x_1, y)\Delta x_1\right)\phi(y)\Delta y \\ &= \frac{1}{3}\tau^2\Big|_{\tau=0}^{\tau=x} + 2\tau\Big|_{\tau=0}^{\tau=x} + \int_0^x h_4(x, y)\phi(y)\Delta y \\ &= \frac{1}{3}x^2 + 2x + \int_0^x h_4(x, y)\phi(y)\Delta y,\end{aligned}$$

from where

$$\phi^\Delta(x) - \phi^\Delta(0) = \frac{1}{3}x^2 + 2x + \int_0^x h_4(x, y)\phi(y)\Delta y$$

or

$$\phi^\Delta(x) = \frac{1}{3}x^2 + 2x - 1 + \int_0^x h_4(x, y)\phi(y)\Delta y.$$

We integrate the last equation from 0 to x and we find

$$\begin{aligned}\int_0^x \phi^\Delta(\tau)\Delta\tau &= \int_0^x \left(\frac{1}{3}\tau^2 + 2\tau - 1\right)\Delta\tau + \int_0^x \int_0^{x_1} h_4(x_1, y)\phi(y)\Delta y\Delta x_1 \\ &= \int_0^x \left(\frac{1}{21}(\tau^3)^\Delta + \frac{2}{3}(\tau^2)^\Delta - 1\right)\Delta\tau \\ &\quad + \int_0^x \left(\int_{\sigma(y)}^x h_4(x_1, y)\Delta x_1\right)\phi(y)\Delta y \\ &= \frac{1}{21}\int_0^x (\tau^3)^\Delta\Delta\tau + \frac{2}{3}\int_0^x (\tau^2)^\Delta\Delta\tau - \int_0^x \Delta\tau \\ &\quad + \int_0^x \left(\int_y^x h_4(x_1, y)\Delta x_1\right)\phi(y)\Delta y\end{aligned}$$

$$= \frac{1}{21}\tau^3\Big|_{\tau=0}^{\tau=x} + \frac{2}{3}\tau^2\Big|_{\tau=0}^{\tau=x} - \tau\Big|_{\tau=0}^{\tau=x} + \int_0^x h_5(x,y)\phi(y)\Delta y$$

$$= \frac{1}{21}x^3 + \frac{2}{3}x^2 - x + \int_0^x h_5(x,y)\phi(y)\Delta y.$$

Hence,

$$\phi(x) - \phi(0) = \frac{1}{21}x^3 + \frac{2}{3}x^2 - x + \int_0^x h_5(x,y)\phi(y)\Delta y$$

or

$$\phi(x) = \frac{1}{21}x^3 + \frac{2}{3}x^2 - x + 1 + \int_0^x h_5(x,y)\phi(y)\Delta y.$$

Example 10 Let $\mathscr{T} = 2^{\mathscr{N}_0} \cup \{0\}$. Consider the equation

$$\phi^{\Delta^3}(x) + \phi^{\Delta^2}(x) + \phi^{\Delta}(x) = x + \int_0^x h_1(x,y)\phi(y)\Delta y,$$

$$\phi(0) = \phi^{\Delta}(0) = \phi^{\Delta^2}(0) = 1.$$

Here $\sigma(x) = 2x, \quad x \in \mathscr{T}$.
We integrate the given equation from 0 to x and we get

$$\int_0^x \left(\phi^{\Delta^3}(\tau) + \phi^{\Delta^2}(\tau) + \phi^{\Delta}(\tau)\right)\Delta\tau = \int_0^x \tau\Delta\tau + \int_0^x\int_0^{x_1} h_1(x_1,y)\phi(y)\Delta y\Delta x_1$$

or

$$\int_0^x \phi^{\Delta^3}(\tau)\Delta\tau + \int_0^x \phi^{\Delta^2}(\tau)\Delta\tau + \int_0^x \phi^{\Delta}(\tau)\Delta\tau$$

$$= \frac{1}{3}\int_0^x (\tau^2)^{\Delta}\Delta\tau + \int_0^x \left(\int_{\sigma(y)}^x h_1(x_1,y)\Delta x_1\right)\phi(y)\Delta y,$$

or

$$\phi^{\Delta^2}(x) - \phi^{\Delta^2}(0) + \phi^{\Delta}(x) - \phi^{\Delta}(0) + \phi(x) - \phi(0)$$

$$= \frac{1}{3}\tau^2\Big|_{\tau=0}^{\tau=x} + \int_0^x \left(\int_y^x h_1(x_1,y)\Delta x_1\right)\Delta y,$$

or

$$\phi^{\Delta^2}(x) + \phi^{\Delta}(x) + \phi(x) = 3 + \frac{1}{3}x^2 + \int_0^x h_2(x,y)\phi(y)\Delta y.$$

We integrate the last equation from 0 to x and we find

$$\int_0^x \left(\phi^{\Delta^2}(\tau) + \phi^{\Delta}(\tau) + \phi(\tau)\right)\Delta\tau = \int_0^x \left(3 + \frac{1}{3}\tau^2\right)\Delta\tau + \int_0^x\int_0^{x_1} h_2(x_1,y)\phi(y)\Delta y\Delta x_1,$$

or

$$\int_0^x \phi^{\Delta^2}(\tau)\Delta\tau + \int_0^x \phi^{\Delta}(\tau)\Delta\tau + \int_0^x \phi(y)\Delta y$$

$$= \int_0^x \left(3 + \tfrac{1}{21}(\tau^3)^{\Delta}\right)\Delta\tau + \int_0^x \left(\int_{\sigma(y)}^x h_2(x_1, y)\Delta x_1\right)\phi(y)\Delta y,$$

or

$$\phi^{\Delta}(x) - \phi^{\Delta}(0) + \phi(x) - \phi(0) + \int_0^x \phi(y)\Delta y$$

$$= 3\int_0^x \Delta\tau + \tfrac{1}{21}\int_0^x (\tau^3)^{\Delta}\Delta\tau + \int_0^x \left(\int_y^x h_2(x_1, y)\Delta x_1\right)\phi(y)\Delta y,$$

or

$$\phi^{\Delta}(x) + \phi(x) + \int_0^x \phi(y)\Delta y = 2 + 3\tau\Big|_{\tau=0}^{\tau=x} + \tfrac{1}{21}\tau^3\Big|_{\tau=0}^{\tau=x} + \int_0^x h_3(x, y)\phi(y)\Delta y,$$

or

$$\phi^{\Delta}(x) + \phi(x) + \int_0^x \phi(y)\Delta y = 2 + 3x + \frac{1}{21}x^3 + \int_0^x h_3(x, y)\phi(y)\Delta y.$$

We integrate the last equation from 0 to x and we obtain

$$\int_0^x \left(\phi^{\Delta}(x_1) + \phi(x_1) + \int_0^{x_1} \phi(y)\Delta y\right)\Delta x_1 = \int_0^x \left(2 + 3\tau + \tfrac{1}{21}\tau^3\right)\Delta\tau$$

$$+ \int_0^x \int_0^{x_1} h_3(x_1, y)\phi(y)\Delta y\Delta x_1$$

or

$$\int_0^x \phi^{\Delta}(\tau)\Delta\tau + \int_0^x \phi(y)\Delta y + \int_0^x \int_0^{x_1} \phi(y)\Delta y\Delta x_1$$

$$= \int_0^x \left(2 + (\tau^2)^{\Delta} + \tfrac{1}{315}(\tau^4)^{\Delta}\right)\Delta\tau + \int_0^x \left(\int_{\sigma(y)}^x h_3(x_1, y)\Delta x_1\right)\phi(y)\Delta y,$$

or

$$\phi(x) - \phi(0) + \int_0^x \phi(y)\Delta y + \int_0^x (x - \sigma(y))\phi(y)\Delta y = 2\int_0^x \Delta\tau$$

$$+ \int_0^x (\tau^2)^{\Delta}\Delta\tau + \tfrac{1}{315}\int_0^x (\tau^4)^{\Delta}\Delta\tau + \int_0^x \left(\int_y^x h_3(x_1, y)\Delta x_1\right)\phi(y)\Delta y,$$

or

$$\phi(x) - 1 + \int_0^x \phi(y)\Delta y + \int_0^x (x - 2y)\phi(y)\Delta y = 2\tau\Big|_{\tau=0}^{\tau=x}$$

$$+\tau^2\Big|_{\tau=0}^{\tau=x} + \tfrac{1}{315}\tau^4\Big|_{\tau=0}^{\tau=x} + \int_0^x h_4(x, y)\phi(y)\Delta y,$$

or

$$\phi(x) - 1 + \int_0^x (x + 1 - 2y)\phi(y)\Delta y = 2x + x^2 + \frac{1}{315}x^4 + \int_0^x h_4(x, y)\phi(y)\Delta y,$$

or

$$\phi(x) = 1 + 2x + x^2 + \frac{1}{315}x^4 + \int_0^x (h_4(x,y) - x - 1 + 2y)\phi(y)\Delta y.$$

Exercise 3 Convert the following equations into generalized Volterra integral equations.

1.
$$\begin{cases} \phi^{\Delta^3}(x) + x^3\phi(x) = x^2 + 2x + 7 + x^4 \int_0^x y^4\phi(y)\Delta y, \\ \phi(0) = \phi^{\Delta}(0) = 1, \quad \phi^{\Delta^2}(0) = -3, \quad \mathscr{T} = 4\mathscr{Z}, \end{cases}$$

2.
$$\begin{cases} \phi^{\Delta^2}(x) + 3x\phi^{\Delta}(x) - 3\phi(x) = 2x - 5 + x^6 \int_0^x \phi(y)\Delta y, \\ \phi(0) = \phi^{\Delta}(0) = 1, \quad \mathscr{T} = 3^{\mathscr{N}_0} \cup \{0\}, \end{cases}$$

3.
$$\begin{cases} \phi^{\Delta^6}(x) + \phi^{\Delta^3}(x) + x^4\phi^{\Delta}(x) = x \int_0^x \phi(y)\Delta y, \\ \phi(0) = \phi^{\Delta}(0) = \phi^{\Delta^2}(0) = \phi^{\Delta^3}(0) = \phi^{\Delta^4}(0) = \phi^{\Delta^5}(0) = 0, \quad \mathscr{T} = \mathscr{N}. \end{cases}$$

4.2 Generalized Volterra Integro-Differential Equations of the First Kind

The standard form of the generalized Volterra integro-differential equations of the first kind is given by

$$\int_a^x K(x,y)\phi(y)\Delta y + \sum_{i=1}^n \lambda_i \int_a^b K_i(x,y)\phi^{\Delta^i}(y)\Delta y = u(x),$$

$$\phi^{\Delta^j}(a) = a_j, \quad j = 0, 1, \ldots, n-1,$$

where $K, K_i : \mathscr{T} \times \mathscr{T} \longmapsto \mathscr{R}$, $i = 1, 2, \ldots, n$, are rd-continuous functions, $u : \mathscr{T} \longmapsto \mathscr{R}$ is given rd-continuous function, λ_i, $i = 1, 2, \ldots, n$, are parameters, a_i, $i = 0, 1, \ldots, n-1$, are given constants.
Using integration by parts, the generalized Volterra integro-differential equations of the first kind can be reduced to generalized Volterra integro-differential equations of the second kind or generalized Volterra integral equations of the second kind.

Example 11 Let $\mathscr{T} = \mathscr{Z}$. Consider the equation

$$\int_0^x (x+y)\phi(y)\Delta y + \int_0^x (2x - y - 1)\phi^{\Delta}(y)\Delta y = x, \quad \phi(0) = 1.$$

Here $\sigma(x) = x + 1, \quad x \in \mathscr{T}$. Then

$$\int_0^x (2x - y - 1)\phi^{\Delta}(y)\Delta y = \int_0^x (2x - y)^{\sigma_y}\phi^{\Delta}(y)\Delta y$$

$$= (2x - y)\phi(y)\Big|_{y=0}^{y=x} - \int_0^x (2x - y)_y^{\Delta}\phi(y)\Delta y$$

$$= x\phi(x) - 2x\phi(0) + \int_0^x \phi(y)\Delta y$$

$$= x\phi(x) - 2x + \int_0^x \phi(y)\Delta y.$$

Then the given equation we can rewrite in the form

$$\int_0^x (x + y)\phi(y)\Delta y + x\phi(x) - 2x + \int_0^x \phi(y)\Delta y = x$$

or

$$x\phi(x) = 3x - \int_0^x (1 + x + y)\phi(y)\Delta y.$$

Example 12 Let $\mathscr{T} = \mathscr{Z}$. Consider the equation

$$2x \int_0^x \phi(y)\Delta y + \int_0^x y^2\phi^{\Delta}(y)\Delta y = 1, \quad \phi(0) = 2.$$

We have

$$\int_0^x y^2\phi^{\Delta}(y)\Delta y = y^2\phi(y)\Big|_{y=0}^{y=x} - \int_0^x (y^2)^{\Delta}\phi(\sigma(y))\Delta y$$

$$= x^2\phi(x) - \int_0^x (\sigma(y) + y)\phi(\sigma(y))\Delta y$$

$$= x^2\phi(x) - \int_0^x (2y + 1)\phi(y + 1)\Delta y.$$

Then

$$2x \int_0^x \phi(y)\Delta y + x^2\phi(x) - \int_0^x (2y + 1)\phi(y + 1)\Delta y = 1.$$

Example 13 Let $\mathscr{T} = 2^{\mathscr{N}_0} \cup \{0\}$. Consider the equation

$$\phi^{\Delta^3}(x) + \phi^{\Delta}(x) = \int_0^x \phi(y)\Delta y,$$

$$\phi(0) = \phi^{\Delta}(0) = \phi^{\Delta^2}(0) = 1.$$

Here $\sigma(x) = 2x, \quad x \in \mathscr{T}$.
We integrate from 0 to x both sides of the given equation and we get

$$\int_0^x \phi^{\Delta^3}(y)\Delta y + \int_0^x \phi^{\Delta}(y)\Delta y = \int_0^x \int_0^{x_1} \phi(y)\Delta y \Delta x_1$$

or

$$\phi^{\Delta^2}(x) - \phi^{\Delta^2}(0) + \phi(x) - \phi(0) = \int_0^x (x - \sigma(y))\phi(y)\Delta y,$$

or

$$\phi^{\Delta^2}(x) - \phi^{\Delta^2}(0) + \phi(x) - \phi(0) = \int_0^x (x - 2y)\phi(y)\Delta y,$$

or

$$\phi^{\Delta^2}(x) + \phi(x) - 2 = \int_0^x (x - 2y)\phi(y)\Delta y,$$

or

$$\phi^{\Delta^2}(x) = 2 - \phi(x) + \int_0^x (x - 2y)\phi(y)\Delta y.$$

Now we integrate from 0 to x both sides of the last equation and we find

$$\int_0^x \phi^{\Delta^2}(y)\Delta y = \int_0^x (2 - \phi(y))\Delta y + \int_0^x \int_0^{x_1} (x_1 - 2y)\phi(y)\Delta y \Delta x_1,$$

or

$$\phi^{\Delta}(x) - \phi^{\Delta}(0) = 2\int_0^x \Delta y - \int_0^x \phi(y)\Delta y$$

$$+ \int_0^x \int_0^{x_1} (h_1(x_1, y) - \mu(y)h_0(x_1, y))\phi(y)\Delta y \Delta x_1$$

$$= 2x - \int_0^x \phi(y)\Delta y$$

$$+ \int_0^x \int_{\sigma(y)}^x (h_1(x_1, y) - \mu(y)h_0(x_1, y))\phi(y)\Delta x_1 \Delta y$$

$$= 2x - \int_0^x \phi(y)\Delta y$$

$$+ \int_0^x \left(\int_y^x (h_1(x_1, y) - \mu(y)h_0(x_1, y))\Delta x_1 \right) \phi(y)\Delta y$$

$$= 2x - \int_0^x \phi(y)\Delta y + \int_0^x (h_2(x, y) - \mu(y)h_1(x, y))\phi(y)\Delta y,$$

or

$$\phi^\Delta(x) = 2x + 1 - \int_0^x \phi(y)\Delta y + \int_0^x (h_2(x, y) - \mu(y)h_1(x, y))\phi(y)\Delta y.$$

Now we integrate from 0 to x both sides of the last equation and we find

$$\int_0^x \phi^\Delta(y)\Delta y = \int_0^x (2y + 1)\Delta y - \int_0^x \int_0^{x_1} \phi(y)\Delta y \Delta x_1$$

$$+ \int_0^x \int_0^{x_1} (h_2(x_1, y) - \mu(y)h_1(x_1, y))\phi(y)\Delta y \Delta x_1$$

$$= \int_0^x \left(\frac{2}{3}(y^2)^\Delta + 1 \right) \Delta y - \int_0^x (x - \sigma(y))\phi(y)\Delta y$$

$$+ \int_0^x \left(\int_{\sigma(y)}^x (h_2(x_1, y) - \mu(y)h_1(x_1, y))\Delta x_1 \right) \phi(y)\Delta y$$

$$= \frac{2}{3} \int_0^x (y^2)^\Delta \Delta y + \int_0^x \Delta y - \int_0^x (x - \sigma(y))\phi(y)\Delta y$$

$$+ \int_0^x \left(\int_y^x (h_2(x_1, y) - \mu(y)h_1(x_1, y))\Delta x_1 \right) \phi(y)\Delta y$$

$$= \frac{2}{3} y^2 \Big|_{y=0}^{y=x} + y \Big|_{y=0}^{y=x} - \int_0^x (x - 2y)\phi(y)\Delta y$$

$$+ \int_0^x (h_3(x, y) - \mu(y)h_2(x, y))\phi(y)\Delta y$$

$$= \frac{2}{3} x^2 + x - \int_0^x (x - 2y)\phi(y)\Delta y$$

$$+\int_0^x (h_3(x,y)-\mu(y)h_2(x,y))\phi(y)\Delta y,$$

or

$$\phi(x)-\phi(0)=\frac{2}{3}x^2+x-\int_0^x (x-2y)\phi(y)\Delta y+\int_0^x (h_3(x,y)-\mu(y)h_2(x,y))\phi(y)\Delta y,$$

or

$$\phi(x)=\frac{2}{3}x^2+x+1-\int_0^x (x-2y)\phi(y)\Delta y+\int_0^x (h_3(x,y)-\mu(y)h_2(x,y))\phi(y)\Delta y.$$

Exercise 4 Convert the following equations into generalized Volterra integro-differential equations of the second kind.

1.
$$\begin{cases} \int_0^x y\phi^{\Delta^2}(y)\Delta y-2x^3\int_0^x y\phi(y)\Delta y=x^2+x^4\int_0^x y^4\phi(y)\Delta y, \\ \phi(0)=\phi^{\Delta}(0)=1, \quad \mathscr{T}=\mathscr{Z}, \end{cases}$$

2.
$$\begin{cases} \int_0^x (y+1)\phi^{\Delta^3}(y)\Delta y=\int_0^x \phi(y)\Delta y, \\ \phi(0)=\phi^{\Delta}(0)=\phi^{\Delta^2}(0)=-2, \quad \mathscr{T}=2^{\mathscr{N}_0}\cup\{0\}, \end{cases}$$

3.
$$\begin{cases} \int_0^x y^2\phi^{\Delta^5}(y)\Delta y=x^2+x\int_0^x \phi(y)\Delta y, \\ \phi(0)=\phi^{\Delta}(0)=\phi^{\Delta^2}(0)=\phi^{\delta^3}(0)=\phi^{\Delta^4}(0)=0, \quad \mathscr{T}=\mathscr{Z}. \end{cases}$$

4.3 Advanced Practical Exercises

Problem 1 Using ADM, find the recurrence relation for $\{\phi_k(x)\}_{k=0}^{\infty}$ for the following equations.

1.
$$\begin{cases} \phi^{\Delta^4}(x)=x^2+2x+x\int_0^x y\phi(y)\Delta y, \\ \phi(0)=\phi^{\Delta}(0)=\phi^{\Delta^2}(0)=\phi^{\Delta^3}(0)=-1, \quad \mathscr{T}=\mathscr{Z}, \end{cases}$$

2.
$$\begin{cases} \phi^{\Delta}(x)+(x^2+2x+7)\phi(x)=1+4x+x^4\int_0^x y^3\phi(y)\Delta y, \\ \phi(0)=1, \quad \mathscr{T}=\mathscr{Z}, \end{cases}$$

3.

$$\begin{cases} \phi^{\Delta^3}(x) = x^2\phi(x) - x - x\int_0^x y\phi(y)\Delta y, \\ \phi(0) = \phi^{\Delta}(0) = \phi^{\Delta^2}(0) = -2, \quad \mathscr{T} = 3^{\mathscr{N}_0} \cup \{0\}. \end{cases}$$

Problem 2 Convert the following equations into initial value problems.

1.

$$\begin{cases} \phi^{\Delta^2}(x) - 2x^3\phi(x) = x^2 + x^4\int_0^x y^4\phi(y)\Delta y, \\ \phi(0) = \phi^{\Delta}(0) = 1, \quad \mathscr{T} = \mathscr{Z}, \end{cases}$$

2.

$$\begin{cases} \phi^{\Delta^3}(x) - 3x^2\phi^{\Delta^2}(x) - 3x\phi^{\Delta}(x) - 3\phi(x) = \int_0^x \phi(y)\Delta y, \\ \phi(0) = \phi^{\Delta}(0) = \phi^{\Delta^2}(0) = -2, \quad \mathscr{T} = 2^{\mathscr{N}_0} \cup \{0\}, \end{cases}$$

3.

$$\begin{cases} \phi^{\Delta^5}(x) + \phi^{\Delta^3}(x) + \phi^{\Delta}(x) = x^2 + x\int_0^x \phi(y)\Delta y, \\ \phi(0) = \phi^{\Delta}(0) = \phi^{\Delta^2}(0) = \phi^{\Delta^3}(0) = \phi^{\Delta^4}(0) = 0, \quad \mathscr{T} = \mathscr{Z}. \end{cases}$$

Problem 3 Convert the following equations into generalized Volterra integral equations.

1.

$$\begin{cases} \phi^{\Delta^3}(x) + x\phi(x) = x^2 + x\int_0^x (x+y)^2\phi(y)\Delta y, \\ \phi(0) = \phi^{\Delta}(0) = 1, \quad \phi^{\Delta^2}(0) = -1, \quad \mathscr{T} = 3\mathscr{Z}, \end{cases}$$

2.

$$\begin{cases} \phi^{\Delta^4}(x) + x^3\phi^{\Delta^3}(x) + 7x^2\phi^{\Delta}(x) = x^2\int_0^x \phi(y)\Delta y, \\ \phi(0) = \phi^{\Delta}(0) = \phi^{\Delta^2}(0) = -2, \quad \phi^{\Delta^3}(0) = 0, \quad \mathscr{T} = 7^{\mathscr{N}_0} \cup \{0\}, \end{cases}$$

3.

$$\begin{cases} \phi^{\Delta^3}(x) + 4\phi^{\Delta^2}(x) + x^5\phi^{\Delta}(x) = x\int_0^x \phi(y)\Delta y, \\ \phi(0) = \phi^{\Delta}(0) = \phi^{\Delta^2}(0) = 0, \quad \mathscr{T} = 2\mathscr{Z}. \end{cases}$$

Problem 4 Convert the following equations into generalized Volterra integro-differential equations of the second kind.

1.

$$\begin{cases} \int_0^x (y^2 + 2y + 1 + x)\phi^{\Delta^2}(y)\Delta y - 2x^3\int_0^x (x+y)\phi^{\Delta}(y)\Delta y = \int_0^x y^4\phi(y)\Delta y, \\ \phi(0) = 1, \quad \phi^{\Delta}(0) = -11, \quad \mathscr{T} = \mathscr{Z}, \end{cases}$$

2.

$$\begin{cases} \int_0^x (y^2 + xy + 1)\phi^{\Delta^3}(y)\Delta y = x - 3\int_0^x y\phi(y)\Delta y, \\ \phi(0) = \phi^{\Delta}(0) = \phi^{\Delta^2}(0) = 0, \quad \mathscr{T} = 3^{\mathscr{N}_0} \cup \{0\}, \end{cases}$$

3.

$$\begin{cases} \int_0^x (x - y)^2\phi^{\Delta^4}(y)\Delta y = x^2 + \int_0^x y\phi(y)\Delta y, \\ \phi(0) = \phi^{\Delta}(0) = \phi^{\Delta^2}(0) = \phi^{\Delta^3}(0) = 0, \quad \mathscr{T} = 7\mathscr{Z}. \end{cases}$$

Chapter 5
Generalized Fredholm Integral Equations

In this chapter we adapt the Adomian decomposition method, the modified decomposition method, the noise term phenomenon, the direct computation method and the successive approximation method for generalized Fredholm integral equations. They are considered generalizations of the Fredholm Alternative theorem, the Smidth expansion theorem and Mercer's expansion theorem.

5.1 Generalized Fredholm Integral Equations of the Second Kind

We will first study the generalized Fredholm integral equations of the second kind given by

$$\phi(x) = u(x) + \lambda \int_a^b K(x, y)\phi(y)\Delta y. \tag{5.1}$$

The unknown function $\phi(x)$, that will be determined, occurs inside and outside the integral sign. The kernel $K(x, y)$ and the source term $u(x)$ are real-valued functions, and λ is a real parameter.

5.1.1 The Adomian Decomposition Method

The Adomian decomposition method (ADM) was introduced in Chap. 3 and it consists of decomposing the unknown function $\phi(x)$ of any equation into a sum of an infinite number of components defined by the decomposition series

S.G. Georgiev, *Integral Equations on Time Scales*, Atlantis Studies in Dynamical Systems 5, DOI 10.2991/978-94-6239-228-1_5

$$\phi(x) = \sum_{l=0}^{\infty} \phi_l(x) \tag{5.2}$$

or equivalently

$$\phi(x) = \phi_0(x) + \phi_1(x) + \phi_2(x) + \cdots,$$

where the components $\phi_l(x)$, $l \geq 0$, will be determined recurrently.

The Adomian decomposition method concerns itself with finding the components $\phi_0(x)$, $\phi_1(x)$, $\phi_2(x)$, $\phi_3(x)$, $\ldots$, individually.

To establish the recurrence relation, we substitute (5.2) into the generalized Fredholm integral equation (5.1) to obtain

$$\phi_0(x) + \phi_1(x) + \phi_2(x) + \cdots = u(x) + \lambda \int_a^b K(x, y) (\phi_0(y) + \phi_1(y) + \phi_2(y) + \cdots) \Delta y.$$

We set

$$\begin{aligned} \phi_0(x) &= u(x), \\ \phi_n(x) &= \lambda \textstyle\int_a^b K(x, y)\phi_{n-1}(y)\Delta y, \quad n \geq 1, \end{aligned} \tag{5.3}$$

or equivalently

$$\begin{aligned} \phi_0(x) &= u(x), \\ \phi_1(x) &= \lambda \textstyle\int_a^b K(x, y)\phi_0(y)\Delta y, \\ \phi_2(x) &= \lambda \textstyle\int_a^b K(x, y)\phi_1(y)\Delta y, \\ \phi_3(x) &= \lambda \textstyle\int_a^b K(x, y)\phi_2(y)\Delta y, \end{aligned}$$

and so on for other components.

In view of (5.3), the components $\phi_0(x)$, $\phi_1(x)$, $\phi_2(x)$, $\phi_3(x)$, $\ldots$, are completely determined. As a result, the solution $\phi(x)$ of the generalized Fredholm integral equation (5.1) is obtained in a series form by using the series assumption in (5.2).

Example 1 Let $\mathscr{T} = \mathscr{Z}$. Consider the equation

$$\phi(x) = x + \int_0^4 y\phi(y)\Delta y.$$

Here

$$\sigma(x) = x + 1, \quad x \in \mathscr{T}, \quad u(x) = x, \quad K(x, y) = y, \quad \lambda = 1.$$

Then

$$
\begin{aligned}
\phi_0(x) &= x,\\
\phi_1(x) &= \int_0^4 y\phi_0(y)\Delta y\\
&= \int_0^4 y^2\Delta y\\
&= \int_0^4 \left(\frac{1}{3}(y^3)^\Delta - \frac{1}{2}(y^2)^\Delta + \frac{1}{6}\right)\Delta y\\
&= \frac{1}{3}\int_0^4 (y^3)^\Delta\Delta y - \frac{1}{2}\int_0^4 (y^2)^\Delta\Delta y + \frac{1}{6}\int_0^4 \Delta y\\
&= \frac{1}{3}y^3\Big|_{y=0}^{y=4} - \frac{1}{2}y^2\Big|_{y=0}^{y=4} + \frac{1}{6}y\Big|_{y=0}^{y=4}\\
&= \frac{64}{3} - 8 + \frac{2}{3}\\
&= 14,
\end{aligned}
$$

$$
\begin{aligned}
\phi_2(x) &= \int_0^4 y\phi_1(y)\Delta y\\
&= 14\int_0^4 y\Delta y\\
&= 14\int_0^4 \left(\frac{1}{2}(y^2)^\Delta - \frac{1}{2}\right)\Delta y\\
&= 7\left(\int_0^4 (y^2)^\Delta\Delta y - \int_0^4 \Delta y\right)\\
&= 7\left(y^2\Big|_{y=0}^{y=4} - y\Big|_{y=0}^{y=4}\right)\\
&= 7(16-4)\\
&= 84,\\
\phi_3(x) &= \int_0^4 y\phi_2(y)\Delta y\\
&= 84\int_0^4 y\Delta y\\
&= 504,\\
\phi_4(x) &= \int_0^4 y\phi_3(y)\Delta y\\
&= 504\int_0^4 y\Delta y\\
&= 3024
\end{aligned}
$$

Example 2 Let $\mathscr{T} = \mathscr{Z}$. Consider the equation

$$\phi(x) = 1 + 2x \int_0^2 \phi(y)\Delta y.$$

Here

$$\sigma(x) = x + 1, \quad x \in \mathscr{T}, \quad u(x) = 1, \quad K(x, y) = x, \quad \lambda = 2.$$

Then

$$\begin{aligned}
\phi_0(x) &= 1,\\
\phi_1(x) &= 2x \int_0^2 \phi_0(y)\Delta y\\
&= 2x \int_0^2 \Delta y\\
&= 4x,
\end{aligned}$$

$$\begin{aligned}
\phi_2(x) &= 2x \int_0^2 \phi_1(y)\Delta y\\
&= 8x \int_0^2 y\Delta y\\
&= 4x \int_0^2 \left((y^2)^{\Delta} - 1\right) \Delta y\\
&= 4x \left(\int_0^2 (y^2)^{\Delta}\Delta y - \int_0^2 \Delta y\right)\\
&= 4x \left(y^2\Big|_{y=0}^{y=2} - y\Big|_{y=0}^{y=2}\right)\\
&= 4x(4-2)\\
&= 8x,\\
\phi_3(x) &= 2x \int_0^2 \phi_2(y)\Delta y\\
&= 16x \int_0^2 y\Delta y\\
&= 16x,\\
\phi_4(x) &= 2x \int_0^2 \phi_3(y)\Delta y\\
&= 32x \int_0^2 y\Delta y\\
&= 32x.
\end{aligned}$$

Example 3 Let $\mathscr{T} = 2^{\mathscr{N}_0} \cup \{0\}$. Consider the equation

$$\phi(x) = x + 1 + x \int_0^4 y\phi(y)\Delta y.$$

Here

$$\sigma(x) = x + 1, \quad x \in \mathscr{T}, \quad u(x) = x + 1, \quad K(x, y) = xy, \quad \lambda = 1.$$

Then

$$\begin{aligned}
\phi_0(x) &= x + 1, \\
\phi_1(x) &= x \int_0^4 y\phi_0(y)\Delta y \\
&= x \int_0^4 y(y + 1)\Delta y \\
&= x \int_0^4 (y^2 + y)\Delta y \\
&= x \int_0^4 \left(\frac{1}{7}(y^3)^\Delta + \frac{1}{3}(y^2)^\Delta\right) \Delta y \\
&= x \left(\frac{1}{7} \int_0^4 (y^3)^\Delta \Delta y + \frac{1}{3} \int_0^4 (y^2)^\Delta \Delta y\right) \\
&= x \left(\frac{1}{7} y^3 \Big|_{y=0}^{y=4} + \frac{1}{3} y^2 \Big|_{y=0}^{y=4}\right) \\
&= x \left(\frac{64}{7} + \frac{16}{3}\right) \\
&= \frac{304}{21} x, \\
\phi_2(x) &= x \int_0^4 y\phi_1(y)\Delta y \\
&= \frac{304}{21} x \int_0^4 y\Delta y \\
&= \frac{304}{21} \frac{1}{3} x \int_0^4 (y^2)^\Delta \Delta y \\
&= \left(\frac{304}{21}\right) \left(\frac{16}{3}\right) x, \\
\phi_3(x) &= x \int_0^4 y\phi_2(y)\Delta y \\
&= \left(\frac{304}{21}\right) \left(\frac{16}{3}\right) x \int_0^4 y^2 \Delta y
\end{aligned}$$

$$= \left(\frac{304}{21}\right)\left(\frac{16}{3}\right)^2 x,$$

$$\phi_4(x) = x\int_0^4 y\phi_3(y)\Delta y$$

$$= \left(\frac{304}{21}\right)\left(\frac{16}{3}\right)^2 x\int_0^4 y^2\Delta y$$

$$= \left(\frac{304}{21}\right)\left(\frac{16}{3}\right)^3 x.$$

Exercise 1 Using ADM, find $\phi_0(x), \phi_1(x), \phi_2(x), \phi_3(x)$ and $\phi_4(x)$ for the equation

$$\phi(x) = 1 + x^2\int_0^2 y\phi(y)\Delta y, \quad \mathscr{T} = 2^{\mathscr{N}_0} \cup \{0\}.$$

Answer

$$\phi_0(x) = 1, \quad \phi_1(x) = \tfrac{4}{3}x^2, \quad \phi_2(x) = \tfrac{4}{3}\left(\tfrac{16}{15}\right)x^2,$$

$$\phi_3(x) = \tfrac{4}{3}\left(\tfrac{16}{15}\right)^2 x^2, \quad \phi_4(x) = \tfrac{4}{3}\left(\tfrac{16}{15}\right)^3 x^2.$$

5.1.2 The Modified Decomposition Method

For many cases, the function $u(x)$ can be set as the sum of two partial functions, namely $u_1(x)$ and $u_2(x)$. In other words, we can set

$$u(x) = u_1(x) + u_2(x).$$

The modified decomposition method(MDM) identifies the zeroth component $\phi_0(x)$ by one part of $u(x)$, namely $u_1(x)$ or $u_2(x)$. The other part of $u(x)$ can be added to the component $\phi_1(x)$ that exists in the standard recurrence relation. The modified decomposition method admits the use of the modified recurrence relation

$$\begin{aligned} \phi_0(x) &= u_1(x), \\ \phi_1(x) &= u_2(x) + \lambda\int_a^b K(x,y)\phi_0(y)\Delta y, \\ \phi_{k+1}(x) &= \lambda\int_a^b K(x,y)\phi_k(y)\Delta y, \quad k \geq 1. \end{aligned}$$

A rule that may help for the proper choice of $u_1(x)$ and $u_2(x)$ could not be found yet. If $u(x)$ consists of one term only, the modified decomposition method cannot be used in this case.

Example 4 Let $\mathscr{T} = \mathscr{Z}$. Consider the equation

$$\phi(x) = 1 + x + \int_{-1}^{1} \phi(y)\Delta y.$$

Here

$$\sigma(x) = x + 1, \quad x \in \mathscr{T}, \quad u(x) = 1 + x, \quad K(x, y) = 1, \quad \lambda = 1.$$

Let

$$u_1(x) = 1, \quad u_2(x) = x.$$

Then

$$\begin{aligned}
\phi_0(x) &= 1,\\
\phi_1(x) &= x + \int_{-1}^{1} \phi_0(y)\Delta y\\
&= x + \int_{-1}^{1} \Delta y\\
&= x + 2,\\
\phi_2(x) &= \int_{-1}^{1} \phi_1(y)\Delta y\\
&= \int_{-1}^{1} (y + 2)\Delta y\\
&= \int_{-1}^{1} \left(\frac{1}{2}(y^2)^{\Delta} - \frac{1}{2} + 2\right) \Delta y\\
&= \frac{1}{2}\int_{-1}^{1} (y^2)^{\Delta}\Delta y + \frac{3}{2}\int_{-1}^{1} \Delta y\\
&= \frac{1}{2}y^2\Big|_{y=-1}^{y=1} + \frac{3}{2}y\Big|_{y=-1}^{y=1}\\
&= 3,\\
\phi_3(x) &= \int_{-1}^{1} \phi_2(y)\Delta y\\
&= 3\int_{-1}^{1} \Delta y\\
&= 3y\Big|_{y=-1}^{y=1}\\
&= 6,
\end{aligned}$$

$$\begin{aligned}
\phi_4(x) &= \int_{-1}^{1} \phi_3(y)\Delta y \\
&= 6\int_{-1}^{1} \Delta y \\
&= 6y\Big|_{y=-1}^{y=1} \\
&= 12.
\end{aligned}$$

Example 5 Let $\mathscr{T} = 2^{\mathscr{N}_0} \cup \{0\}$. Consider the equation

$$\phi(x) = x + x^2 + 2\int_0^2 y\phi(y)\Delta y.$$

Here

$$\sigma(x) = 2x, \quad x \in \mathscr{T}, \quad u(x) = x + x^2, \quad K(x, y) = y, \quad \lambda = 2.$$

Let

$$u_1(x) = x, \quad u_2(x) = x^2.$$

Then

$$\begin{aligned}
\phi_0(x) &= x, \\
\phi_1(x) &= x^2 + 2\int_0^2 y\phi_0(y)\Delta y \\
&= x^2 + 2\int_0^2 y^2\Delta y \\
&= x^2 + \frac{2}{7}\int_0^2 (y^3)^\Delta \Delta y \\
&= x^2 + \frac{2}{7}y^3\Big|_{y=0}^{y=2} \\
&= x^2 + \frac{16}{7}, \\
\phi_2(x) &= 2\int_0^2 y\phi_1(y)\Delta y \\
&= 2\int_0^2 y\left(y^2 + \frac{16}{7}\right)\Delta y \\
&= 2\int_0^2 \left(y^3 + \frac{16}{7}y\right)\Delta y \\
&= 2\int_0^2 \left(\frac{1}{15}(y^4)^\Delta + \frac{16}{21}(y^2)^\Delta\right)\Delta y
\end{aligned}$$

$$
\begin{aligned}
&= \frac{2}{15}\int_0^2 (y^4)^{\Delta}\Delta y + \frac{32}{21}\int_0^2 (y^2)^{\Delta}\Delta y \\
&= \frac{2}{15}y^4\Big|_{y=0}^{y=2} + \frac{32}{21}y^2\Big|_{y=0}^{y=2} \\
&= \frac{288}{35},
\end{aligned}
$$

$$
\begin{aligned}
\phi_3(x) &= 2\int_0^2 y\phi_2(y)\Delta y \\
&= \frac{576}{35}\int_0^2 y\Delta y \\
&= \frac{192}{35}y^2\Big|_{y=0}^{y=2} \\
&= \frac{768}{35}, \\
\phi_4(x) &= 2\int_0^2 y\phi_3(y)\Delta y \\
&= \frac{1536}{35}\int_0^2 y\Delta y \\
&= \frac{512}{35}y^2\Big|_{y=0}^{y=2} \\
&= \frac{2048}{35}.
\end{aligned}
$$

Example 6 Let $\mathscr{T} = 2^{\mathscr{N}_0} \cup \{0\}$. Consider the equation

$$
\phi(x) = x^2 + x^3 + x\int_0^2 \phi(y)\Delta y.
$$

Here

$$
\sigma(x) = 2x, \quad x \in \mathscr{T}, \quad u(x) = x^2 + x^3, \quad K(x, y) = x, \quad \lambda = 1.
$$

Let

$$
u_1(x) = x^3, \quad u_2(x) = x^2.
$$

Then

$$\begin{aligned}
\phi_0(x) &= x^3, \\
\phi_1(x) &= x^2 + x \int_0^2 \phi_0(y) \Delta y \\
&= x^2 + x \int_0^2 y^3 \Delta y \\
&= x^2 + \frac{1}{15} x \int_0^2 (y^4)^{\Delta} \Delta y \\
&= x^2 + \frac{1}{15} x y^4 \Big|_{y=0}^{y=2} \\
&= x^2 + \frac{16}{15} x,
\end{aligned}$$

$$\begin{aligned}
\phi_2(x) &= x \int_0^2 \phi_1(y) \Delta y \\
&= x \int_0^2 \left(y^2 + \frac{16}{15} y \right) \Delta y \\
&= x \int_0^2 \left(\frac{1}{7} (y^3)^{\Delta} + \frac{16}{45} (y^2)^{\Delta} \right) \\
&= x \left(\frac{1}{7} \int_0^2 (y^3)^{\Delta} \Delta y + \frac{16}{45} \int_0^2 (y^2)^{\Delta} \Delta y \right) \\
&= x \left(\frac{1}{7} y^3 \Big|_{y=0}^{y=2} + \frac{16}{45} y^2 \Big|_{y=0}^{y=2} \right) \\
&= x \left(\frac{8}{7} + \frac{64}{45} \right) \\
&= \frac{808}{315} x, \\
\phi_3(x) &= x \int_0^2 \phi_2(y) \Delta y \\
&= \frac{808}{315} x \int_0^2 y \Delta y \\
&= \frac{808}{945} x \int_0^2 (y^2)^{\Delta} \Delta y \\
&= \frac{808}{945} x y^2 \Big|_{y=0}^{y=2} \\
&= \frac{3232}{945} x,
\end{aligned}$$

$$\begin{aligned}\phi_4(x) &= x\int_0^2 \phi_3(y)\Delta y\\ &= \frac{3232}{945}x\int_0^2 y\Delta y\\ &= \frac{3232}{2835}xy^2\Big|_{y=0}^{y=2}\\ &= \frac{12928}{2835}x.\end{aligned}$$

Exercise 2 Let $\mathscr{T} = 3^{\mathscr{N}_0} \cup \{0\}$. Consider the equation

$$\phi(x) = 1 + x + x\int_0^3 y\phi(y)\Delta y.$$

Set $u_1(x) = x$, $u_2(x) = 1$. Using MDM, find $\phi_0(x)$, $\phi_1(x)$, $\phi_2(x)$, $\phi_3(x)$, $\phi_4(x)$.

Answer

$$\begin{aligned}&\phi_0(x) = x,\quad \phi_1(x) = 1 + \tfrac{27}{13}x,\quad \phi_2(x) = \tfrac{4437}{676}x,\\ &\phi_3(x) = \tfrac{4437}{676}\left(\tfrac{27}{13}\right)x,\quad \phi_4(x) = \tfrac{4437}{676}\left(\tfrac{27}{13}\right)^2 x.\end{aligned}$$

5.1.3 The Noise Terms Phenomenon

It was shown that a proper choice of $u_1(x)$ and $u_2(x)$ is essential to use the modified decomposition method. In Chap. 3 the noise terms phenomenon was introduced and it demonstrates a fast convergence of the solution. The noise terms as defined before are the identical terms with opposite signs that appear between the components $\phi_0(x)$ and $\phi_1(x)$. Other noise terms may appear between other components. By canceling the noise terms between $\phi_0(x)$ and $\phi_1(x)$, even though $\phi_1(x)$ contains further terms, the remaining non-canceled terms of $\phi_0(x)$ may give the exact solution of the given generalized Fredholm integral equation of the second kind. The appearance of the noise terms between $\phi_0(x)$ and $\phi_1(x)$ is not always sufficient to obtain the exact solution by cancelling these noise terms. Therefore, it is necessary to show that the non-cancelled terms of $\phi_0(x)$ satisfy the given generalized Fredholm integral equation of the second kind.

Example 7 Let $\mathscr{T} = \mathscr{Z}$. Consider the equation

$$\phi(x) = x - 2 + 2\int_0^2 \phi(y)\Delta y.$$

Here

$$\sigma(x) = x + 1,\quad x \in \mathscr{T},\quad u(x) = x - 2,\quad K(x, y) = 2.$$

Then

$$\begin{aligned}
\phi_0(x) &= u(x) \\
&= x - 2, \\
\phi_1(x) &= 2 \int_0^2 \phi_0(y) \Delta y \\
&= 2 \int_0^2 (y - 2) \Delta y
\end{aligned}$$

$$\begin{aligned}
&= 2 \int_0^2 \left(\frac{1}{2} (y^2)^{\Delta} - \frac{1}{2} - 2 \right) \Delta y \\
&= \int_0^2 ((y^2)^{\Delta} - 5) \Delta y \\
&= \int_0^2 (y^2)^{\Delta} \Delta y - 5 \int_0^2 \Delta y \\
&= y^2 \Big|_{y=0}^{y=2} - 5y \Big|_{y=0}^{y=2} \\
&= 4 - 10 \\
&= -6 \\
&= 2 - 8.
\end{aligned}$$

The noise term here is 2. Therefore we have to show that $\phi(x) = x$ is a solution to the considered equation.

Really, we have

$$\begin{aligned}
x - 2 + 2 \int_0^2 y \Delta y &= x - 2 + 2 \int_0^2 \left(\frac{1}{2} (y^2)^{\Delta} - \frac{1}{2} \right) \Delta y \\
&= x - 2 + \int_0^2 ((y^2)^{\Delta} - 1) \Delta y \\
&= x - 2 + \int_0^2 (y^2)^{\Delta} \Delta y - \int_0^2 \Delta y \\
&= x - 2 + y^2 \Big|_{y=0}^{y=2} - y \Big|_{y=0}^{y=2} \\
&= x - 2 + 4 - 2 \\
&= x \\
&= \phi(x).
\end{aligned}$$

Consequently, $\phi(x) = x$ is a solution to the considered equation.

Example 8 Let $\mathscr{T} = 2^{\mathscr{N}_0} \cup \{0\}$. Consider the equation

$$\phi(x) = x^2 - 256 + 15 \int_0^4 y\phi(y)\Delta y.$$

Here

$$\sigma(x) = 2x, \quad x \in \mathscr{T}, \quad u(x) = x^2 - 256, \quad K(x, y) = 15y.$$

Then

$$\begin{aligned}
\phi_0(x) &= u(x) \\
&= x^2 - 256, \\
\phi_1(x) &= 15 \int_0^4 y\phi_0(y)\Delta y \\
&= 15 \int_0^4 y(y^2 - 256)\Delta y \\
&= 15 \int_0^4 (y^3 - 256y)\Delta y \\
&= 15 \int_0^4 \left(\frac{1}{15}(y^4)^\Delta - \frac{256}{3}(y^2)^\Delta\right)\Delta y \\
&= \int_0^4 (y^4)^\Delta \Delta y - 1280 \int_0^4 (y^2)^\Delta \Delta y \\
&= y^4\Big|_{y=0}^{y=4} - 1280y^2\Big|_{y=0}^{y=4} \\
&= 256 - 20480.
\end{aligned}$$

The noise term is 256. Therefore we have to show that $\phi(x) = x^2$ is a solution to the given equation.

Really,

$$\begin{aligned}
x^2 - 256 + 15 \int_0^4 y\phi(y)\Delta y &= x^2 - 256 + 15 \int_0^4 y^3 \Delta y \\
&= x^2 - 256 + \int_0^4 (y^4)^\Delta \Delta y \\
&= x^2 - 256 + y^4\Big|_{y=0}^{y=4} \\
&= x^2 - 256 + 256 \\
&= x^2 \\
&= \phi(x).
\end{aligned}$$

Consequently $\phi(x) = x^2$ is a solution to the given equation.

Example 9 Let $\mathscr{T} = \mathscr{Z}$. Consider the equation

$$\phi(x) = 2x^2 - x + x\int_0^2 \phi(y)\Delta y.$$

Here

$$\sigma(x) = x + 1, \quad x \in \mathscr{T}, \quad u(x) = 2x^2 - x, \quad K(x, y) = x.$$

Then

$$\begin{aligned}
\phi_0(x) &= u(x)\\
&= 2x^2 - x,\\
\phi_1(x) &= x\int_0^2 \phi_0(y)\Delta y\\
&= x\int_0^2 (2y^2 - y)\Delta y\\
&= x\int_0^2 \left(\frac{2}{3}(y^3)^\Delta - (y^2)^\Delta + \frac{1}{3} - \frac{1}{2}(y^2)^\Delta + \frac{1}{2}\right)\Delta y\\
&= x\int_0^2 \left(\frac{2}{3}(y^3)^\Delta - \frac{3}{2}(y^2)^\Delta + \frac{5}{6}\right)\Delta y\\
&= x\left(\frac{2}{3}\int_0^2 (y^3)^\Delta\Delta y - \frac{3}{2}\int_0^2 (y^2)^\Delta\Delta y + \frac{5}{6}\int_0^2 \Delta y\right)\\
&= x\left(\frac{2}{3}y^3\Big|_{y=0}^{y=2} - \frac{3}{2}y^2\Big|_{y=0}^{y=2} + \frac{5}{6}y\Big|_{y=0}^{y=2}\right)\\
&= x\left(\frac{16}{3} - 6 + \frac{5}{3}\right)\\
&= x.
\end{aligned}$$

The noise term here is x. We have to check if $\phi(x) = 2x^2$ is a solution of the considered equation.

We have

$$\begin{aligned}
2x^2 - x + x\int_0^2 \phi(y)\Delta y &= 2x^2 - x + 2x\int_0^2 y^2\Delta y\\
&= 2x^2 - x + 2x\int_0^2 \left(\frac{1}{3}(y^3)^\Delta - \frac{1}{2}(y^2)^\Delta + \frac{1}{6}\right)\Delta y\\
&= 2x^2 - x + x\int_0^2 \left(\frac{2}{3}(y^3)^\Delta - (y^2)^\Delta + \frac{1}{3}\right)\Delta y\\
&= 2x^2 - x + x\left(\frac{2}{3}\int_0^2 (y^3)^\Delta\Delta y - \int_0^2 (y^2)^\Delta\Delta y + \frac{1}{3}\int_0^2 \Delta y\right)
\end{aligned}$$

$$
\begin{aligned}
&= 2x^2 - x + x\left(\frac{2}{3}y^3\Big|_{y=0}^{y=2} - y^2\Big|_{y=0}^{y=2} + \frac{1}{3}y\Big|_{y=0}^{y=2}\right) \\
&= 2x^2 - x + x\left(\frac{16}{3} - 4 + \frac{2}{3}\right) \\
&= 2x^2 - x + 2x \\
&= 2x^2 + x \\
&\neq \phi(x).
\end{aligned}
$$

Consequently $\phi(x) = 2x^2$ is not a solution of the considered equation.

Exercise 3 Let $\mathscr{T} = \mathscr{Z}$. Using the noise terms phenomenon, find a solution of the following generalized Fredholm integral equation of the second kind.

$$\phi(x) = x^2 + x - 20 + \int_0^4 \phi(y)\Delta y.$$

Answer

$$\phi(x) = x^2 + x.$$

5.1.4 The Direct Computation Method

The direct computation method(DCM) will be applied for solving the generalized Fredholm integral equations. The method gives the solution in an exact form and not in a series form. Note that this method will be applied for the degenerate or separable kernels of the form

$$K(x, y) = \sum_{l=1}^{n} f_l(x)g_l(y). \tag{5.4}$$

We substitute (5.4) in (5.1) and we get

$$\phi(x) = u(x) + \lambda \int_a^b \sum_{l=1}^{n} f_l(x)g_l(y)\phi(y)\Delta y,$$

or equivalently

$$\phi(x) = u(x) + \lambda \sum_{l=1}^{n} f_l(x) \int_a^b g_l(y)\phi(y)\Delta y.$$

We set

$$\alpha_l = \int_a^b g_l(y)\phi(y)\Delta y, \quad l = 1, 2, \ldots, n. \tag{5.5}$$

Then

$$\phi(x) = u(x) + \lambda \sum_{l=1}^{n} \alpha_l f_l(x). \tag{5.6}$$

Using (5.6), the solution $\phi(x)$ will be determined if the constants α_l, $1 \le l \le n$, are determined. We substitute (5.6) in (5.5) and we obtain

$$\begin{aligned}
\alpha_l &= \int_a^b g_l(y)\left(u(y) + \lambda \sum_{m=1}^{n} \alpha_m f_m(y)\right)\Delta y \\
&= \int_a^b g_l(y)u(y)\Delta y + \lambda \sum_{m=1}^{n} \alpha_m \int_a^b g_l(y) f_m(y)\Delta y \\
&= \int_a^b g_l(y)u(y)\Delta y + \lambda \sum_{m=1, m\neq l}^{n} \alpha_m \int_a^b g_m(y) f_m(y)\Delta y \\
&\quad + \lambda\alpha_l \int_a^b g_l(y) f_l(y)\Delta y, \quad l = 1, 2, \ldots, n,
\end{aligned}$$

whereupon

$$\begin{aligned}
\int_a^b g_l(y)u(y)\Delta y = \alpha_l &\left(1 - \lambda \int_a^b g_l(y) f_l(y)\Delta y\right) \\
&- \lambda \sum_{m=1, m\neq l}^{n} \alpha_m \int_a^b g_m(y) f_m(y)\Delta y, \quad l = 1, 2, \ldots, n.
\end{aligned}$$

Consequently we go to a system of n algebraic equations that can be solved to determine the constants α_l, $1 \le l \le n$. Using the obtained numerical values of α_l into (5.6), the solution $\phi(x)$ of the generalized Fredholm integral equation of the second kind (5.1) is readily obtained.

Example 10 Let $\mathscr{T} = \mathscr{Z}$. Consider the equation

$$\phi(x) = 1 + \int_0^2 (x + y + xy)\phi(y)\Delta y.$$

We can rewrite the given equation in the form

$$\phi(x) = 1 + x\int_0^2 \phi(y)\Delta y + (x+1)\int_0^2 y\phi(y)\Delta y.$$

Let

$$\alpha_1 = \int_0^2 \phi(y)\Delta y, \quad \alpha_2 = \int_0^2 y\phi(y)\Delta y. \tag{5.7}$$

Then

$$\phi(x) = 1 + \alpha_1 x + \alpha_2(x+1). \tag{5.8}$$

We substitute (5.8) into (5.7) and we find

$$\begin{aligned}
\alpha_1 &= \int_0^2 (1+\alpha_1 y + \alpha_2(y+1))\Delta y \\
&= \int_0^2 \Delta y + \alpha_1 \int_0^2 y\Delta y + \alpha_2 \int_0^2 (y+1)\Delta y \\
&= 2 + \alpha_1 \int_0^2 \left(\frac{1}{2}(y^2)^\Delta - \frac{1}{2}\right)\Delta y + \alpha_2 \int_0^2 \left(\frac{1}{2}(y^2)^\Delta - \frac{1}{2} + 1\right)\Delta y \\
&= 2 + \alpha_1 \left(\frac{1}{2}\int_0^2 (y^2)^\Delta \Delta y - \frac{1}{2}\int_0^2 \Delta y\right) + \alpha_2 \left(\frac{1}{2}\int_0^2 (y^2)^\Delta \Delta y + \frac{1}{2}\int_0^2 \Delta y\right) \\
&= 2 + \alpha_1 \left(\frac{1}{2}y^2\Big|_{y=0}^{y=2} - \frac{1}{2}y\Big|_{y=0}^{y=2}\right) + \alpha_2 \left(\frac{1}{2}y^2\Big|_{y=0}^{y=2} + \frac{1}{2}y\Big|_{y=0}^{y=2}\right) \\
&= 2 + \alpha_1 + 3\alpha_2,
\end{aligned}$$

whereupon

$$\alpha_2 = -\frac{2}{3}, \tag{5.9}$$

and

$$\begin{aligned}
\alpha_2 &= \int_0^2 y(1+\alpha_1 y + \alpha_2(y+1))\Delta y \\
&= \int_0^2 y\Delta y + \alpha_1 \int_0^2 y^2\Delta y + \alpha_2 \int_0^2 (y^2+y)\Delta y \\
&= \int_0^2 \left(\frac{1}{2}(y^2)^\Delta - \frac{1}{2}\right)\Delta y + \alpha_1 \int_0^2 \left(\frac{1}{3}(y^3)^\Delta - \frac{1}{2}(y^2)^\Delta + \frac{1}{6}\right)\Delta y \\
&\quad + \alpha_2 \int_0^2 \left(\frac{1}{3}(y^3)^\Delta - \frac{1}{2}(y^2)^\Delta + \frac{1}{6} + \frac{1}{2}(y^2)^\Delta - \frac{1}{2}\right)\Delta y \\
&= \frac{1}{2}\int_0^2 (y^2)^\Delta \Delta y - \frac{1}{2}\int_0^2 \Delta y + \alpha_1 \left(\frac{1}{3}\int_0^2 (y^3)^\Delta \Delta y - \frac{1}{2}\int_0^2 (y^2)^\Delta \Delta y + \frac{1}{6}\int_0^2 \Delta y\right) \\
&\quad + \alpha_2 \left(\frac{1}{3}\int_0^2 (y^3)^\Delta \Delta y - \frac{1}{3}\int_0^2 \Delta y\right) \\
&= \frac{1}{2}y^2\Big|_{y=0}^{y=2} - \frac{1}{2}y\Big|_{y=0}^{y=2} + \alpha_1 \left(\frac{1}{3}y^3\Big|_{y=0}^{y=2} - \frac{1}{2}y^2\Big|_{y=0}^{y=2} + \frac{1}{6}y\Big|_{y=0}^{y=2}\right) \\
&\quad + \alpha_2 \left(\frac{1}{3}y^3\Big|_{y=0}^{y=2} - \frac{1}{3}y\Big|_{y=0}^{y=2}\right) \\
&= 1 + \alpha_1 + 2\alpha_2,
\end{aligned}$$

from where, using (5.9),

$$\alpha_1 = -\frac{1}{3}.$$

Consequently

$$\begin{aligned}\phi(x) &= 1 + \left(-\frac{1}{3}x\right) + \left(-\frac{2}{3}\right)(x+1)\\ &= 1 - \frac{1}{3}x - \frac{2}{3}x - \frac{2}{3}\\ &= \frac{1}{3} - x.\end{aligned}$$

Example 11 Let $\mathscr{T} = 2^{\mathscr{N}_0} \cup \{0\}$. Consider the equation

$$\phi(x) = x + \int_0^2 (x+y)\phi(y)\Delta y.$$

We can rewrite the given equation in the form

$$\phi(x) = x + x\int_0^2 \phi(y)\Delta y + \int_0^2 y\phi(y)\Delta y.$$

Let

$$\alpha_1 = \int_0^2 \phi(y)\Delta y, \quad \alpha_2 = \int_0^2 y\phi(y)\Delta y. \tag{5.10}$$

Then

$$\phi(x) = \alpha_2 + (1+\alpha_1)x. \tag{5.11}$$

We substitute (5.11) into (5.10) and we get

$$\begin{aligned}\alpha_1 &= \int_0^2 (\alpha_2 + (1+\alpha_1)y)\Delta y\\ &= \alpha_2 \int_0^2 \Delta y + (1+\alpha_1)\int_0^2 y\Delta y\\ &= \alpha_2 y\Big|_{y=0}^{y=2} + \frac{1}{3}(1+\alpha_1)\int_0^2 (y^2)^{\Delta}\Delta y\\ &= 2\alpha_2 + \frac{1}{3}(1+\alpha_1)y^2\Big|_{y=0}^{y=2}\\ &= 2\alpha_2 + \frac{4}{3}(1+\alpha_1)\\ &= 2\alpha_2 + \frac{4}{3} + \frac{4}{3}\alpha_1,\end{aligned}$$

whereupon

$$-\frac{1}{3}\alpha_1 - 2\alpha_2 = \frac{4}{3}, \tag{5.12}$$

$$\begin{aligned}
\alpha_2 &= \int_0^2 y(\alpha_2 + (1+\alpha_1)y)\Delta y \\
&= \alpha_2 \int_0^2 y\Delta y + (1+\alpha_1)\int_0^2 y^2 \Delta y \\
&= \frac{1}{3}\alpha_2 \int_0^2 (y^2)^\Delta \Delta y + \frac{1}{7}(1+\alpha_1)\int_0^2 (y^3)^\Delta \Delta y \\
&= \frac{1}{3}\alpha_2 y^2\Big|_{y=0}^{y=2} + \frac{1}{7}(1+\alpha_1)y^3\Big|_{y=0}^{y=2} \\
&= \frac{4}{3}\alpha_2 + \frac{8}{7}(1+\alpha_1) \\
&= \frac{4}{3}\alpha_2 + \frac{8}{7} + \frac{8}{7}\alpha_1,
\end{aligned}$$

from where

$$-\frac{8}{7}\alpha_1 - \frac{1}{3}\alpha_2 = \frac{8}{7}.$$

From the last equation and from (5.12) we obtain the system

$$\begin{cases} -\frac{1}{3}\alpha_1 - 2\alpha_2 = \frac{4}{3} \\ -\frac{8}{7}\alpha_1 - \frac{1}{3}\alpha_2 = \frac{8}{7} \end{cases} \Longrightarrow \begin{cases} -\alpha_1 - 6\alpha_2 = 4 \\ -24\alpha_1 - 7\alpha_2 = 24 \end{cases} \Longrightarrow$$

$$\begin{cases} \alpha_1 = -4 - 6\alpha_2 \\ \alpha_2 = -\frac{72}{137} \end{cases} \Longrightarrow \begin{cases} \alpha_1 = -\frac{116}{137} \\ \alpha_2 = -\frac{72}{137}. \end{cases}$$

Consequently

$$\begin{aligned}
\phi(x) &= -\frac{72}{137} + \left(1 - \frac{116}{137}\right)x \\
&= -\frac{72}{137} + \frac{21}{137}x.
\end{aligned}$$

Exercise 4 Let $\mathscr{T} = \mathscr{Z}$. Using DCM, find a solution to the equation

$$\phi(x) = -1 + \int_0^2 (x+y)\phi(y)\Delta y.$$

Answer $\phi(x) = x$.

5.1.5 The Successive Approximations Method

The successive approximations method(SAM) or Picard iteration method was introduced in Chap. 3. This method provides a scheme for solving initial value problems or generalized integral equations.
For given generalized Fredholm integral equation (5.1), the successive approximations method introduce the recurrence relation

$$\begin{aligned} \phi_0(x) &= any \quad selective \quad real - valued \quad function, \\ \phi_{n+1}(x) &= u(x) + \lambda \int_a^b K(x, y)\phi_n(y)\Delta y, \quad n \in \mathcal{N}_0. \end{aligned}$$

The question of convergence of $\phi_n(x)$ is justified by Theorem 7 in Chap. 3. At the limit, the solution is determined by using the limit

$$\phi)(x) = \lim_{n\to\infty} \phi_n(x).$$

We will point out the difference between the successive approximations method and the Adomian decomposition method.

1. The successive approximations method gives successive approximations of the solution $\phi(x)$, whereas the Adomian decomposition method gives successive components of the solution $\phi(x)$.
2. The successive approximations method admits to use of a selective real-valued function for the zeroth approximation $\phi_0(x)$, whereas the Adomian decomposition method uses all terms that are not inside the integral sign. Recall that this assignment was modified when we use the modified decomposition method.
3. The successive approximations method gives the exact solution, if it exists by

   $$\lim_{n\to\infty} \phi_n(x).$$

   The Adomian decomposition method gives the solution as infinite series of components by

   $$\phi(x) = \sum_{n=0}^{\infty} \phi_n(x).$$

   This series converges rapidly to the exact solution if such a solution exists.

Example 12 Let $\mathcal{T} = \mathcal{Z}$. Consider the equation

$$\phi(x) = x^2 + \int_0^2 xy\phi(y)\Delta y.$$

Let $\phi_0(x) = 0$. Then

$$\begin{aligned}
\phi_1(x) &= x^2,\\
\phi_2(x) &= x^2 + x\int_0^2 y\phi_1(y)\Delta y\\
&= x^2 + x\int_0^2 y^3\Delta y\\
&= x^2 + x\int_0^2 \left(\frac{1}{4}(y^4)^\Delta - \frac{1}{2}(y^3)^\Delta + \frac{1}{4}(y^2)^\Delta\right)\Delta y\\
&= x^2 + x\left(\frac{1}{4}\int_0^2 (y^4)^\Delta\Delta y - \frac{1}{2}\int_0^2 (y^3)^\Delta\Delta y + \frac{1}{4}\int_0^2 (y^2)^\Delta\Delta y\right)\\
&= x^2 + x\left(\frac{1}{4}y^4\Big|_{y=0}^{y=2} - \frac{1}{2}y^3\Big|_{y=0}^{y=2} + \frac{1}{4}y^2\Big|_{y=0}^{y=2}\right)\\
&= x^2 + x,
\end{aligned}$$

$$\begin{aligned}
\phi_3(x) &= x^2 + x\int_0^2 y\phi_2(y)\Delta y\\
&= x^2 + x\int_0^2 y(y^2+y)\Delta y\\
&= x^2 + x\int_0^2 (y^3+y^2)\Delta y\\
&= x^2 + x\int_0^2 \left(\frac{1}{4}(y^4)^\Delta - \frac{1}{2}(y^3)^\Delta + \frac{1}{4}(y^2)^\Delta + \frac{1}{3}(y^3)^\Delta - \frac{1}{2}(y^2)^\Delta + \frac{1}{6}\right)\Delta y\\
&= x^2 + x\int_0^2 \left(\frac{1}{4}(y^4)^\Delta - \frac{1}{6}(y^3)^\Delta - \frac{1}{4}(y^2)^\Delta + \frac{1}{6}\right)\Delta y\\
&= x^2 + x\left(\frac{1}{4}\int_0^2 (y^4)^\Delta\Delta y - \frac{1}{6}\int_0^2 (y^3)^\Delta\Delta y - \frac{1}{4}\int_0^2 (y^2)^\Delta\Delta y + \frac{1}{6}\int_0^2 \Delta y\right)\\
&= x^2 + x\left(\frac{1}{4}y^4\Big|_{y=0}^{y=2} - \frac{1}{6}y^3\Big|_{y=0}^{y=2} - \frac{1}{4}y^2\Big|_{y=0}^{y=2} + \frac{1}{6}y\Big|_{y=0}^{y=2}\right)\\
&= x^2 + x\left(4 - \frac{4}{3} - 1 + \frac{1}{3}\right)\\
&= x^2 + 2x,\\
\phi_4(x) &= x^2 + x\int_0^2 y\phi_3(y)\Delta y\\
&= x^2 + x\int_0^2 y(y^2+2y)\Delta y\\
&= x^2 + x\int_0^2 (y^3+2y^2)\Delta y
\end{aligned}$$

$$= x^2 + x\int_0^2 \left(\frac{1}{4}(y^4)^\Delta - \frac{1}{2}(y^3)^\Delta + \frac{1}{4}(y^2)^\Delta + \frac{2}{3}(y^3)^\Delta - (y^2)^\Delta + \frac{1}{3}\right)\Delta y$$
$$= x^2 + x\int_0^2 \left(\frac{1}{4}(y^4)^\Delta + \frac{1}{6}(y^3)^\Delta - \frac{3}{4}(y^2)^\Delta + \frac{1}{3}\right)\Delta y$$
$$= x^2 + x\left(\frac{1}{4}\int_0^2 (y^4)^\Delta \Delta y + \frac{1}{6}\int_0^2 (y^3)^\Delta \Delta y - \frac{3}{4}\int_0^2 (y^2)^\Delta \Delta y + \frac{1}{3}\int_0^2 \Delta y\right)$$
$$= x^2 + x\left(\frac{1}{4}y^4\Big|_{y=0}^{y=2} + \frac{1}{6}y^3\Big|_{y=0}^{y=2} - \frac{3}{4}y^2\Big|_{y=0}^{y=2} + \frac{1}{3}y\Big|_{y=0}^{y=2}\right)$$
$$= x^2 + x\left(4 + \frac{4}{3} - 3 + \frac{2}{3}\right)$$
$$= x^2 + 3x.$$

Example 13 Let $\mathscr{T} = 2^{\mathscr{N}_0} \cup \{0\}$. Consider the equation

$$\phi(x) = x + x^2 + \int_0^2 y\phi(y)\Delta y.$$

Let $\phi_0(x) = 0$. Then

$$\begin{aligned}
\phi_1(x) &= x + x^2,\\
\phi_2(x) &= x + x^2 + \int_0^2 y\phi_1(y)\Delta y\\
&= x + x^2 + \int_0^2 y(y + y^2)\Delta y\\
&= x + x^2 + \int_0^2 (y^2 + y^3)\Delta y\\
&= x + x^2 + \int_0^2 \left(\frac{1}{7}(y^3)^\Delta + \frac{1}{15}(y^4)^\Delta\right)\Delta y\\
&= x + x^2 + \frac{1}{7}\int_0^2 (y^3)^\Delta \Delta y + \frac{1}{15}\int_0^2 (y^4)^\Delta \Delta y\\
&= x + x^2 + \frac{1}{7}y^3\Big|_{y=0}^{y=2} + \frac{1}{15}y^4\Big|_{y=0}^{y=2}\\
&= x + x^2 + \frac{8}{7} + \frac{16}{15}\\
&= x + x^2 + \frac{232}{105},
\end{aligned}$$

$$\begin{aligned}
\phi_3(x) &= x + x^2 + \int_0^2 y\phi_2(y)\Delta y \\
&= x + x^2 + \int_0^2 y\left(y + y^2 + \frac{232}{105}\right)\Delta y \\
&= x + x^2 + \int_0^2 \left(y^2 + y^3 + \frac{232}{105}y\right)\Delta y \\
&= x + x^2 + \int_0^2 \left(\frac{1}{7}(y^3)^\Delta + \frac{1}{15}(y^4)^\Delta + \frac{232}{315}(y^2)^\Delta\right)\Delta y \\
&= x + x^2 + \frac{1}{7}\int_0^2 (y^3)^\Delta \Delta y + \frac{1}{15}\int_0^2 (y^4)^\Delta \Delta y + \frac{232}{315}\int_0^2 (y^2)^\Delta \Delta y \\
&= x + x^2 + \frac{1}{7}y^3\Big|_{y=0}^{y=2} + \frac{1}{15}y^4\Big|_{y=0}^{y=2} + \frac{232}{315}y^2\Big|_{y=0}^{y=2} \\
&= x + x^2 + \frac{8}{7} + \frac{16}{15} + \frac{928}{315} \\
&= x + x^2 + \frac{1624}{315}, \\
\phi_4(x) &= x + x^2 + \int_0^2 y\phi_3(y)\Delta y \\
&= x + x^2 + \int_0^2 y\left(y + y^2 + \frac{1624}{315}\right)\Delta y \\
&= x + x^2 + \int_0^2 \left(y^2 + y^3 + \frac{1624}{315}y\right)\Delta y \\
&= x + x^2 + \int_0^2 \left(\frac{1}{7}(y^3)^\Delta + \frac{1}{15}(y^4)^\Delta + \frac{1624}{945}(y^2)^\Delta\right)\Delta y \\
&= x + x^2 + \frac{1}{7}\int_0^2 (y^3)^\Delta \Delta y + \frac{1}{15}\int_0^2 (y^4)^\Delta \Delta y + \frac{1624}{945}\int_0^2 (y^2)^\Delta \Delta y \\
&= x + x^2 + \frac{1}{7}y^3\Big|_{y=0}^{y=2} + \frac{1}{15}y^4\Big|_{y=0}^{y=2} + \frac{1624}{945}y^2\Big|_{y=0}^{y=2} \\
&= x + x^2 + \frac{8}{7} + \frac{16}{15} + \frac{6496}{945} \\
&= x + x^2 + \frac{8584}{945}.
\end{aligned}$$

Example 14 Let $\mathscr{T} = 4\mathscr{Z}$. Consider the equation

$$\phi(x) = 1 + x\int_0^8 \phi(y)\Delta y.$$

Let $\phi_0(x) = 1$. Then

$$\begin{aligned}
\phi_1(x) &= 1 + x \int_0^8 \phi_0(y)\Delta y \\
&= 1 + x \int_0^8 \Delta y \\
&= 1 + 8x, \\
\phi_2(x) &= 1 + x \int_0^8 \phi_1(y)\Delta y \\
&= 1 + x \int_0^8 (1 + 8y)\Delta y \\
&= 1 + x \int_0^8 \Delta y + 8x \int_0^8 y\Delta y \\
&= 1 + 8x + 8x \int_0^8 \left(\frac{1}{2}(y^2)^{\Delta} - 2\right) \Delta y \\
&= 1 + 8x + 4x \int_0^8 (y^2)^{\Delta}\Delta y - 16x \int_0^8 \Delta y \\
&= 1 + 8x + 4xy^2\Big|_{y=0}^{y=8} - 16xy\Big|_{y=0}^{y=8} \\
&= 1 + 8x + 256x - 128x \\
&= 1 + 136x, \\
\phi_3(x) &= 1 + x \int_0^8 \phi_2(y)\Delta y \\
&= 1 + x \int_0^8 (1 + 136y)\Delta y \\
&= 1 + x \int_0^8 \Delta y + 136x \int_0^8 y\Delta y
\end{aligned}$$

$$\begin{aligned}
&= 1 + 8x + 136x \int_0^8 \left(\frac{1}{2}(y^2)^{\Delta} - 2\right) \Delta y \\
&= 1 + 8x + 68x \int_0^8 (y^2)^{\Delta}\Delta y - 272x \int_0^8 \Delta y \\
&= 1 + 8x + 4352x - 2176x \\
&= 1 + 2184x, \\
\phi_4(x) &= 1 + x \int_0^8 \phi_3(y)\Delta y \\
&= 1 + x \int_0^8 (1 + 2184y)\Delta y
\end{aligned}$$

$$
\begin{aligned}
&= 1 + x\int_0^8 \Delta y + 2184x\int_0^8 y\Delta y \\
&= 1 + 8x + 2184x\int_0^8 \left(\frac{1}{2}(y^2)^\Delta - 2\right)\Delta y \\
&= 1 + 8x + 1092x\int_0^8 (y^2)^\Delta \Delta y - 4368x\int_0^8 \Delta y \\
&= 1 + 8x + 1092xy^2\Big|_{y=0}^{y=8} - 34944x \\
&= 1 + 8x + 69888x - 34944x \\
&= 1 + 34952x.
\end{aligned}
$$

Exercise 5 Using SAM, find $\phi_0(x)$, $\phi_1(x)$, $\phi_2(x)$, $\phi_3(x)$ and $\phi_4(x)$ for the following equations.

1. $\phi(x) = 1 + x\int_0^2 y^2\phi(y)\Delta y, \quad \mathscr{T} = \mathscr{Z},$
2. $\phi(x) = -3 + 2\int_0^4 y\phi(y)\Delta y, \quad \mathscr{T} = \mathscr{Z},$
3. $\phi(x) = 1 + x + \int_0^4 y\phi(y)\Delta y, \quad \mathscr{T} = \mathscr{Z},$
4. $\phi(x) = x + \int_0^4 y^2\phi(y)\Delta y, \quad \mathscr{T} = 2^{\mathscr{N}_0} \cup \{0\},$
5. $\phi(x) = 3x^3 + (x-1)\int_0^9 y\phi(y)\Delta y, \quad \mathscr{T} = 3^{\mathscr{N}_0} \cup \{0\},$
6. $\phi(x) = 1 - x + \int_0^4 y\phi(y)\Delta y, \quad \mathscr{T} = \mathscr{N}_0^2,$
7. $\phi(x) = 1 + x + x^2 + \int_0^4 (x+y)\phi(y)\Delta y, \quad \mathscr{T} = \mathscr{Z},$
8. $\phi(x) = 1 - 2x + \int_{-2}^4 y\phi(y)\Delta y, \quad \mathscr{T} = \mathscr{Z},$
9. $\phi(x) = x^2 + x\int_0^9 \phi(y)\Delta y, \quad \mathscr{T} = 3^{\mathscr{N}_0} \cup \{0\}.$

5.2 Homogeneous Generalized Fredholm Integral Equations of the Second Kind

Substituting $u(x) = 0$ into the generalized Fredholm integral equation of the second kind (5.1) we get

$$
\phi(x) = \lambda\int_a^b K(x, y)\phi(y)\Delta y. \tag{5.13}
$$

Definition 1 The Eq. (5.13) is called homogeneous generalized Fredholm integral equation of the second kind.

In this section we will focus our attention on the Eq. (5.13) for separable kernel $K(x, y)$ only. The main goal is to find nontrivial solution, because the trivial solution $\phi(x)$ is a solution to this equation. Moreover, the Adomian decomposition method is not applicable here because it depends on assigning a non-zero value for the zeroth component $\phi_0(x)$, and in this kind of equations $u(x) = 0$. Based on this, the direct computation method will be used for this kind of equations.

Assume that

$$K(x, y) = \sum_{l=1}^{n} f_l(x) g_l(y). \tag{5.14}$$

The direct computation method can be applied as follows.

1. We first substitute (5.14) into the Eq. (5.13).
2. This substitution leads to

$$\begin{aligned} \phi(x) &= \lambda \int_a^b \sum_{l=1}^{n} f_l(x) g_l(y) \phi(y) \Delta y \\ &= \lambda \sum_{l=1}^{n} f_l(x) \int_a^b g_l(y) \phi(y) \Delta y \end{aligned}$$

or

$$\begin{aligned} \phi(x) = \lambda f_1(x) \int_a^b g_1(y)\phi(y)\Delta y + \lambda f_2(x) \int_a^b g_2(y)\phi(y)\Delta y \\ + \cdots + \lambda f_n(x) \int_a^b g_n(y)\phi(y)\Delta y. \end{aligned}$$

3. Each integral at the right side is equivalent to a constant. Setting

$$\alpha_l = \int_a^b g_l(y)\phi(y)\Delta y, \quad l = 1, 2, \ldots, n, \tag{5.15}$$

we obtain the equation

$$\phi(x) = \lambda \alpha_1 f_1(x) + \lambda \alpha_2 f_2(x) + \cdots + \lambda \alpha_n f_n(x). \tag{5.16}$$

4. Substituting (5.16) in (5.15) gives

$$\alpha_l = \lambda \int_a^b g_l(y) \sum_{m=1}^{n} \alpha_m f_m(y) \Delta y$$

or

$$\alpha_l\left(1-\lambda\int_a^b g_l(y)f_l(y)\Delta y\right)-\lambda\sum_{m=1,m\neq l}^{n}\alpha_m\int_a^b g_l(y)f_m(y)\Delta y=0,$$

$l=1,2,\ldots,n$. In this way we obtain a system of $n-$simultaneous algebraic equations that can be solved to determine the constants α_l, $1\le l\le n$. Using the obtained numerical values of α_l, $1\le l\le n$, into (5.16), the solution $\phi(x)$ of the homogeneous generalized Fredholm integral equation of the second kind (5.13) follows immediately.

Example 15 Let $\mathscr{T}=\mathscr{Z}$. Consider the equation

$$\phi(x)=\lambda x\int_0^4 y\phi(y)\Delta y.$$

Set

$$\alpha=\int_0^4 y\phi(y)\Delta y. \tag{5.17}$$

Then

$$\phi(x)=\lambda\alpha x. \tag{5.18}$$

Substituting (5.18) into the Eq. (5.17) we get

$$\begin{aligned}
\alpha&=\int_0^4 y(\lambda\alpha y)\Delta y\\
&=\lambda\alpha\int_0^4 y^2\Delta y\\
&=\lambda\alpha\int_0^4\left(\frac{1}{3}(y^3)^\Delta-\frac{1}{2}(y^2)^\Delta+\frac{1}{6}\right)\Delta y\\
&=\lambda\alpha\left(\frac{1}{3}\int_0^4(y^3)^\Delta\Delta y-\frac{1}{2}\int_0^4(y^2)^\Delta\Delta y+\frac{1}{6}\int_0^4\Delta y\right)\\
&=\lambda\alpha\left(\frac{1}{3}y^3\Big|_{y=0}^{y=4}-\frac{1}{2}y^2\Big|_{y=0}^{y=4}+\frac{1}{6}y\Big|_{y=0}^{y=4}\right)\\
&=\lambda\alpha\left(\frac{64}{3}-8+\frac{2}{3}\right)\\
&=\lambda\alpha(22-8)\\
&=14\lambda\alpha,
\end{aligned}$$

whereupon

$$\alpha(1 - 14\lambda) = 0.$$

Then, if $\lambda = \frac{1}{14}$, we obtain that

$$\phi(x) = \frac{\alpha}{14}x, \quad \alpha = \text{const},$$

is a solution to the considered equation.

Example 16 Let $\mathscr{T} = \mathscr{Z}$. Consider the equation

$$\phi(x) = \lambda \int_0^2 (1 + y)\phi(y)\Delta y.$$

The given equation we can rewrite in the following form

$$\phi(x) = \lambda \left(\int_0^2 \phi(y)\Delta y + \int_0^2 y\phi(y)\Delta y \right).$$

Let

$$\alpha_1 = \int_0^2 \phi(y)\Delta y, \quad \alpha_2 = \int_0^2 y\phi(y)\Delta y. \tag{5.19}$$

Then

$$\phi(x) = \lambda(\alpha_1 + \alpha_2). \tag{5.20}$$

We set (5.20) into (5.19) and we obtain

$$\begin{aligned} \alpha_1 &= \int_0^2 \lambda(\alpha_1 + \alpha_2)\Delta y \\ &= \lambda(\alpha_1 + \alpha_2) \int_0^2 \Delta y \\ &= 2\lambda(\alpha_1 + \alpha_2), \end{aligned}$$

whereupon

$$(2\lambda - 1)\alpha_1 + 2\lambda\alpha_2 = 0,$$

$$\begin{aligned} \alpha_2 &= \int_0^2 y(\lambda(\alpha_1 + \alpha_2))\Delta y \\ &= \lambda(\alpha_1 + \alpha_2) \int_0^2 y\Delta y \end{aligned}$$

$$
\begin{aligned}
&= \lambda(\alpha_1+\alpha_2)\int_0^2\left(\frac{1}{2}(y^2)^{\Delta}-\frac{1}{2}\right)\Delta y\\
&= \lambda(\alpha_1+\alpha_2)\left(\frac{1}{2}\int_0^2(y^2)^{\Delta}\Delta y-\frac{1}{2}\int_0^2\Delta y\right)\\
&= \lambda(\alpha_1+\alpha_2)\left(\frac{1}{2}y^2\Big|_{y=0}^{y=2}-\frac{1}{2}y\Big|_{y=0}^{y=2}\right)\\
&= \lambda(\alpha_1+\alpha_2)(2-1)\\
&= \lambda(\alpha_1+\alpha_2),
\end{aligned}
$$

from where

$$\lambda\alpha_1+(\lambda-1)\alpha_2=0.$$

For α_1, α_2 and λ we get the following system

$$\begin{cases}(2\lambda-1)\alpha_1+2\lambda\alpha_2=0\\ \lambda\alpha_1+(\lambda-1)\alpha_2=0.\end{cases}$$

If $\lambda=1$, then

$$\begin{cases}\alpha_1+2\alpha_2=0\\ \alpha_1=0\end{cases}\quad or\quad \begin{cases}\alpha_1=0\\ \alpha_2=0,\end{cases}$$

from here we obtain the trivial solution $\phi(x)=0$.

Let $\lambda\neq 1$. Then

$$\begin{cases}(2\lambda-1)\alpha_1+2\lambda\alpha_2=0\\ \alpha_2=-\frac{\lambda}{\lambda-1}\alpha_1\end{cases}\implies\begin{cases}(2\lambda-1)\alpha_1-\frac{2\lambda^2}{\lambda-1}\alpha_1=0\\ \alpha_2=-\frac{\lambda}{\lambda-1}\alpha_1\end{cases}\implies$$

$$\begin{cases}\left(2\lambda-1-\frac{2\lambda^2}{\lambda-1}\right)\alpha_1=0\\ \alpha_2=-\frac{\lambda}{\lambda-1}\alpha_1.\end{cases}$$

Hence,

$$2\lambda-1-\frac{2\lambda^2}{\lambda-1}=0\implies$$

$$(2\lambda-1)(\lambda-1)-2\lambda^2=0\implies$$

$$-3\lambda+1=0\implies$$

$$\lambda=\frac{1}{3}.$$

Consequently

$$\begin{aligned}\phi(x) &= \lambda(\alpha_1+\alpha_2)\\ &= \lambda\left(\alpha_1-\frac{\lambda}{\lambda-1}\alpha_1\right)\\ &= \frac{-\lambda}{\lambda-1}\alpha_1,\\ &= \frac{1}{2}\alpha_1 \quad \alpha_1=\text{const},\end{aligned}$$

is a solution to the given equation.

Example 17 Let $\mathscr{T}=2^{\mathscr{N}_0}\cup\{0\}$. Consider the equation

$$\phi(x)=\lambda\int_0^2(x+y)\phi(y)\Delta y.$$

The considered equation we can rewrite in the following form.

$$\phi(x)=\lambda x\int_0^2\phi(y)\Delta y+\lambda\int_0^2 y\phi(y)\Delta y.$$

We set

$$\alpha_1=\int_0^2\phi(y)\Delta y,\quad \alpha_2=\int_0^2 y\phi(y)\Delta y. \tag{5.21}$$

Then

$$\phi(x)=\lambda\alpha_1 x+\lambda\alpha_2. \tag{5.22}$$

We substitute (5.22) in (5.21) and we find

$$\begin{aligned}\alpha_1 &= \int_0^2(\lambda\alpha_1 y+\lambda\alpha_2)\Delta y\\ &= \lambda\alpha_1\int_0^2 y\Delta y+\lambda\alpha_2\int_0^2\Delta y\\ &= \frac{1}{3}\alpha_1\lambda\int_0^2(y^2)^{\Delta}\Delta y+\lambda\alpha_2 y\Big|_{y=0}^{y=2}\\ &= \frac{1}{3}\alpha_1\lambda y^2\Big|_{y=0}^{y=2}+2\lambda\alpha_2\\ &= \frac{4}{3}\alpha_1\lambda+2\lambda\alpha_2,\end{aligned}$$

whereupon

$$\left(\frac{4}{3}\lambda - 1\right)\alpha_1 + 2\lambda\alpha_2 = 0,$$

$$\begin{aligned}\alpha_2 &= \int_0^2 y(\lambda\alpha_1 y + \lambda\alpha_2)\Delta y \\ &= \lambda\alpha_1 \int_0^2 y^2 \Delta y + \lambda\alpha_2 \int_0^2 y\Delta y \\ &= \frac{1}{7}\lambda\alpha_1 \int_0^2 (y^3)^\Delta \Delta y + \frac{1}{3}\lambda\alpha_2 \int_0^2 (y^2)^\Delta \Delta y \\ &= \frac{1}{7}\lambda\alpha_1 y^3\Big|_{y=0}^{y=2} + \frac{1}{3}\lambda\alpha_2 y^2\Big|_{y=0}^{y=2} \\ &= \frac{8}{7}\lambda\alpha_1 + \frac{4}{3}\lambda\alpha_2\end{aligned}$$

or

$$\frac{8}{7}\lambda\alpha_1 + \left(\frac{4}{3}\lambda - 1\right)\alpha_2 = 0.$$

We obtain for α_1, α_2 and λ the following system

$$\begin{cases} \left(\frac{4}{3}\lambda - 1\right)\alpha_1 + 2\lambda\alpha_2 = 0 \\ \frac{8}{7}\lambda\alpha_1 + \left(\frac{4}{3}\lambda - 1\right)\alpha_2 = 0. \end{cases}$$

If $\lambda = \dfrac{3}{4}$, then

$$\begin{cases} \alpha_1 = 0 \\ \alpha_2 = 0, \end{cases}$$

i.e., we get the trivial solution $\phi(x) = 0$.

Let $\lambda \neq \dfrac{3}{4}$. Then

$$\begin{cases} \left(\frac{4}{3}\lambda - 1\right)\alpha_1 + 2\lambda\alpha_2 = 0 \\ \alpha_2 = -\frac{24\lambda}{7(4\lambda-3)}\alpha_1, \end{cases}$$

whereupon

$$\frac{4\lambda - 3}{3}\alpha_1 - \frac{48\lambda^2}{7(4\lambda - 3)}\alpha_1 = 0 \implies$$

$$\frac{4\lambda - 3}{3} - \frac{48\lambda^2}{7(4\lambda - 3)} = 0 \implies$$

$$7(4\lambda - 3)^2 - 144\lambda^2 = 0 \implies$$

$$32\lambda^2 + 168\lambda - 63 = 0 \implies$$

$$\lambda_{1,2} = \frac{-21 \pm 9\sqrt{7}}{8}.$$

Consequently

$$\begin{aligned}\phi(x) &= \lambda\alpha_1 x + \lambda\alpha_2 \\ &= \lambda\alpha_1 x - \frac{24\lambda^2}{7(4\lambda - 3)}\alpha_1 \\ &= \lambda\left(x - \frac{24\lambda}{7(4\lambda - 3)}\right)\alpha_1, \quad \lambda = \frac{-21 \pm 9\sqrt{7}}{8},\end{aligned}$$

$\alpha_1 = \text{const}$, is a solution to the given equation.

Exercise 6 Find a solution to the following equation

$$\phi(x) = \lambda x^2 \int_0^2 \phi(y)\Delta y, \quad \mathscr{T} = \mathscr{Z}.$$

Answer $\lambda = 1, \quad \phi(x) = \alpha x^2, \alpha = \text{const}.$

5.3 Fredholm Alternative Theorem

5.3.1 *The Case When* $\int_a^b \int_a^b |K(x, Y)|^2 \Delta X \Delta Y < 1$

For the sake of simplicity, we take $\lambda = 1$ in (5.1) and we consider the equation

$$\phi(x) - \int_a^b K(x, y)\phi(y)\Delta y = u(x), \tag{5.23}$$

where $u(x)$ is a continuous function on the interval $[a, b]$.

Definition 2 An equation, in the unknown $\psi(y)$, of the form

$$\psi(y) - \int_a^b K(x, y)\psi(x)\Delta x = v(y), \tag{5.24}$$

$v(y)$ being a given continuous function on the interval $[a, b]$, is called the conjugate equation of the Eq. (5.23).

Lemma 1 *Let* $\alpha, \beta > 0$. *For* $p > 1$ *and* $\dfrac{1}{p} + \dfrac{1}{q} = 1$, *we have*

$$\alpha^{\frac{1}{p}} \beta^{\frac{1}{q}} \leq \frac{\alpha}{p} + \frac{\beta}{q}. \tag{5.25}$$

Proof Since

$$e^{\frac{1}{p}x + \frac{1}{q}y} \leq \frac{1}{p} e^x + \frac{1}{q} e^y \quad for \quad all \quad x, y \in \mathscr{R},$$

for $x = \log \alpha$ and $y = \log \beta$, we get the inequality (5.25).

Theorem 1 (Hölder's Inequality) *Let* $a, b \in \mathscr{T}$, $a < b$. *For rd-continuous functions* $f, g : [a, b] \longmapsto \mathscr{R}$ *we have*

$$\int_a^b |f(t)g(t)| \Delta t \leq \left(\int_a^b |f(t)|^p \Delta t \right)^{\frac{1}{p}} \left(\int_a^b |g(t)|^q \Delta t \right)^{\frac{1}{q}},$$

where $p > 1$ *and* $\dfrac{1}{p} + \dfrac{1}{q} = 1$.

Proof Without loss of generality we suppose that

$$\left(\int_a^b |f(t)|^p \Delta t \right) \left(\int_a^b |g(t)|^q \Delta t \right) \neq 0.$$

Applying (5.25) to

$$\alpha(t) = \frac{|f(t)|^p}{\int_a^b |f(t)|^p \Delta t} \quad and \quad \beta(t) = \frac{|g(t)|^q}{\int_a^b |g(t)|^q \Delta t},$$

we get

$$\frac{|f(t)g(t)|}{\left(\int_a^b |f(t)|^p \Delta t \right)^{\frac{1}{p}} \left(\int_a^b |g(t)|^q \Delta t \right)^{\frac{1}{q}}} \leq \frac{1}{p} \frac{|f(t)|^p}{\int_a^b |f(t)|^p \Delta t} + \frac{1}{q} \frac{|g(t)|^q}{\int_a^b |g(t)|^q \Delta t}.$$

The obtained inequality we integrate between a and b, and we obtain

$$\int_a^b \frac{|f(t)g(t)|}{\left(\int_a^b |f(t)|^p \Delta t \right)^{\frac{1}{p}} \left(\int_a^b |g(t)|^q \Delta t \right)^{\frac{1}{q}}} \Delta t \leq \frac{1}{p} \int_a^b \frac{|f(t)|^p}{\int_a^b |f(t)|^p \Delta t} \Delta t + \frac{1}{q} \int_a^b \frac{|g(t)|^q}{\int_a^b |g(t)|^q \Delta t} \Delta t$$

or

$$\begin{aligned}\frac{\int_a^b |f(t)g(t)|\Delta t}{\left(\int_a^b |f(t)|^p \Delta t\right)^{\frac{1}{p}} \left(\int_a^b |g(t)|^q \Delta t\right)^{\frac{1}{q}}} &\le \frac{1}{p}\frac{\int_a^b |f(t)|^p \Delta t}{\int_a^b |f(t)|^p \Delta t} + \frac{1}{q}\frac{\int_a^b |g(t)|^q \Delta t}{\int_a^b |g(t)|^q \Delta t} \\ &= \frac{1}{p} + \frac{1}{q} \\ &= 1.\end{aligned}$$

This directly gives Hölder's inequality.

The special case $p = q = 2$ yields the Cauchy-Schwartz inequality.

Theorem 2 (Cauchy-Schwartz Inequality) *Let $a, b \in \mathscr{T}$, $a < b$. For rd-continuous functions $f, g : \mathscr{T} \longmapsto \mathscr{R}$ we have*

$$\int_a^b |f(t)g(t)|\Delta t \le \sqrt{\left(\int_a^b |f(t)|^2 \Delta t\right)\left(\int_a^b |g(t)|^2 \Delta t\right)}.$$

Theorem 3 *Let $K : [a, b] \times [a, b] \longmapsto \mathscr{R}$ be continuous. Under the assumption*

$$\int_a^b \int_a^b |K(x, y)|^2 \Delta x \Delta y < 1, \tag{5.26}$$

the Eqs. (5.23) (5.24) admits one and only one solution $\phi(x)$ $(\psi(y))$ for any continuous function $u(x)(v(y))$ on $[a, b]$. In particular, $\phi(x) \equiv 0$ $(\psi(y) \equiv 0)$ for the corresponding homogeneous equation

$$\phi(x) - \int_a^b K(x, y)\phi(y)\Delta y = 0 \quad \left(\psi(y) - \int_a^b K(x, y)\psi(x)\Delta x = 0\right).$$

Proof Starting with the kernel $K(x, y)$, we define the iterated kernels $K^{(1)}(x, y)$, $K^{(2)}(x, y), \ldots, K^{(n)}(x, y), \ldots$, as follows

$$\begin{aligned}&K^{(1)}(x, y) = K(x, y), \\ &K^{(2)}(x, y) = \int_a^b K(x, t)K^{(1)}(t, y)\Delta t, \\ &\ldots \\ &K^{(n)}(x, y) = \int_a^b K(x, t)K^{(n-1)}(t, y)\Delta t, \quad n \ge 3.\end{aligned}$$

Observe that

$$\begin{aligned}
K^{(2)}(x, y) &= \int_a^b K(x, t)K^{(1)}(t, y)\Delta t \\
&= \int_a^b K(x, t)K(t, y)\Delta t \\
&= \int_a^b K^{(1)}(x, t)K(t, y)\Delta t, \\
K^{(3)}(x, y) &= \int_a^b K(x, t)K^{(2)}(t, y)\Delta t \\
&= \int_a^b K(x, t)\int_a^b K^{(1)}(t, z)K(z, y)\Delta z\Delta t \\
&= \int_a^b \left(\int_a^b K(x, t)K^{(1)}(t, z)\Delta t\right) K(z, y)\Delta z \\
&= \int_a^b K^{(2)}(x, z)K(z, y)\Delta z.
\end{aligned}$$

Let for some $n \in \mathcal{N}$, $n \geq 3$, we have

$$K^{(n-1)}(x, y) = \int_a^b K^{(n-2)}(x, t)K(t, y)\Delta t.$$

Then

$$\begin{aligned}
K^{(n)}(x, y) &= \int_a^b K(x, t)K^{(n-1)}(t, y)\Delta t \\
&= \int_a^b K(x, t)\int_a^b K^{(n-2)}(t, z)K(z, y)\Delta z\Delta t \\
&= \int_a^b \left(\int_a^b K(x, t)K^{(n-2)}(t, z)\Delta t\right) K(z, y)\Delta z \\
&= \int_a^b K^{(n-1)}(x, z)K(z, y)\Delta z.
\end{aligned}$$

By the Cauchy-Schwartz inequality we have (for $n \geq 2$)

$$\begin{aligned}
|K^{(n)}(x, y)| &= \left|\int_a^b K(x, t)K^{(n-1)}(t, y)\Delta t\right| \\
&\leq \int_a^b |K(x, t)||K^{(n-1)}(t, y)|\Delta t \\
&\leq \left(\int_a^b |K(x, t)|^2\Delta t\right)^{\frac{1}{2}} \left(\int_a^b |K^{(n-1)}(t, y)|^2\Delta t\right)^{\frac{1}{2}},
\end{aligned}$$

whereupon

$$|K^{(n)}(x,y)|^2 \le \left(\int_a^b |K(x,t)|^2 \Delta t\right)\left(\int_a^b |K^{(n-1)}(t,y)|^2 \Delta t\right).$$

The obtained inequality we integrate over $[a,b] \times [a,b]$ and we find

$$\int_a^b \int_a^b |K^{(n)}(x,y)|^2 \Delta x \Delta y \le \left(\int_a^b \int_a^b |K(x,t)|^2 \Delta t \Delta x\right)\left(\int_a^b \int_a^b |K^{(n-1)}(t,y)|^2 \Delta t \Delta y\right).$$

Repeating this procedure, we finally obtain

$$\int_a^b \int_a^b |K^{(n)}(x,y)|^2 \Delta x \Delta y \le \left(\int_a^b \int_a^b |K(x,y)|^2 \Delta x \Delta y\right)^n. \tag{5.27}$$

Note that for $n \ge 3$ we have

$$\begin{aligned} K^{(n)}(x,y) &= \int_a^b K(x,t) K^{(n-1)}(t,y) \Delta y \\ &= \int_a^b K(x,t) \int_a^b K(t,r) K^{(n-2)}(r,y) \Delta r \Delta t \\ &= \int_a^b \int_a^b K(x,t) K^{(n-2)}(t,r) K(r,y) \Delta r \Delta t. \end{aligned}$$

Hence by the Cauchy-Schwartz inequality we have

$$|K^{(n)}(x,y)|^2 \le \left(\int_a^b \int_a^b |K^{(n-2)}(t,r)|^2 \Delta t \Delta r\right)\left(\int_a^b \int_a^b |K(x,t)K(r,y)|^2 \Delta r \Delta t\right).$$

From here and from (5.27) we obtain

$$|K^{(n)}(x,y)|^2 \le \left(\int_a^b \int_a^b |K(t,r)|^2 \Delta r \Delta t\right)^{n-2} \left(\int_a^b |K(x,t)|^2 \Delta t\right)\left(\int_a^b |K(r,y)|^2 \Delta r\right).$$

Note that $\left(\int_a^b |K(x,t)|^2 \Delta t\right)\left(\int_a^b |K(r,y)|^2 \Delta r\right)$ is continuous on the domain $a \le x \le b$, $a \le y \le b$. Hence this term is bounded. Therefore, according to the assumption (5.26), the series

$$\Gamma(x,y) = \sum_{n=1}^{\infty} K^{(n)}(x,y) \tag{5.28}$$

converges uniformly on the domain $a \le x \le b$, $a \le y \le b$.

Also,

$$\begin{aligned}
\Gamma(x, y) &= K(x, y) + \sum_{n=2}^{\infty} K^{(n)}(x, y) \\
&= K(x, y) + \sum_{n=2}^{\infty} \int_a^b K(x, t) K^{(n-1)}(t, y) \Delta t \\
&= K(x, y) + \int_a^b K(x, t) \sum_{n=2}^{\infty} K^{(n-1)}(t, y) \Delta y \\
&= K(x, y) + \int_a^b K(x, t) \Gamma(t, y) \Delta t,
\end{aligned}$$

i.e.,

$$\Gamma(x, y) = K(x, y) + \int_a^b K(x, t) \Gamma(t, y) \Delta t. \tag{5.29}$$

As in above, we obtain the equation

$$\Gamma(x, y) = K(x, y) + \int_a^b \Gamma(x, t) K(t, y) \Delta t.$$

Now we can prove that

$$\phi(x) = u(x) + \int_a^b \Gamma(x, y) u(y) \Delta y \tag{5.30}$$

satisfies (5.23).

In fact, substituting (5.30) into (5.23), we have

$$\begin{aligned}
&\phi(x) - \textstyle\int_a^b K(x, y)\phi(y)\Delta y = u(x) + \int_a^b \Gamma(x, y) u(y) \Delta y \\
&- \textstyle\int_a^b K(x, y) \left(u(y) + \int_a^b \Gamma(y, s) u(s) \Delta s \right) \Delta y \\
&= u(x) + \textstyle\int_a^b \Gamma(x, y) u(y) \Delta y - \int_a^b K(x, y) u(y) \Delta y - \int_a^b \int_a^b K(x, y) \Gamma(y, s) u(s) \Delta s \Delta y \\
&= u(x) + \textstyle\int_a^b \Gamma(x, s) u(s) \Delta s - \int_a^b K(x, s) u(s) \Delta s - \int_a^b \left(\int_a^b K(x, y) \Gamma(y, s) \Delta y \right) u(s) \Delta s \\
&= u(x) + \textstyle\int_a^b \left(\Gamma(x, s) - K(x, s) - \int_a^b K(x, y) \Gamma(y, s) \Delta y \right) u(s) \Delta s \\
&= u(x).
\end{aligned}$$

Remark 1 Accordingly, we see that (5.23) is equivalent to (5.30). Similarly, we can prove that the conjugate equation (5.24) is equivalent to the equation

$$\psi(y) = v(y) + \int_a^b \Gamma(x, y) v(x) \Delta x.$$

Definition 3 The series (5.28) is called the Neumann series for the kernel $K(x, y)$.

5.3.2 The General Case

Theorem 4 *Let $K(x, y)$ be continuous on $[a, b] \times [a, b]$. Then there exist two sets of linearly independent continuous functions*

$$f_1(x), \quad f_2(x), \quad \ldots, \quad f_m(x), g_1(y), \quad g_2(y), \quad \ldots, \quad g_m(y),$$

defined on the interval $[a, b]$, such that

$$\int_a^b \int_a^b \left| K(x, y) - \sum_{l=1}^{m} f_l(x) g_l(y) \right|^2 \Delta x \Delta y < 1. \tag{5.31}$$

Proof Let $\varepsilon > 0$ be arbitrarily chosen. Then, using that $K(x, y)$ is continuous on $[a, b] \times [a, b]$, there exists a division $\{I_\nu\}_{\nu=1}^{m}$ of the interval $[a, b]$ such that

$$\sup_{a \le x \le b} |K(x, y_1) - K(x, y_2)| < \varepsilon$$

for any pair of points y_1 and y_2 in each I_ν, $\nu = 1, 2, \ldots, m$.
Let $y_\nu \in I_\nu$, $\nu = 1, 2, \ldots, m$, and $I'_\nu \subset I_\nu$ be such that $y_\nu \in I'_\nu$.
Define the functions $g_\nu(y)$, $\nu = 1, 2, \ldots, m$, as follows

$$g_\nu(y) = \begin{cases} 0 & outside \quad of \quad I_\nu \\ 1 & on \quad I'_\nu \end{cases}$$

such that the functions $g_\nu(y)$ are continuous and $|g_\nu(y)| \le 1$ on $[a, b]$.

We now set

$$f_\nu(x) = K(x, y_\nu), \quad \nu = 1, 2, \ldots, m.$$

Let also,

$$L(x, y) = \left| K(x, y) - \sum_{\nu=1}^{m} f_\nu(x) g_\nu(y) \right|.$$

We have

$$L(x, y) = \left| K(x, y) - \sum_{\nu=1}^{m} K(x, y_\nu) g_\nu(y) \right|.$$

For $y \in I'_\nu$ we have that $g_l(y) = 0$, $l = 1, 2, \ldots, m$, $l \neq \nu$, $g_\nu(y) = 1$ and hence, using the definition of I_ν,

$$L(x, y) = |K(x, y) - K(x, y_\nu)| < \varepsilon.$$

For $y \in I_\nu \backslash I'_\nu$ we have that $g_l(y) = 0$, $l = 1, 2, \ldots, m$, $l \neq \nu$, and $|g_\nu(y)| \le 1$, and hence

$$\begin{aligned}|L(x,y)| &= |K(x,y) - K(x,y_\nu)g_\nu(y)| \\ &\le |K(x,y)| + |K(x,y_\nu)g_\nu(y)| \\ &= |K(x,y)| + |K(x,y_\nu)||g_\nu(y)| \\ &\le |K(x,y)| + |K(x,y_\nu)| \\ &\le 2M,\end{aligned}$$

where

$$M = \sup_{a\le x,y\le b} |K(x,y)|.$$

Since $\varepsilon > 0$ and the sum of lengths of $I_\nu \backslash I'_\nu$ are both arbitrary, we can choose the values of them so small that

$$\int_a^b \int_a^b \left| K(x,y) - \sum_{\nu=1}^m f_\nu(x)g_\nu(y) \right|^2 \Delta x \Delta y < 1.$$

Note that $g_1(y), g_2(y), \ldots, g_m(y)$ are linearly independent. Hence, if $f_1(x), f_2(x), \ldots, f_m(x)$ are linearly independent, then our proof is completed. Assume that $f_1(x), f_2(x), \ldots, f_m(x)$ are linearly dependent. Without loss of generality we suppose that $f_m(x)$ is written as a linear combination of $f_1(x), f_2(x), \ldots, f_{m-1}(x)$, i.e.,

$$f_m(x) = \alpha_1 f_1(x) + \alpha_2 f_2(x) + \cdots + \alpha_{m-1} f_{m-1}(x), \quad x \in [a,b],$$

where α_l, $l = 1, 2, \ldots, m-1$, are constants and

$$(\alpha_1, \alpha_2, \ldots, \alpha_{m-1}) \ne (0, 0, \ldots, 0).$$

Hence,

$$\begin{aligned}R(x,y) &= \sum_{l=1}^m f_l(x)g_l(y) \\ &= \sum_{l=1}^{m-1} f_l(x)g_l(y) + f_m(x)g_m(y) \\ &= \sum_{l=1}^{m-1} f_l(x)g_l(y) + (\alpha_1 f_1(x) + \alpha_2 f_2(x) + \cdots + \alpha_{m-1} f_{m-1}(x))g_m(y) \\ &= \sum_{l=1}^{m-1} f_l(x)(g_l(y) + \alpha_l g_m(y)).\end{aligned}$$

We set

$$g_l^{(1)}(y) = g_l(y) + \alpha_l g_m(y), \quad l = 1, 2, \ldots, m-1.$$

Then

$$R(x, y) = \sum_{l=1}^{m-1} f_l(x) g_l^{(1)}(y).$$

If $g_1^{(1)}(y), g_2^{(1)}(y), \ldots, g_{m-1}^{(1)}(y)$ are linearly independent, then by setting $g_l^{(1)}(y) = g_l(y)$, the number m is diminished. If otherwise, say $g_{m-1}^{(1)}(y)$ is a linear combination of $g_1^{(1)}(y), g_2^{(1)}(y), \ldots, g_{m-2}^{(1)}(y)$, i.e.,

$$g_{m-1}^{(1)}(y) = \beta_1 g_1^{(1)}(y) + \beta_2 g_2^{(1)}(y) + \cdots + \beta_{m-2} g_{m-2}^{(1)}(y),$$

where β_l, $l = 1, \ldots, m-2$, are constants and

$$(\beta_1, \ldots, \beta_{m-2}) \neq (0, \ldots, 0).$$

Then

$$\begin{aligned}
R(x, y) &= \sum_{l=1}^{m} f_l(x) g_l(y) \\
&= \sum_{l=1}^{m-1} f_l(x) g_l^{(1)}(y) \\
&= \sum_{l=1}^{m-2} f_l(x) g_l^{(1)}(y) + f_{m-1}(x) g_{m-1}^{(1)}(y) \\
&= \sum_{l=1}^{m-2} f_l(x) g_l^{(1)}(y) + f_{m-1}(x)(\beta_1 g_1^{(1)}(y) + \beta_2 g_2^{(1)}(y) + \cdots + \beta_{m-2} g_{m-2}^{(1)}(y)) \\
&= \sum_{l=1}^{m-2} (f_l(x) + \beta_l f_{m-1}(x)) g_l^{(1)}(y) \\
&= \sum_{l=1}^{m-2} f_l^{(1)}(x) g_l^{(1)}(y),
\end{aligned}$$

where

$$f_l^{(1)}(x) = f_l(x) + \beta_l f_{m-1}(x), \quad l = 1, 2, \ldots, m-1.$$

Repeating this argument alternatively for f and g, we finally obtain two sets of linearly independent functions $f_l^{(k)}(x)$ and $g_l^{(k)}(y)$ in terms of which $R(x, y)$ is written as

$$R(x, y) = \sum_{l=1}^{m_k} f_l^{(k)}(x) g_l^{(k)}(y)$$

provided that $K(x, y) \neq 0$ and $R(x, y) \neq 0$. Then by setting $f_l(x) = f_l^{(k)}(x)$ and $g_l(y) = g_l^{(k)}(y)$, the proof is completed.

Below we suppose that $K(x, y)$ is continuous on $[a, b] \times [a, b]$. Let also, $f_1(x)$, $f_2(x), \ldots, f_m(x)$, $g_1(y), g_2(y), \ldots, g_m(y)$ be two sets of linearly independent continuous functions on $[a, b]$ such that (5.31) holds.

We now set

$$K_1(x, y) = K(x, y) - \sum_{l=1}^{m} f_l(x) g_l(y)$$

and denote the resolvent kernel of $K_1(x, y)$ by

$$\Gamma_1(x, y) = \sum_{n=1}^{\infty} K_1^{(n)}(x, y).$$

Then the Eq. (5.23) takes the form

$$\phi(x) - \int_a^b \left(K_1(x, y) + \sum_{l=1}^{m} f_l(x) g_l(y) \right) \phi(y) \Delta y = u(x)$$

or

$$\phi(x) - \int_a^b K_1(x, y) \phi(y) \Delta y = u(x) + \int_a^b \sum_{l=1}^{m} f_l(x) g_l(y) \phi(y) \Delta y.$$

Hence, using (5.30), we get

$$\begin{aligned}
&\phi(x) = u(x) + \textstyle\int_a^b \sum_{l=1}^m f_l(x) g_l(y) \phi(y) \Delta y \\
&+ \textstyle\int_a^b \Gamma_1(x, y) \left(u(y) + \int_a^b \sum_{l=1}^m f_l(y) g_l(s) \phi(s) \Delta s \right) \Delta y \\
&= u(x) + \textstyle\int_a^b \sum_{l=1}^m f_l(x) g_l(y) \phi(y) \Delta y \\
&+ \textstyle\int_a^b \Gamma_1(x, y) u(y) \Delta y + \int_a^b \Gamma_1(x, y) \int_a^b \sum_{l=1}^m f_l(y) g_l(s) \phi(s) \Delta s \Delta y \\
&= u(x) + \textstyle\int_a^b \sum_{l=1}^m f_l(x) g_l(y) \phi(y) \Delta y \\
&+ \textstyle\int_a^b \Gamma_1(x, y) u(y) \Delta y + \int_a^b \int_a^b \Gamma_1(x, y) \sum_{l=1}^m f_l(y) g_l(s) \phi(s) \Delta s \Delta y \\
&= u(x) + \textstyle\int_a^b \sum_{l=1}^m f_l(x) g_l(y) \phi(y) \Delta y + \int_a^b \Gamma_1(x, y) u(y) \Delta y \\
&+ \textstyle\int_a^b \int_a^b \Gamma_1(x, y) \sum_{l=1}^m f_l(y) g_l(s) \phi(s) \Delta y \Delta s \\
&= u(x) + \textstyle\int_a^b \sum_{l=1}^m f_l(x) g_l(y) \phi(y) \Delta y + \int_a^b \Gamma_1(x, y) u(y) \Delta y \\
&+ \textstyle\int_a^b \int_a^b \Gamma_1(x, s) \sum_{l=1}^m f_l(s) g_l(y) \phi(y) \Delta s \Delta y \\
&= u(x) + \textstyle\int_a^b \sum_{l=1}^m f_l(x) g_l(y) \phi(y) \Delta y + \int_a^b \Gamma_1(x, y) u(y) \Delta y \\
&+ \textstyle\int_a^b \sum_{l=1}^m \left(\int_a^b \Gamma_1(x, s) f_l(s) \Delta s \right) g_l(y) \phi(y) \Delta y \\
&= u(x) + \textstyle\int_a^b \sum_{l=1}^m \left(f_l(x) + \int_a^b \Gamma_1(x, s) f_l(s) \Delta s \right) g_l(y) \phi(y) \Delta y \\
&+ \textstyle\int_a^b \Gamma_1(x, y) u(y) \Delta y
\end{aligned}$$

or

$$\begin{aligned}&\phi(x) - \int_a^b \sum_{l=1}^m \left(f_l(x) + \int_a^b \Gamma_1(x,s) f_l(s)\Delta s \right) g_l(y)\phi(y)\Delta y \\ &= u(x) + \int_a^b \Gamma_1(x,y)u(y)\Delta y.\end{aligned} \tag{5.32}$$

From this follows the fact that to solve the Eq. (5.23) is equivalent to finding a solution $\phi(x)$ of the Eq. (5.32) with the kernel given by

$$K'(x,y) = \sum_{l=1}^m \left(f_l(x) + \int_a^b \Gamma_1(x,s) f_l(s)\Delta s \right) g_l(y)$$

and the right side the given function

$$u(x) + \int_a^b \Gamma_1(x,y)u(y)\Delta y.$$

Theorem 5 *The functions*

$$f_l(x) + \int_a^b \Gamma_1(x,y) f_l(y)\Delta y, \quad l = 1, 2, \ldots, m, \tag{5.33}$$

are linearly independent on $[a, b]$.

Proof Assume that (5.33) are linearly dependent on $[a, b]$. Then there exist constants $c_1, c_2, \ldots, c_m$ such that

$$(c_1, c_2, \ldots, c_m) \neq (0, 0, \ldots, 0)$$

and

$$\sum_{l=1}^m c_l \left(f_l(x) + \int_a^b \Gamma_1(x,y) f_l(y)\Delta y \right) = 0. \tag{5.34}$$

Let

$$\phi_1(x) = \sum_{l=1}^m c_l f_l(x), \quad x \in [a,b].$$

Hence, using (5.34), we obtain

$$\sum_{l=1}^m c_l f_l(x) + \sum_{l=1}^m c_l \int_a^b \Gamma_1(x,y) f_l(y)\Delta y = 0$$

or

$$\sum_{l=1}^m c_l f_l(x) + \int_a^b \Gamma_1(x,y) \sum_{l=1}^m c_l f_l(y)\Delta y = 0,$$

or

$$\phi_1(x) + \int_a^b \Gamma_1(x, y)\phi_1(y)\Delta y = 0,$$

or

$$\phi_1(x) = -\int_a^b \Gamma_1(x, y)\phi_1(y)\Delta y.$$

From here, using (5.30), we obtain

$$\phi_1(x) = 0 - \int_a^b \Gamma_1(x, y).0\Delta y = 0,$$

i.e.,

$$\sum_{l=1}^{m} c_l f_l(x) = 0 \quad on \quad [a, b],$$

which is a contradiction because $\{f_l(x)\}_{l=1}^{m}$ are linearly independent.

Let

$$\alpha_l = \int_a^b g_l(y)\phi(y)\Delta y, \quad l = 1, 2, \ldots, m. \tag{5.35}$$

Then the Eq. (5.32) takes the form

$$\phi(x) = \sum_{l=1}^{m} \alpha_l \left(f_l(x) + \int_a^b \Gamma_1(x, s) f_l(s)\Delta s \right) + u(x) + \int_a^b \Gamma_1(x, y)u(y)\Delta y. \tag{5.36}$$

We substitute (5.36) in (5.35) and we get

$$\begin{aligned} \alpha_l &= \int_a^b g_l(y)\Big(\sum_{l=1}^{m} \alpha_i \Big(f_i(y) + \int_a^b \Gamma_1(y, s) f_i(s)\Delta s\Big) \\ &\quad + u(y) + \int_a^b \Gamma_i(y, s)u(s)\Delta s\Big)\Delta y \\ &= \sum_{i=1}^{m} \alpha_i \int_a^b g_l(y)\Big(f_i(y) + \int_a^b \Gamma_1(y, s) f_i(s)\Delta s\Big)\Delta y \\ &\quad + \int_a^b g_l(y)\Big(u(y) + \int_a^b \Gamma_1(y, s)u(s)\Delta s\Big)\Delta y \end{aligned}$$

or

$$\begin{aligned}&\left(1-\int_a^b g_l(y)\Big(f_l(y)+\int_a^b \Gamma_l(y,s)f_l(s)\Delta s\Big)\Delta y\right)\alpha_l\\&-\sum_{i=1,i\neq l}^m \alpha_i\int_a^b g_l(y)\Big(f_i(y)+\int_a^b \Gamma_1(y,s)f_i(s)\Delta s\Big)\Delta y\\&=\int_a^b g_l(y)\left(u(y)+\int_a^b \Gamma_1(y,s)u(s)\Delta s\right)\Delta y,\quad l=1,2,\ldots,m.\end{aligned}\tag{5.37}$$

Substituting the solutions $(\alpha_1,\alpha_2,\ldots,\alpha_m)$ of (5.37) in (5.36) we obtain the solutions of (5.23).

Similarly, we see that to solve the Eq. (5.24) is equivalent to solving the following linear equations in the unknowns $(\alpha_1',\alpha_2',\ldots,\alpha_m')$,

$$\begin{aligned}&\left(1-\int_a^b g_l(y)f_l(y)\Delta y-\int_a^b\int_a^b \Gamma_1(y,s)g_l(y)f_l(s)\Delta y\Delta s\right)\alpha_l'\\&-\sum_{i=1,i\neq l}\alpha_i'\left(\int_a^b g_i(y)f_l(y)\Delta y+\int_a^b\int_a^b \Gamma_1(y,s)g_i(y)f_l(s)\Delta y\Delta s\right)\\&=\int_a^b f_l(y)v(y)\Delta y+\int_a^b\int_a^b \Gamma_1(y,s)f_l(y)v(s)\Delta y\Delta s,\quad l=1,2,\ldots,m,\end{aligned}\tag{5.38}$$

and the solution $\psi(y)$ of the system (5.24) is given by

$$\psi(y)=v(y)+\int_a^b \Gamma_1(x,y)v(x)\Delta x+\sum_{l=1}^m \alpha_l'\left(g_l(y)+\int_a^b \Gamma_1(x,y)g_l(x)\Delta x\right),$$

where $(\alpha_1',\alpha_2',\ldots,\alpha_m')$ are the solutions of (5.38).

Let A be the matrix of the equations (5.37), in the unknowns α, and A' be the matrix of the equations (5.38), in the unknowns α'. Then

$$A' \ \textit{is the transposed matrix of}\ A.$$

Hence,

$$\det A'\neq 0\ \textit{if and only if}\ \det A\neq 0.$$

We first consider the case when $\det A\neq 0$, and hence, $\det A'\neq 0$. In this case the system (5.37) (5.38) admits a unique solution $\phi(x)(\psi(y))$. In particular, if $u(x)\equiv 0$ $(v(y)\equiv 0)$, then

$$\begin{aligned}\alpha&=(\alpha_1,\alpha_2,\ldots,\alpha_m)=(0,0,\ldots,0),\\(\alpha'&=(\alpha_1',\alpha_2',\ldots\alpha_m')=(0,0,\ldots,0)),\end{aligned}$$

hence $\phi(x)\equiv 0(\psi(y)\equiv 0)$.

We next consider the case when $\det A=0$, and hence $\det A'=0$. For the sake of simplicity, we write (5.37) and (5.38) as follows

$$\alpha_l-\sum_{i=1}^m c_{li}\alpha_i=u_l,\quad l=1,2,\ldots,m,\tag{5.39}$$

$$\alpha_l'-\sum_{i=1}^m c_{il}\alpha_i'=v_l,\quad l=1,2,\ldots,m,\tag{5.40}$$

respectively. Here

$$\begin{aligned}
c_{ll} &= 1 - \int_a^b g_l(y)\left(f_l(y) + \int_a^b \Gamma_1(y,s) f_l(s)\Delta s\right)\Delta y,\\
c_{li} &= \int_a^b g_l(y)\left(f_i(y) + \int_a^b \Gamma_1(y,s) f_i(s)\Delta s\right)\Delta y, \quad i = 1, \dots, m, i \neq l,\\
u_l &= \int_a^b g_l(y)\left(u(y) + \int_a^b \Gamma_1(y,s) u(s)\Delta s\right)\Delta y,\\
v_l &= \int_a^b f_l(y) v(y)\Delta y + \int_a^b \int_a^b \Gamma_1(y,s) f_l(y) v(s)\Delta y \Delta s, \quad l = 1, \dots, m.
\end{aligned}$$

The matrices A and A' are written as

$$A = (\delta_{li} - c_{li}), \quad A' = (\delta_{il} - c_{il}),$$

where $\delta_{il} = 0$ for $i \neq l$ and $\delta_{ll} = 1$, $i, l = 1, 2, \dots, m$.

For the case when $\det A = \det A' = 0$, the associated systems of linear homogeneous equations

$$\alpha_l - \sum_{i=1}^{m} c_{li}\alpha_i = 0, \quad i = 1, 2, \dots, m, \tag{5.41}$$

$$\alpha_l' - \sum_{i=1}^{m} c_{il}\alpha_i' = 0, \quad l = 1, 2, \dots, m, \tag{5.42}$$

admit a number r, $r \geq 1$, of linearly independent solutions

$$\begin{aligned}
&\alpha(1) = (\alpha_{11}, \alpha_{12}, \dots, \alpha_{1m}),\\
&\dots\\
&\alpha(r) = (\alpha_{r1}, \alpha_{r2}, \dots, \alpha_{rm}),
\end{aligned}$$

and

$$\begin{aligned}
&\alpha'(1) = (\alpha_{11}', \alpha_{12}', \dots, \alpha_{1m}'),\\
&\dots\\
&\alpha'(r) = (\alpha_{r1}', \alpha_{r2}', \dots, \alpha_{rm}'),
\end{aligned}$$

respectively. The inhomogeneous system (5.39) admits a solution for given u_1, u_2, $\dots$, u_m if and only if

$$\sum_{i=1}^{m} u_i\alpha_{li}' = 0, \quad l = 1, 2, \dots, m. \tag{5.43}$$

In other words, for the general solution of (5.42),

$$\sum_{j=1}^{r} c_j\alpha'(j) = \left(\sum_{j=1}^{r} c_j\alpha_{j1}', \sum_{j=1}^{r} c_j\alpha_{j2}', \dots, \sum_{j=1}^{r} c_j\alpha_{jm}'\right),$$

which contains a number r of arbitrary constants $c_1, c_2, \ldots, c_r$, there hold the following relation

$$\sum_{l=1}^{m} u_l \left(\sum_{j=1}^{r} c_j \alpha'_{jl} \right) = \sum_{j=1}^{r} c_j \left(\sum_{l=1}^{m} u_l \alpha'_{jl} \right) = 0. \tag{5.44}$$

If the condition (5.44) is satisfied, then the general solution of the Eq. (5.39) is given by the sum of a particular solution

$$\overline{\alpha} = (\overline{\alpha}_1, \overline{\alpha}_2, \ldots, \overline{\alpha}_m)$$

of (5.39) and the general solution $\sum_{j=1}^{r} c_j \alpha(j)$ of (5.41), that is by the following expression containing r arbitrary constants $c_1, c_2, \ldots, c_r$,

$$\begin{aligned} \alpha &= \overline{\alpha} + \textstyle\sum_{j=1}^{r} c_j \alpha(j) \\ &= \left(\overline{\alpha}_1 + \textstyle\sum_{j=1}^{r} c_j \alpha_{j1}, \overline{\alpha}_2 + \sum_{j=1}^{r} c_j \alpha_{j2}, \ldots, \overline{\alpha}_m + \sum_{j=1}^{r} c_j \alpha_{jm} \right). \end{aligned} \tag{5.45}$$

Similarly, the equations (5.40) admit a solution for given $v_1, v_2, \ldots, v_m$ if and only if the following relation

$$\sum_{\mu=1}^{m} v_\mu \left(\sum_{j=1}^{r} c_j \alpha_{j\mu} \right) = 0 \tag{5.46}$$

holds, and, under the condition (5.46), the general solution of (5.40) is given by the sum of a particular solution

$$\overline{\alpha}' = (\overline{\alpha}'_1, \overline{\alpha}'_2, \ldots, \overline{\alpha}'_m)$$

of the Eq. (5.40) and the general solution $\sum_{j=1}^{m} c_j \alpha'(j)$ of the equations (5.42), that is by the following expression containing r arbitrary constants $c_1, c_2, \ldots, c_r$,

$$\begin{aligned} \alpha' &= \overline{\alpha}' + \textstyle\sum_{j=1}^{r} c_j \alpha'(j) \\ &= \left(\overline{\alpha}'_1 + \textstyle\sum_{j=1}^{r} c_j \alpha'_{j1}, \overline{\alpha}'_2 + \sum_{j=1}^{r} c_j \alpha'_{j2}, \ldots, \overline{\alpha}'_m + \sum_{j=1}^{r} c_j \alpha'_{jm} \right). \end{aligned} \tag{5.47}$$

Accordingly, substituting the solution α given by (5.46) in the equations (5.36) we obtain the general solution $\phi(x)$ of (5.23). The solution $\phi(x)$ contains r arbitrary constants. In fact, if

$$0 = \sum_{l=1}^{m} \left(f_l(x) + \int_a^b \Gamma_1(x, y) f_l(y) \Delta y \right) \left(\sum_{j=1}^{r} c_j \alpha_{jl} \right),$$

then, by Theorem 5,

$$0 = \sum_{j=1}^{r} c_j \alpha_{jl}, \quad l = 1, 2, \ldots, m,$$

which is a contradiction since $\alpha(1), \alpha(2), \ldots, \alpha(r)$ are linearly independent solutions of (5.41). We can also substitute (5.47) in (5.38), to obtain the general solution $\psi(y)$ of (5.24) which contains a number r of arbitrary constants.

Finally, we shall reduce the solvability condition (5.43) to a more readable and useful form as follows. Using the definition of the functions u_l, we can rewrite (5.43) in the following form

$$\begin{aligned} 0 &= \sum_{i=1}^{m} \alpha'_{li} u_i \\ &= \sum_{i=1}^{m} \alpha'_{li} \int_a^b \left(g_i(y) + \int_a^b \Gamma_1(s, y) g_i(s) \Delta s \right) u(y) \Delta y \\ &= \int_a^b \left[\sum_{i=1}^{m} \alpha'_{li} \left(g_i(y) + \int_a^b \Gamma_1(s, y) g_i(s) \Delta s \right) \right] u(y) \Delta y. \end{aligned}$$

We observe that

$$\sum_{i=1}^{m} \alpha'_{li} \left(g_i(y) + \int_a^b \Gamma_1(s, y) g_i(s) \Delta s \right) \tag{5.48}$$

is a solution to the equation

$$\psi(y) - \int_a^b K(x, y) \psi(x) \Delta x = 0. \tag{5.49}$$

On the other hand the general solution of (5.49) is given by a linear combinations of the functions (5.48). Therefore the condition (5.43) is equivalent to the following. For every solution $\psi(y)$ of (5.49),

$$\int_a^b u(y) \psi(y) \Delta y = 0.$$

Similarly, we see that the condition (5.46) is equivalent to the following. For every solution $\phi(x)$ of the equation

$$\begin{aligned} \phi(x) - \textstyle\int_a^b K(x, y)\phi(y)\Delta y &= 0, \\ \textstyle\int_a^b v(y)\phi(y)\Delta y &= 0. \end{aligned}$$

5.3.3 Fredholm's Alternative Theorem

The results obtained in the previous two sections are known as Fredholm's alternative theorem concerning a continuous kernel $K(x, y)$. The theorem reads as follows.

Theorem 6 (Fredholm's Alternative Theorem) *Either the generalized integral equation of the second kind*

$$u(x) = \phi(x) - \lambda \int_a^b K(x, y)\phi(y)\Delta y \tag{5.50}$$

with fixed λ, admits a unique continuous solution $\phi(x)$ for any continuous function $u(x)$, in particular $\phi(x) \equiv 0$ for $u(x) \equiv 0$, or the associative homogeneous equation

$$\phi(x) = \lambda \int_a^b K(x, y)\phi(y)\Delta y \tag{5.51}$$

admits a number r, $r \geq 1$, of linearly independent continuous solutions $\phi_l(x)$, $l = 1, 2, \dots, r$. In the first case the conjugate equation

$$v(y) = \psi(y) - \lambda \int_a^b K(x, y)\psi(x)\Delta x \tag{5.52}$$

also admits a unique continuous solution $\psi(y)$ for any continuous function $v(y)$. In the second case, the associated homogeneous equation

$$\psi(y) = \lambda \int_a^b K(x, y)\psi(x)\Delta x \tag{5.53}$$

admits a number r of linearly independent continuous solutions $\psi_1(y)$, $\psi_2(y)$, $\dots$, $\psi_r(y)$. In the second case, Eq. (5.50) admits solution if and only if

$$\int_a^b u(x)\psi_l(x)\Delta x = 0, \quad l = 1, 2, \dots, r. \tag{5.54}$$

If the condition (5.54) is satisfied, the general solution $\phi(x)$ of the Eq. (5.50) is written as

$$\phi(x) = \phi^{(1)}(x) + \sum_{j=1}^{r} c_j \phi_j(x)$$

by means of a particular solution $\phi^{(1)}(x)$ *of (5.50) and r arbitrary constants* c_1, c_2, $\ldots$, c_r. *Similarly, the conjugate equation (5.52) admits a solution if and only if*

$$\int_a^b v(y)\phi_i(y)\Delta y = 0, \quad i = 1, 2, \ldots, r. \tag{5.55}$$

If the condition (5.55) is satisfied, the general solution of the Eq. (5.52) is written as

$$\psi(y) = \psi^{(1)}(y) + \sum_{j=1}^{r} c_j \psi_j(y)$$

by means of a particular solution $\psi^{(1)}(y)$ *of (5.51) and r arbitrary constants* c_1, c_2, $\ldots$, c_r.

Definition 4 Every $\lambda \in \mathscr{C}$ for which the Eq. (5.51) or (5.53) has a nontrivial solution $\phi(x)$ or $\psi(x)$, respectively, will be called eigenvalue corresponding to the kernel $K(x, y)$. The solution $\phi(x)$ is called eigenfunction belonging to the eigenvalue λ. If the maximal number of linearly independent eigenfunctions belonging to the eigenvalue λ is k, we will say that the multiplicity of λ is k.

Remark 2 Fredholm's alternative theorem shows that the unique solution of (5.50) exists for any continuous function $u(x)$ if and only if λ is not an eigenvalue.

5.4 The Schmidth Expansion Theorem and the Mercer Expansion Theorem

5.4.1 *Operator-Theoretical Notations*

Assume that $K(x, y)$ is a complex-valued continuous function on the region $a \leq x \leq b$, $a \leq y \leq b$. By K we denote the operator

$$Kf(x) = \int_a^b K(x, y) f(y)\Delta y$$

which transforms every continuous function $f(x)$ on the interval $[a, b]$ into a continuous function $(Kf)(x)$.

Definition 5 Let $L(x, y)$ be a complex-valued continuous function. Then the kernel $M(x, y)$ defined by

$$M(x, y) = \int_a^b K(x, r)L(r, y)\Delta r$$

is said to be the composition of the kernels $K(x, y)$ and $L(x, y)$.

If the operator M is defined by $M(x, y)$, then

$$\begin{aligned} Mf(x) &= \int_a^b \left(\int_a^b K(x, r)L(r, y)\Delta r \right) f(y)\Delta y \\ &= \int_a^b K(x, r) \left(\int_a^b L(r, y)f(y)\Delta y \right) \Delta r \\ &= \int_a^b K(x, r)L(f)(r)\Delta r \\ &= K(Lf)(x) \end{aligned}$$

for any continuous function $f(x)$ on $[a, b]$. Consequently

$$M = KL.$$

Definition 6 For a pair of continuous functions $f(x)$ and $g(x)$ on $[a, b]$ we define inner product by

$$\langle f, g \rangle = \int_a^b f(x)\overline{g(x)}\Delta x. \tag{5.56}$$

Exercise 7 Prove that (5.56) satisfies all axioms for inner product.

Definition 7 For a continuous function $f(x)$ on $[a, b]$ we define norm by

$$||f|| = \langle f, f \rangle^{\frac{1}{2}}. \tag{5.57}$$

Exercise 8 Prove that (5.57) satisfies the axioms for norm.

Definition 8 The number

$$||K|| = \sup_{||f||=1} ||Kf||$$

will be called the norm of the operator K.

Exercise 9 Prove

$$||Kf|| \le ||K|| ||f||$$

for every $f \in \mathscr{C}([a, b])$.

Remark 3 In fact we have

$$||K|| = \sup_{||f||=1} |\langle Kf, f \rangle|.$$

Definition 9 We define the transposed conjugate kernel $K^*(x, y)$ of the kernel $K(x, y)$ by

$$K^*(x, y) = \overline{K(y, x)}.$$

Theorem 7 *For every continuous functions f, g on $[a, b]$ we have*

$$\langle Kf, g\rangle = \langle f, K^*g\rangle.$$

Proof We have

$$\begin{aligned}
\langle Kf, g\rangle &= \int_a^b Kf(x)\overline{g(x)}\Delta x \\
&= \int_a^b \left(\int_a^b K(x, y)f(y)\Delta y\right)\overline{g(x)}\Delta x \\
&= \int_a^b f(y)\left(\int_a^b K(x, y)\overline{g(x)}\Delta x\right)\Delta y \\
&= \int_a^b f(y)\left(\int_a^b \overline{K^*(y, x)g(x)}\Delta x\right)\Delta y \\
&= \int_a^b f(y)\overline{\left(\int_a^b K^*(y, x)g(x)\Delta x\right)}\Delta y \\
&= \int_a^b f(y)\overline{(K^*g)(y)}\Delta y \\
&= \langle f, K^*g\rangle.
\end{aligned}$$

Theorem 8 *We have the following relation*

$$(KL)^* = L^*K^*. \tag{5.58}$$

Proof For every $f, g \in \mathscr{C}([a, b])$ we have

$$\langle KLf, g\rangle = \langle f, (KL)^*g\rangle. \tag{5.59}$$

Also,

$$\begin{aligned}
\langle KLf, g\rangle &= \langle Lf, K^*g\rangle \\
&= \langle f, L^*K^*g\rangle.
\end{aligned}$$

Hence and (5.59) we obtain (5.58).

Theorem 9 *We have*

$$(K^*)^* = K.$$

Proof For every $f, g \in \mathscr{C}([a, b])$ we have

$$\begin{aligned}\langle (K^*)^* f, g\rangle &= \langle f, K^* g\rangle \\ &= \langle Kf, g\rangle,\end{aligned}$$

which completes the proof.

Definition 10 The complex-valued function $K(x, y)$, defined on $a \le x, y \le b$, will be called Hermitian symmetric if

$$K(x, y) = \overline{K(y, x)}$$

for all $x, y \in [a, b]$. In other words, $K(x, y)$ is Hermitian symmetric if $K(x, y) = K^*(x, y)$.

Exercise 10 Let $K(x, y)$ be Hermitian symmetric, defined and symmetric on $a \le x, y \le b$. Prove that

$$\langle Kf, g\rangle = \langle f, Kg\rangle$$

for all $f, g \in \mathscr{C}([a, b])$.

Lemma 2 *Let $K(x, y)$ be continuous and Hermitian symmetric on $a \le x, y \le b$. Then*

$$\langle Kf, f\rangle \in \mathscr{R}$$

for every $f \in \mathscr{C}([a, b])$.

Proof Since $K(x, y)$ is continuous and Hermitian symmetric on $a \le x, y \le b$, we have that $K = K^*$. Hence, for $f \in \mathscr{C}([a, b])$, we obtain

$$\begin{aligned}\langle Kf, f\rangle &= \langle f, K^* f\rangle \\ &= \langle f, Kf\rangle \\ &= \overline{\langle Kf, f\rangle},\end{aligned}$$

which completes the proof.

Theorem 10 *Let $K(x, y)$ be continuous and Hermitian symmetric on $[a, b] \times [a, b]$. Let also, $||K|| \ne 0$. Then either $\dfrac{1}{||K||}$ or $-\dfrac{1}{||K||}$ is an eigenvalue of the operator K.*

Proof By Lemma 2 we have that $\langle Kf, f\rangle \in \mathscr{R}$ for every $f \in \mathscr{C}([a, b])$. Since

$$||K|| = \sup_{||f||=1} |\langle Kf, f\rangle|,$$

there exists a sequence $\{f_m\}_{m=1}^{\infty}$ of elements of $\mathscr{C}([a, b])$ such that

$$\begin{cases} ||f_m|| = 1 \quad and \\ \langle Kf_m, f_m\rangle \longrightarrow_{m\to\infty} ||K|| \quad or \quad \langle Kf_m, f_m\rangle \longrightarrow_{m\to\infty} -||K||. \end{cases}$$

Suppose that

$$\langle Kf_m, f_m\rangle \longrightarrow_{m\to\infty} ||K||.$$

By Ascoli-Arzelà theorem it follows that the sequence $\{Kf_m(x)\}_{m=1}^{\infty}$ contains a subsequence which converges to a continuous function $\phi(x)$ uniformly on the interval $[a, b]$. We may assume without loss of generality that $\{(Kf_m)(x)\}_{m=1}^{\infty}$ itself converges uniformly to the function $\phi(x)$. Then

$$\begin{aligned} &||Kf_m - ||K||f_m||^2 = \langle Kf_m - ||K||f_m, Kf_m - ||f_m||\rangle \\ &= \langle Kf_m, Kf_m - ||K||f_m\rangle - \langle ||K||f_m, Kf_m - ||K||f_m\rangle \\ &= \langle Kf_m, Kf_m\rangle - \langle Kf_m, ||K||f_m\rangle - \langle ||K||f_m, Kf_m\rangle + \langle ||K||f_m, ||K||f_m\rangle \\ &= ||Kf_m||^2 - ||K||\langle Kf_m, f_m\rangle - ||K||\langle f_m, Kf_m\rangle + ||K||^2\langle f_m, f_m\rangle \\ &= ||Kf_m||^2 - 2||K||\langle Kf_m, f_m\rangle + ||K||^2||f_m||^2. \end{aligned} \tag{5.60}$$

Hence, using that $Kf_m(x) \longrightarrow_{m\to\infty} \phi(x)$, we get

$$0 \le ||\phi||^2 - 2||K||^2 + ||K||^2 = ||\phi||^2 - ||K||^2,$$

whereupon

$$||\phi|| \ge ||K|| > 0.$$

Now, from (5.60), using that

$$||Kf_m|| \le ||K||||f_m||,$$

we get

$$\begin{aligned} &||Kf_m - ||K||f_m||^2 \le ||K||^2||f_m||^2 - 2||K||\langle Kf_m, f_m\rangle + ||K||^2 \\ &= 2||K||^2 - 2||K||\langle Kf_m, f_m\rangle \longrightarrow_{m\to\infty} 2||K||^2 - 2||K||^2 = 0. \end{aligned}$$

Thus

$$Kf_m - ||K||f_m \longrightarrow_{m\to\infty} 0.$$

Hence, using that

$$||K(Kf_m) - ||K||Kf_m|| \le ||K||||Kf_m - ||K||f_m|| \longrightarrow_{m\to\infty} 0,$$

we obtain

$$||K(Kf_m) - ||K||Kf_m|| \longrightarrow_{m\to\infty} 0$$

or

$$||K\phi - ||K||\phi|| = 0,$$

i.e.,

$$\int_a^b |K\phi(x) - ||K||\phi(x)|^2 \Delta x = 0,$$

whereupon

$$K\phi(x) = ||K||\phi(x).$$

If $\langle K f_m, f_m \rangle \longrightarrow_{m\to\infty} -||K||$, applying the same arguments to the operator $-K$, we see that $\frac{1}{||K||}$ is an eigenvalue of $-K$. Hence, in this case, $-\frac{1}{||K||}$ is an eigenvalue of K.

Definition 11 The method of the proof of Theorem 10 is called maximal method.

Definition 12 Let $f, g \in \mathscr{C}([a, b])$. When

$$\langle f, g \rangle = 0$$

$f(x)$ is said to be orthogonal to $g(x)$, and this fact is indicated by writing $f \perp g$. Clearly, the orthogonality relation is reflexive, that is, if $f \perp g$, then $g \perp f$.

Definition 13 We say that the system

$$f_1(x), \quad f_2(x), \quad \ldots, \quad f_m(x), \quad \ldots,$$

of continuous functions on $[a, b]$, satisfy the orthonormality relations if

$$\langle f_l, f_m \rangle = \delta_{lm} = \begin{cases} 1 & for \quad l = m \\ 0 & for \quad l \neq m. \end{cases}$$

Definition 14 We say that a continuous function $f(x)$ on $[a, b]$ is normalized if $||f|| = 1$.

Theorem 11 (Bessel Inequality) *Let* $f \in \mathscr{C}([a, b])$ *and* $\{\phi_m\}_{m=1}^{\infty}$ *be an orthonormal system. Then*

$$||f||^2 \geq \sum_{m=1}^{\infty} |\langle f, \phi_m \rangle|^2. \tag{5.61}$$

Proof Let $l \in \mathscr{N}$ be arbitrarily chosen. Set

$$f_l = f - \sum_{m=1}^{l} \langle f, \phi_m \rangle \phi_m.$$

Then, from the orthonormality of the system $\{\phi_m\}_{m=1}^{\infty}$, we have

$$\begin{aligned}
0 &\le ||f_l||^2 = \langle f_l, f_l \rangle \\
&= \langle f - \textstyle\sum_{m=1}^{l} \langle f, \phi_m \rangle \phi_m, f - \sum_{k=1}^{l} \langle f, \phi_k \rangle \phi_k \rangle \\
&= \langle f, f \rangle - \langle f, \textstyle\sum_{k=1}^{l} \langle f, \phi_k \rangle \phi_k \rangle - \langle \sum_{m=1}^{l} \langle f, \phi_m \rangle \phi_m, f \rangle + \sum_{m=1}^{l} \sum_{k=1}^{l} \langle f, \phi_m \rangle \langle f, \phi_k \rangle \langle \phi_m, \phi_k \rangle \\
&= ||f||^2 - \textstyle\sum_{k=1}^{l} |\langle f, \phi_k \rangle|^2 - \sum_{k=1}^{l} |\langle f, \phi_k \rangle|^2 + \sum_{k=1}^{l} |\langle f, \phi_k \rangle|^2 \\
&= ||f||^2 - \textstyle\sum_{k=1}^{l} |\langle f, \phi_k \rangle|^2,
\end{aligned}$$

i.e.,

$$||f||^2 \ge \sum_{k=1}^{l} |\langle f, \phi_k \rangle|^2.$$

Because $l \in \mathcal{N}$ was arbitrarily chosen, we get (5.61).

Definition 15 The numbers

$$\langle f, \phi_k \rangle, \quad k = 1, 2, \ldots,$$

in (5.61) are called the Fourier coefficients of $f(x)$ with respect to the orthonormal system $\{\phi_m\}_{m=1}^{\infty}$.

Theorem 12 *Let $K(x, y)$ be continuous and Hermitian symmetric on $a \le x, y \le b$.*

1. *All eigenvalues of the operator K are real.*
2. *Two eigenfunctions, corresponding to different eigenvalues, are orthogonal.*

Proof 1. Let λ be an eigenvalue of the operator K and ϕ be the corresponding eigenfunction, that is, $K\phi = \frac{1}{\lambda}\phi$. Then

$$\langle K\phi, \phi \rangle = \langle \frac{1}{\lambda}\phi, \phi \rangle = \frac{1}{\lambda}\langle \phi, \phi \rangle. \tag{5.62}$$

By Lemma 2 we have that $\langle K\phi, \phi \rangle \in \mathcal{R}$. Hence, using that $\langle \phi, \phi \rangle \in \mathcal{R}$ and (5.62), we conclude that $\lambda \in \mathcal{R}$.

2. Let λ and μ are two eigenvalues of the operator K and ϕ and ψ are the corresponding eigenfunctions, respectively. Then

$$K\phi = \frac{1}{\lambda}\phi \quad \textit{and} \quad K\psi = \frac{1}{\mu}\psi.$$

Assume that $\lambda \ne \mu$. Then $\frac{\lambda}{\mu} \ne 1$ and

$$\begin{aligned}\langle \phi, \psi \rangle &= \langle \lambda K \phi, \psi \rangle \\ &= \lambda \langle K\phi, \psi \rangle \\ &= \lambda \langle \phi, K\psi \rangle \\ &= \lambda \langle \phi, \frac{1}{\mu}\psi \rangle \\ &= \frac{\lambda}{\mu} \langle \phi, \psi \rangle.\end{aligned}$$

From here we obtain that

$$\langle \phi, \psi \rangle = 0,$$

which completes the proof.

Definition 16 An orthonormal system $\{\phi_m(x)\}_{m=1}^{\infty}$ of elements of $\mathscr{C}([a, b])$ will be called complete orthonormal system in $\mathscr{C}([a, b])$ if for every $f \in \mathscr{C}([a, b])$ the following representation

$$f(x) = \sum_{\alpha} c_\alpha \phi_\alpha(x), \quad c_\alpha = \text{const},$$

holds.

Definition 17 Let $f(x)$ be continuous function on $[a, b]$ and $K(x, y)$ be continuous on the rectangle $a \le x, y \le b$. Let also, $\{\phi_j\}$ be complete orthonormal system of eigenfunctions of K. Then

$$\sum_{j=1}^{\infty} \langle Kf, \phi_j \rangle \phi_j(x)$$

will be called the Fourier expansion of $(Kf)(x)$ with respect to the system $\{\phi_j\}$.

5.4.2 *The Schmidt Expansion Theorem*

We start with the following theorem.

Theorem 13 *Let $K(x, y) \not\equiv 0$ be continuous on $a \le x, y \le b$. Then the operators*

$$A = KK^* \quad \textit{and} \quad B = K^*K$$

are Hermitian symmetric and have the same eigenvalues with the same multiplicity. Further, their eigenvalues are all positive.

Proof Note that

$$A^* = (KK^*)^* = (K^*)^*K^* = KK^* = A,$$
$$B^* = (K^*K)^* = K^*(K^*)^* = K^*K = B,$$

i.e., the operators A and B are Hermitian symmetric.
For $x \in [a, b]$ we have

$$\begin{aligned} KK^*(x, x) &= \int_a^b K(x, y)K^*(y, x)\Delta y \\ &= \int_a^b K(x, y)\overline{K(x, y)}\Delta y \\ &= \int_a^b |K(x, y)|^2 \Delta y \\ &\neq 0. \end{aligned}$$

Therefore $(KK^*)(x, y) = A(x, y) \not\equiv 0$, and similarly $B(x, y) \not\equiv 0$. By Theorem 10 the operators A and B have eigenvalues.
Let λ be an eigenvalue of A, that is,

$$\lambda A\phi = \phi, \quad \phi(x) \not\equiv 0.$$

Then

$$\begin{aligned} 0 &< \langle \phi, \phi \rangle \\ &= \langle \lambda KK^*\phi, \phi \rangle \\ &= \lambda \langle KK^*\phi, \phi \rangle \\ &= \lambda \langle K^*\phi, K^*\phi \rangle. \end{aligned}$$

Since $\langle K^*\phi, K^*\phi \rangle \geq 0$, we conclude that $\lambda \geq 0$, i.e., all the eigenvalues of A are positive. The proof for the operator B is carried out in the same way.
We write the eigenvalues of A and the corresponding orthonormal system of eigenfunctions as follows.

$$\begin{array}{ccccccc} \lambda_1^2 \leq \lambda_2^2 \leq \ldots \leq \lambda_n^2 \leq \ldots \\ \phi_1, \quad \phi_2, \quad \ldots, \quad \phi_n, \quad \ldots. \end{array}$$

Let us set

$$\lambda_j K^*\phi_j = \psi_j.$$

Then

$$\begin{aligned}
\lambda_j^2 B\psi_j &= \lambda_j^2 K^* K \psi_j \\
&= \lambda_j^2 K^* K(\lambda_j K^* \phi_j) \\
&= \lambda_j K^* (\lambda_j^2 K K^* \phi_j) \\
&= \lambda_j K^* (\lambda_j^2 A\phi_j) \\
&= \lambda_j K^* \phi_j \\
&= \psi_j,
\end{aligned}$$

i.e.,

$$\lambda_j^2 B\psi_j = \psi_j.$$

Thus λ_j^2 is also an eigenvalue of B, and ψ_j the corresponding eigenfunction.

Note that $(\phi_i, \phi_j) = \delta_{ij}$ and

$$\begin{aligned}
\langle \psi_i, \psi_j \rangle &= \langle \lambda_i K^* \phi_i, \lambda_j K^* \phi_j \rangle \\
&= \langle \lambda_i \lambda_j K^* \phi_i, K^* \phi_j \rangle \\
&= \langle \lambda_i \lambda_j K K^* \phi_i, \phi_j \rangle \\
&= \frac{\lambda_j}{\lambda_i} \langle \lambda_i^2 K K^* \phi_i, \phi_j \rangle \\
&= \frac{\lambda_j}{\lambda_i} \langle \lambda_i^2 A\phi_i, \phi_j \rangle \\
&= \frac{\lambda_j}{\lambda_i} \langle \phi_i, \phi_j \rangle \\
&= \delta_{ij}.
\end{aligned}$$

Hence, we see that the eigenfunctions $\{\psi_j\}$ of the operator B satisfy the orthonormality relations.

Now we will prove that the system $\{\psi_j\}$ exhaust all eigenfunctions of the operator B. To see this we will show that

$$\begin{aligned}
&if \quad \lambda^2 B\psi = \psi, \quad \psi(x) \not\equiv 0, \quad then \\
&\lambda K\psi = \phi \quad satisfies \quad \lambda^2 A\phi = \phi.
\end{aligned}$$

Let the eigenvalues of B and the corresponding orthonormal system of eigenfunctions are given by

$$\begin{matrix}
\lambda_1^2 \le \lambda_2^2 \le \ldots \le \lambda_n^2 \le \ldots \\
\psi_1, \quad \psi_2, \quad \ldots, \quad \psi_n, \quad \ldots.
\end{matrix}$$

Then we have

$$\lambda_j^2 B\psi_j = \psi_j$$

or

$$\lambda_j^2 K^* K \psi_j = \psi_j,$$

or

$$\lambda_j K^*(\lambda_j K \psi_j) = \psi_j,$$

or

$$\lambda_j K^* \phi_j = \psi_j.$$

Hence,

$$\begin{aligned}
\lambda_j^2 A\phi_j &= \lambda_j^2 K K^* \phi_j \\
&= \lambda_j K(\lambda_j K^* \phi_j) \\
&= \lambda_j K \psi_j \\
&= \phi_j.
\end{aligned}$$

Thus the proof is completed.

The Schmidt expansion theorem reads as follows.

Theorem 14 *Let $f(x)$ be continuous function on $[a, b]$ and $K(x, y)$ be continuous function on $[a, b] \times [a, b]$. Then $(Kf)(x)$ and $(K^* f)(x)$ can be expanded in Fourier series with respect to $\{\phi_j\}$ and $\{\psi_j\}$, respectively, that converge absolutely and uniformly on the interval $a \leq x \leq b$.*

Proof We have

$$\begin{aligned}
\langle Kf, \phi_j \rangle &= \langle Kf, \lambda_j^2 A\phi_j \rangle \\
&= \langle Kf, \lambda_j^2 K K^* \phi_j \rangle \\
&= \langle K^* K f, \lambda_j^2 K^* \phi_j \rangle \\
&= \langle Bf, \lambda_j^2 K^* \phi_j \rangle \\
&= \langle \lambda_j Bf, \lambda_j K^* \phi_j \rangle
\end{aligned}$$

$$\begin{aligned}
&= \lambda_j \langle Bf, \psi_j \rangle \\
&= \lambda_j \langle f, B\psi_j \rangle \\
&= \lambda_j \langle f, \frac{1}{\lambda_j^2} \psi_j \rangle \\
&= \frac{1}{\lambda_j} \langle f, \psi_j \rangle.
\end{aligned}$$

By the Cauchy-Schwartz inequality we obtain

$$\left(\sum_{j=n}^{m}\left|\langle f,\psi_j\rangle\frac{\phi_j(x)}{\lambda_j}\right|\right)^2 \le \sum_{j=n}^{m}|\langle f,\psi_j\rangle|^2\sum_{j=n}^{m}\left|\frac{\phi_j(x)}{\lambda_j}\right|^2$$

for $n < m$.
By the Bessel inequality we have

$$\int_a^b |K(x,y)|^2\Delta y \ge \sum_{j=1}^{\infty}\frac{|\phi_j(x)|^2}{\lambda_j^2}, \quad ||f||^2 \ge \sum_{j=1}^{\infty}|\langle f,\psi_j\rangle|^2.$$

Hence, the series

$$\sum_{j=1}^{\infty}\lambda_j^{-1}\langle f,\psi_j\rangle\phi_j(x)$$

converges absolutely and uniformly on the interval $a \le x \le b$. In other words, the Fourier expansion of $(Kf)(x)$ is absolutely and uniformly convergent on $[a,b]$.

Let now

$$(K_n f)(x) = (Kf)(x) - \sum_{j=1}^{n}\langle f,\psi_j\rangle\frac{\phi_j}{\lambda_j}.$$

We have

$$\begin{aligned}(K_n f)(x) &= (Kf)(x) - \textstyle\sum_{j=1}^{n}\langle f,\frac{1}{\lambda_j}\psi_j\rangle\phi_j\\ &= (Kf)(x) - \textstyle\sum_{j=1}^{n}\langle f,K^*\phi_j\rangle\phi_j\\ &= (Kf)(x) - \textstyle\sum_{j=1}^{n}\langle Kf,\phi_j\rangle\phi_j.\end{aligned} \tag{5.63}$$

Since the system $\{\phi_n\}$ is an orthonormal system, we have

$$\begin{aligned}\langle K_n f,\phi_m\rangle &= \langle Kf - \sum_{j=1}^{n}\langle Kf,\phi_j\rangle\phi_j,\phi_m\rangle\\ &= \langle Kf,\phi_m\rangle - \langle Kf,\phi_m\rangle\\ &= 0\end{aligned}$$

for $n \le m$.
Therefore, if f is an eigenfunction of the operator K_n corresponding to an eigenvalue μ, then for $m \le n$ we have

$$\langle K_n f,\phi_m\rangle = \frac{1}{\mu}\langle f,\phi_m\rangle = 0,$$

and hence, using (5.63), we obtain

$$K_n f = Kf = \frac{1}{\mu} f.$$

This means that μ is also an eigenvalue of the operator K and the corresponding eigenfunction is f.

Accordingly, f must be written as a linear combination of eigenfunctions ϕ_j corresponding to the eigenvalue $\lambda_j = \mu$, namely

$$f = \sum_{\lambda_j = \mu} \beta_j \phi_j,$$

where $\beta_j = \langle f, \phi_j \rangle$, from which we derive that $\beta_j = 0$ for $j < n$. Hence we see that every eigenvalue μ of the operator K_n satisfies

$$|\mu| \geq |\lambda_n|.$$

Therefore

$$||K_n|| \leq \frac{1}{|\lambda_n|}$$

and then $||K_n|| \to 0$ as $n \to \infty$ because $|\lambda_n| \to \infty$ as $n \to \infty$. Thus we obtain that the uniform limit

$$\lim_{n \to \infty} (K_n f)(x),$$

which as was shown does exist, satisfies

$$|| \lim_{n \to \infty} K_n f|| = 0. \tag{5.64}$$

On the other hand, $(Kf)(x)$ is, together with $f(x)$, continuous, and

$$\phi_n(x) = \lambda_n \int_a^b K(x, \xi) \phi_n(\xi) \Delta\xi$$

is, together with $K(x, \xi)$, continuous. Hence, $(K_n f)(x)$ is continuous, so that its uniform limit $\lim_{n \to \infty} (K_n f)(x)$ is also continuous. From here and (5.64), we obtain that

$$\lim_{n \to \infty} (K_n f)(x) = 0.$$

The proof for $(K^* f)(x)$ is carried out in analogous manner.

5.4.3 *Application to Generalized Fredholm Integral Equation of the First Kind*

Here we consider the equation

$$u(x) = \int_a^b K(x, y)\phi(y)\Delta y \tag{5.65}$$

with a continuous kernel $K(x, y)$ on the rectangle $[a, b] \times [a, b]$, $u(x)$ being given continuous function on $[a, b]$.

Theorem 15 *If the Eq. (5.65) admits a continuous solution $\phi(x)$ for given $u(x)$, then $u(x)$ can be expanded in a Fourier series*

$$u(x) = \sum_{j=1}^{\infty} \beta_j \phi_j(x) \tag{5.66}$$

with respect to the orthonormal system of the eigenfunctions $\{\phi_j\}$ of $A = KK^*$, which converges uniformly and absolutely on the interval $[a, b]$.
Conversely, if $u(x)$ is of the form of (5.66), and if the series

$$\sum_{j=1}^{\infty} \beta_j \lambda_j \psi_j(x), \tag{5.67}$$

where $\psi_j(x)$ are the eigenfunctions of $B = K^*K$ so that $B^2\psi_j = \lambda_j^{-2}\psi_j$, converges uniformly and absolutely on the interval $a \le x \le b$, then Eq. (5.65) admits a solution $\phi(x)$ which is given by (5.67).

Proof The proof follows immediately from Theorem 14 for $u(x) = (K\phi)(x)$.

Remark 4 In general, Eq. (5.65) can not be solved. Furthermore, the uniqueness of the solution $\phi(x)$ of the Eq. (5.65), if any, is equivalent to the following property of the kernel: if a continuous function $v(x)$ satisfies

$$\int_a^b K(x, y)v(y)\Delta y = 0,$$

then $v(y) \equiv 0$. A kernel $K(x, y)$ which possesses such a property is called a closed kernel.

5.4.4 Positive Definite Kernels. Mercer's Expansion Theorem

Definition 18 A complex-valued continuous kernel $K(x, y)$ defined on $[a, b] \times [a, b]$ is called a positive definite kernel, when $K(x, y)$ satisfies

$$\int_a^b \int_a^b K(x, y) f(x) \overline{f(y)} \Delta x \Delta y = \langle Kf, f \rangle \geq 0 \tag{5.68}$$

for any continuous function $f(x)$ on $[a, b]$.

Example 18 Consider $A = KK^*$. Then

$$A(x, y) = \int_a^b K(x, t) K^*(t, y) \Delta t = \int_a^b K(x, y) \overline{K(y, t)} \Delta t.$$

Hence,

$$\begin{aligned}
&\int_a^b \int_a^b A(x, y) f(x) \overline{f(y)} \Delta y \Delta x = \int_a^b \int_a^b \left(\int_a^b K(x, t) \overline{K(y, t)} \Delta t \right) f(x) \overline{f(y)} \Delta y \Delta x \\
&= \int_a^b \left(\int_a^b K(x, t) f(x) \Delta x \right) \left(\int_a^b \overline{K(x, t) f(x) \Delta x} \right) \Delta t \\
&= \int_a^b \left(\int_a^b K(x, t) f(x) \Delta x \right) \overline{\left(\int_a^b K(x, t) f(x) \Delta x \right)} \Delta t \\
&= \int_a^b \left| \int_a^b K(x, y) f(x) \Delta x \right|^2 \Delta y \geq 0
\end{aligned}$$

for any continuous function $f(x)$ on $[a, b]$. Therefore $A(x, y)$ is a positive definite kernel.

Exercise 11 Prove that $B = K^* K$ is a positive definite kernel.

Theorem 16 *In order that $K(x, y)$ is positive definite, it is necessary and sufficient that, for any finite number of arbitrary points $\{x_j\}$, $a < x_1 < x_2 < \ldots < x_n < b$, and arbitrary complex numbers $\xi_1, \xi_2, \ldots, \xi_n$, the following relation*

$$\sum_{i,j=1}^{n} K(x_i, x_j) \xi_i \overline{\xi_j} \geq 0$$

holds.

Proof The sufficiency follows from the definition of the integral (5.68).

Now we will prove the necessity. Suppose that there exist $\{x_j\}$ and $\{\xi_j\}$ so that

$$\sum_{i,j=1}^{n} K(x_i, x_j) \xi_i \overline{\xi_j} \leq -\alpha < 0$$

for some positive α. If (a, b) contains only right-scattered points, the assertion is trivial. Otherwise, we divide the interval $[a, b]$ of subintervals every one of which contains only right-dense points. Without loss of generality we assume that (a, b) contains only right-dense points of $\mathscr{T}_0$. Let $a < c < b, \quad c \in \mathscr{T}_0$. We choose sufficiently small numbers ε and η such that

$$a < c - \varepsilon - \eta < c + \varepsilon + \eta < b.$$

For such c, ε and η, we define $\theta_{\varepsilon,\eta}(x, c)$ as follows.

$$\theta_{\varepsilon,\eta} = \begin{cases} 0 & for \quad a \le x \le c - \varepsilon - \eta, \quad c + \varepsilon + \eta \le x \le b \\ 1 & for \quad c - \eta \le x \le c + \eta \end{cases}$$

and in the rest of the interval $[a, b]$, the function $\theta_{\varepsilon,\eta}(x, c)$ is linear of x. Clearly, $\theta_{\varepsilon,\eta}(x, c)$ is a continuous function of x on the interval $[a, b]$.

Now we choose $\varepsilon > 0$ and $\eta > 0$ so small that no pair of the intervals

$$(-\varepsilon - \eta + x_i, x_i + \varepsilon + \eta)$$

have a common point and

$$a < -\varepsilon - \eta + x_i < x_i + \varepsilon + \eta < b.$$

For such ε and η, we define $\theta_{\varepsilon,\eta}(x, x_i)$ as in above and we set

$$\theta(x) = \sum_{i=1}^{n} \xi_i \theta_{\varepsilon,\eta}(x, x_i).$$

Set further

$$L_n(x, y) = \sum_{i,j=1}^{n} K(x_i + x, y_i + y)\xi_i\overline{\xi_j}.$$

Then, using the definition of the integral, we have

$$\langle K\theta, \theta \rangle = \int_{-\eta}^{\eta} \int_{-\eta}^{\eta} L_n(x, y)\Delta x \Delta y + \sum_{i,j=1}^{n} \int\int_{q_{ij}} K(x, y)\theta(x)\theta(y)\Delta x \Delta y,$$

where q_{ij} is the region between the square

$$x_i - \varepsilon - \eta \le x \le x_i + \varepsilon + \eta, \quad x_j - \varepsilon - \eta \le y \le x_j + \varepsilon + \eta$$

and the square

$$x_i - \eta \le x \le x_i + \eta, \quad x_j - \eta \le y \le x_j + \eta.$$

Furthermore, for any point (x, y) in the region q_{ij}, we have

$$|\theta(x)\overline{\theta}(y)| \le |\xi_i \xi_j|.$$

Hence,

$$\left|\sum_{i,j=1}^{n} \int\int_{q_{ij}} K(x, y)\theta(x)\theta(y)\Delta x \Delta y\right| \le 4M\varepsilon(2\eta + \varepsilon)\left(\sum_{i=1}^{n} |\xi_i|\right)^2,$$

where

$$M = \sup_{a \le x, y \le b} |K(x, y)|.$$

Using our assumption, we have $L_n(0, 0) = -\alpha < 0$, hence we can choose $\eta > 0$ so small that

$$L_n(x, y) < -\frac{1}{2}\alpha$$

whenever $|x| < \eta$ and $|y| < \eta$.

Then for such an η, we have

$$\int_{-\eta}^{\eta}\int_{-\eta}^{\eta} L_n(x, y)\Delta x \Delta y < -2\alpha\eta^2.$$

Hence, we get that

$$\langle K\theta, \theta\rangle < -2\alpha\eta^2 + 4M\varepsilon(2\eta + \varepsilon)\left(\sum_{i=1}^{n} |\xi_i|\right)^2.$$

Since

$$-2\alpha\eta^2 + 4M\varepsilon(2\eta + \varepsilon)\left(\sum_{i=1}^{n} |\xi_i|\right)^2 \longrightarrow_{\varepsilon \to 0} -2\alpha\eta^2,$$

we can choose η and ε small so that

$$\langle K\theta, \theta\rangle < 0$$

holds. This is a contradiction.

Theorem 17 *If a kernel $K(x, y)$ is positive definite, then $K(x, x) \ge 0$ for any $x \in [a, b]$.*

Proof By Theorem 16 for any point $x \in (a, b)$ and any complex number ξ we have

$$K(x, x)||\xi|^2 \ge 0,$$

whereupon $K(x, x) \ge 0$.

If a is right-dense and b is left-dense, using the continuity of $K(x, y)$, we conclude that

$$K(x, x) \geq 0$$

for any $x \in [a, b]$.

Let a be right-scattered. Then we take a continuous function $f(x)$, defined on $[a, b]$ such that $0 \leq f(x) \leq 1$ for any $x \in [a, \sigma(a)]$ and $f(x) \equiv 0$ for any $x \in [\sigma(a), b]$. Hence,

$$\begin{aligned}\int_a^b \int_a^b K(x, y) f(x) \overline{f(y)} \Delta y \Delta x &= \int_a^{\sigma(a)} \int_a^{\sigma(a)} K(x, y) f(x) \overline{f(y)} \Delta x \Delta y \\ &= K(a, a) f(a) \overline{f(a)} (\mu(a))^2 \\ &= K(a, a) |f(a)|^2 (\mu(a)|^2 \geq 0,\end{aligned}$$

from where $K(a, a) \geq 0$.

Similarly, $K(b, b) \geq 0$ when b is left-scattered.

Theorem 18 *If a kernel $K(x, y)$ is positive definite, then*

1. *it is Hermitian symmetric on $[a, b]$ whenever a is right-dense and b is left-dense,*
2. *it is Hermitian symmetric on (a, b) whenever a is right-scattered and b is left-scattered,*
3. *it is Hermitian symmetric on $[a, b)$ whenever a is right-dense and b is left-scattered,*
4. *it is Hermitian symmetric on $(a, b]$ whenever a is right-scattered and b is left-dense.*

Proof By Theorem 16, for any points $x_0, x_1 \in (a, b)$ and two complex numbers ξ_0 and ξ_1, we have

$$K(x_0, x_0)\xi_0\overline{\xi_0} + K(x_1, x_0)\overline{\xi_0}\xi_1 + K(x_0, x_1)\xi_0\overline{\xi_1} + K(x_1, x_1)\xi_1\overline{\xi_1} \geq 0.$$

Because $K(x_0, x_0)|\xi_0|^2 + K(x_1, x_1)|\xi_1|^2 \geq 0$, we conclude that

$$K(x_0, x_1)\xi_0\overline{\xi_1} + K(x_1, x_0)\overline{\xi_0}\xi_1 \in \mathscr{R}.$$

Then, setting $\xi_0 = \xi_1 = 1$, we obtain

$$K(x_0, x_1) + K(x_1, x_0) \in \mathscr{R}.$$

By setting $\xi_0 = 1$ and $\xi_1 = i$, we get

$$-K(x_0, x_1)i + K(x_1, x_0)i \in \mathscr{R}.$$

In this way we get the system

$$\begin{cases} K(x_0, x_1) + K(x_1, x_0) \in \mathscr{R} \\ -K(x_0, x_1)i + K(x_1, x_0)i \in \mathscr{R}. \end{cases}$$

Therefore

$$K(x_0, x_1) = \overline{K(x_1, x_0)}.$$

Therefore $K(x, y)$ is Hermitian symmetric on (a, b). Also,

1. if a is right-dense and b is left-dense, using the continuity of $K(x, y)$, we conclude that $K(x, y)$ is Hermitian symmetric on $[a, b]$.
2. if a is right-dense and b is left-scattered, using the continuity of $K(x, y)$, we conclude that $K(x, y)$ is Hermitian symmetric on $[a, b)$.
3. if a is right-scattered and b is left-dense, using the continuity of $K(x, y)$, we conclude that $K(x, y)$ is Hermitian symmetric on $(a, b]$.

Theorem 19 (Mercer Expansion Theorem) *A positive definite and Hermitian symmetric kernel $K(x, y)$ on $[a, b] \times [a, b]$ can be expanded in a series*

$$K(x, y) = \sum_{j=1}^{\infty} \lambda_j^{-1} \phi_j(x) \overline{\phi_j(y)}$$

which converges absolutely and uniformly on the domain $a \leq x, y \leq b$.

Proof Let $f \in \mathscr{C}([a, b])$. The Fourier expansion of $(Kf)(x)$ is given by

$$\sum_{j=1}^{\infty} \langle Kf, \phi_j \rangle \phi_j(x) = \sum_{j=1}^{\infty} \langle f, K\phi_j \rangle \phi_j(x)$$
$$= \sum_{j=1}^{\infty} \lambda_j^{-1} \langle f, \phi_j \rangle \phi_j(x).$$

Let

$$K_n(x, y) = K(x, y) - \sum_{j=1}^{n} \lambda_j^{-1} \phi_j(x) \overline{\phi_j(y)}.$$

Then for each $n \in \mathscr{N}$ we have

$$\begin{aligned}
(K_n f)(x) &= \int_a^b K_n(x,y) f(y) \Delta y \\
&= \int_a^b \left(K(x,y) - \sum_{j=1}^{n} \lambda_j^{-1} \phi_j(x) \overline{\phi_j(y)} \right) f(y) \Delta y \\
&= \int_a^b K(x,y) f(y) \Delta y - \sum_{j=1}^{n} \lambda_j^{-1} \phi_j(x) \int_a^b f(y) \overline{\phi_j(y)} \Delta y \\
&= (Kf)(x) - \sum_{j=1}^{n} \lambda_j^{-1} \phi_j(x) \langle f, \phi_j \rangle \\
&= \sum_{j=1}^{\infty} \lambda_j^{-1} \phi_j(x) \langle f, \phi_j \rangle - \sum_{j=1}^{n} \lambda_j^{-1} \phi_j(x) \langle f, \phi_j \rangle \\
&= \sum_{j=n+1}^{\infty} \lambda_j^{-1} \langle f, \phi_j \rangle \phi_j(x).
\end{aligned}$$

Hence,

$$\begin{aligned}
\langle K_n f, f \rangle &= \langle \sum_{j=n+1}^{\infty} \lambda_j^{-1} \langle f, \phi_j \rangle \phi_j, f \rangle \\
&= \sum_{j=n+1}^{\infty} \lambda_j^{-1} \langle f, \phi_j \rangle \langle \phi_j, f \rangle \\
&= \sum_{j=n+1}^{\infty} \lambda_j^{-1} \langle f, \phi_j \rangle \overline{\langle f, \phi_j \rangle} \\
&= \sum_{j=n+1}^{\infty} \lambda_j^{-1} |\langle f, \phi_j \rangle|^2 \geq 0.
\end{aligned}$$

Because f was arbitrarily chosen continuous function on $[a, b]$, we conclude that $K_n(x, y)$ is positive definite on $[a, b]$.

By Theorem 17 we have that

$$K_n(x,x) = K(x,x) - \sum_{j=1}^{n} \lambda_j^{-1} \phi_j(x) \overline{\phi_j(x)} \geq 0.$$

Hence, the series

$$\sum_{j=1}^{\infty} \lambda_j^{-1} |\phi_j(x)|^2$$

converges and

$$\sum_{j=1}^{\infty} \lambda_j^{-1} |\phi_j(x)|^2 \leq K(x, x). \tag{5.69}$$

Let

$$S(x, y) = \sum_{j=1}^{\infty} \lambda_j^{-1} \phi_j(x) \overline{\phi_j(y)}.$$

Then, using the Cauchy-Schwartz and Bessel inequalities, we obtain

$$\begin{aligned}\left(\sum_{j=n}^{m} |\lambda_j^{-1} \phi_j(x) \overline{\phi_j(y)}| \right)^2 &\leq \sum_{j=n}^{m} \lambda_j^{-1} |\phi_j(x)|^2 \sum_{j=n}^{m} \lambda_j^{-1} |\phi_j(y)|^2 \\ &\leq \sum_{j=n}^{m} \lambda_j^{-1} |\phi_j(x)|^2 K(y, y), \quad n < m.\end{aligned}$$

From here, the series $S(x, y)$ converges absolutely and uniformly with respect to y for x fixed and also with respect to x for y fixed.

Now we set

$$R(x, y) = K(x, y) - S(x, y).$$

Then for any continuous function $f(x)$ on $[a, b]$, we have

$$\int_a^b K(x, y) f(y) \Delta y = \int_a^b S(x, y) f(y) \Delta y + \int_a^b R(x, y) f(y) \Delta y \tag{5.70}$$

or

$$Kf(x) = Sf(x) + Rf(x).$$

Accordingly to the Schmidt expansion theorem, we have that $(Kf)(x)$ can be expanded in a series

$$\begin{aligned}&\textstyle\sum_{j=1}^{\infty} \langle Kf, \phi_j \rangle \phi_j(x) = \sum_{j=1}^{\infty} \langle f, K\phi_j \rangle \phi_j(x) \\ &= \textstyle\sum_{j=1}^{\infty} \lambda_j^{-1} \langle f, \phi_j \rangle \phi_j(x).\end{aligned} \tag{5.71}$$

On the other hand, since the series $S(x, y)$ converges uniformly with respect to y, we have

$$(Sf)(x) = \sum_{j=1}^{\infty} \lambda_j^{-1} \langle f, \phi_j \rangle \phi_j(x). \tag{5.72}$$

Hence, using (5.70), (5.71) and (5.72), we get

$$\sum_{j=1}^{\infty} \lambda_j^{-1} \langle f, \phi_j \rangle \phi_j(x) = \sum_{j=1}^{\infty} \lambda_j^{-1} \langle f, \phi_j \rangle \phi_j(x) + (Rf)(x),$$

whereupon $(Rf)(x) = 0$.

Since the series $S(x, y)$ converges uniformly with respect to y for x fixed, then $R(x, y)$ is a continuous function of y for any fixed x. Hence, using that $(Rf)(x) = 0$ and by setting $f(y) = \overline{R(x, y)}$, we obtain that $R(x, y)$ is identically zero as a function of y for any fixed x. Therefore $R(x, y) \equiv 0$. Thus we obtain that

$$K(x, y) = \sum_{j=1}^{\infty} \lambda_j^{-1} \phi_j(x) \overline{\phi_j(y)}.$$

We will prove that this series converges uniformly. We note that $K(x, x) = \sum_{j=1}^{\infty} \lambda_j^{-1} |\phi_j(x)|^2$.

Note that for any $\varepsilon > 0$ and any given x_0, there is a $n = n(x_0)$ such that

$$\varepsilon > K(x_0, x_0) - \sum_{j=1}^{n} \lambda_j^{-1} |\phi_j(x_0)|^2 \geq 0.$$

Furthermore, since $K(x, x)$ and $\sum_{j=1}^{n} \lambda_j^{-1} |\phi_j(x)|^2$ are both continuous, we can find an open set $U(x_0)$ containing the point x_0 such that

$$2\varepsilon \geq K(x, x) - \sum_{j=1}^{n(x_0)} \lambda_j^{-1} |\phi_j(x)|^2 \geq 0 \tag{5.73}$$

whenever $x \in U(x_0)$.

On the other hand, the interval $[a, b]$ can be covered completely by a finite number of open sets $U(x_1), U(x_2), \ldots, U(x_k)$. Therefore every point $x \in [a, b]$ is contained in some $U(x_j)$. Now we set

$$n_0 = \max_{1 \leq j \leq k} n(x_j).$$

Hence and (5.73), we obtain that for every $x \in [a, b]$

$$2\varepsilon \geq K(x, x) - \sum_{j=1}^{n} \lambda_j^{-1} |\phi_j(x)|^2 \geq 0$$

whenever $n \geq n_0$.

Therefore the series $\sum_{j=1}^{\infty} \lambda_j^{-1} |\phi_j(x)|^2$ converges absolutely to $K(x, x)$ and uniformly in x. By making use of the Cauchy-Schwartz inequality, we have

$$\left(\sum_{j=n}^{m} \lambda_j^{-1} |\phi_j(x) \overline{\phi_j(y)}| \right)^2 \leq \sum_{j=n}^{m} \lambda_j^{-1} |\phi_j(x)| \sum_{j=n}^{m} \lambda_j^{-1} |\phi_j(x)|^2, \quad n < m.$$

Therefore $\sum_{j=1}^{\infty} \lambda_j^{-1} \phi_j(x) \overline{\phi_j(y)}$ converges absolutely and uniformly for x and y.

Remark 5 For any continuous kernel $K(x, y)$, the kernels $A(x, y)$ and $B(x, y)$,

$$A(x, y) = \int_a^b K(x, t) K^*(t, y) \Delta t, \quad B(x, y) = \int_a^b K^*(x, t) K(t, y) \Delta t$$

are both positive definite and Hermitian symmetric. Hence and Theorem 19, the kernels $A(x, y)$ and $B(x, y)$ can be expanded in absolutely and uniformly convergent series

$$A(x, y) = \sum_{j=1}^{\infty} \lambda_j^{-2} \phi_j(x) \overline{\phi_j(y)},$$
$$B(x, y) = \sum_{j=1}^{\infty} \lambda_j^{-2} \psi_j(x) \overline{\psi_j(y)},$$

with respect to the eigenfunctions $\{\phi_j\}$ and $\{\psi_j\}$.

As a consequence we have the following theorem.

Theorem 20 *The necessary and sufficient conditions that A and B have the same eigenvalues, together with the corresponding eigenfunctions, is that $A = B$.*

Definition 19 A kernel $K(x, y)$ defined on $[a, b] \times [a.b]$ will be called a normal kernel if

$$KK^* = K^*K.$$

Definition 20 A kernel $K(x, y)$ defined on $[a, b] \times [a, b]$ will be called Hermitian skew-symmetric kernel if

$$K(x, y) = -\overline{K(y, x)}.$$

Evidently every Hermitian symmetric kernels and Hermitian skew-symmetric kernels are normal kernels.

Theorem 21 *Let $K(x, y)$ is continuous and Hermitian skew-symmetric on $[a, b] \times [a, b]$. Let also $K(x, y) \not\equiv 0$ on $[a, b] \times [a, b]$. Then the eigenvalues of K are all purely imaginary numbers.*

Proof Let $L(x, y) = iK(x, y)$.
Then

$$\begin{aligned} K^*(x, y) &= \overline{iK(y, x)} \\ &= -i\overline{K(y, x)} \\ &= iK(x, y) \\ &= L(x, y). \end{aligned}$$

Therefore $L(x, y)$ is Hermitian symmetric and continuous on $[a, b] \times [a, b]$. Hence and Theorem 12 we have that all eigenvalues of L are real. Therefore all eigenvalues of K are purely imaginary numbers.

5.5 Advanced Practical Exercises

Problem 1 Using ADM, find $\phi_0(x)$, $\phi_1(x)$, $\phi_2(x)$, $\phi_3(x)$ and $\phi_4(x)$ for the equation

$$\phi(x) = x - 2\int_0^3 y\phi(y)\Delta y, \quad \mathscr{T} = 3^{\mathscr{N}_0} \cup \{0\}.$$

Answer

$$\begin{aligned} &\phi_0(x) = x, \quad \phi_1(x) = -\tfrac{54}{13}, \\ &\phi_2(x) = \tfrac{108}{13}\left(\tfrac{9}{4}\right), \quad \phi_3(x) = (-2)\tfrac{108}{13}\left(\tfrac{9}{4}\right)^2, \quad \phi_4(x) = (-2)^2\tfrac{108}{13}\left(\tfrac{9}{4}\right)^3. \end{aligned}$$

Problem 2 Let $\mathscr{T} = 2^{\mathscr{N}_0} \cup \{0\}$. Consider the equation

$$\phi(x) = 1 + x^3 + \int_0^2 \phi(y)\Delta y.$$

Set $u_1(x) = x^3$, $u_2(x) = 1$. Using MDM, find $\phi_0(x)$, $\phi_1(x)$, $\phi_2(x)$, $\phi_3(x)$, $\phi_4(x)$.

Answer

$$\phi_0(x) = x^3, \quad \phi_1(x) = \tfrac{31}{15}, \quad \phi_2(x) = \tfrac{62}{15}, \quad \phi_3(x) = \tfrac{124}{15}, \quad \phi_4(x) = \tfrac{248}{15}.$$

Problem 3 Let $\mathscr{T} = 3^{\mathscr{N}_0} \cup \{0\}$. Using the noise terms phenomenon, find a solution of the following generalized Fredholm integral equation of the second kind.

$$\phi(x) = -\frac{79}{2}x + 2x\int_0^9 \phi(y)\Delta y.$$

Answer $\phi(x) = x$.

Problem 4 Let $\mathscr{T} = 2^{\mathscr{N}_0} \cup \{0\}$. Using DCM, find a solution to the equation

$$\phi(x) = x^2 - \frac{8}{7}x - \frac{16}{15} + \int_0^2 (x+y)\phi(y)\Delta y.$$

Answer $\phi(x) = x^2$.

Problem 5 Using SAM, find $\phi_0(x)$, $\phi_1(x)$, $\phi_2(x)$, $\phi_3(x)$ and $\phi_4(x)$ for the following equations.

1. $\phi(x) = 1 + 2x + x^2 + x\int_0^2 y^2\phi(y)\Delta y, \quad \mathscr{T} = 3\mathscr{Z},$
2. $\phi(x) = 2x - 3 + 2\int_0^4 y\phi(y)\Delta y, \quad \mathscr{T} = 2\mathscr{Z},$
3. $\phi(x) = 1 + \int_0^4 y\phi(y)\Delta y, \quad \mathscr{T} = 4\mathscr{Z},$
4. $\phi(x) = x + x^2 - x^3 + \int_0^4 y^2\phi(y)\Delta y, \quad \mathscr{T} = 2^{\mathscr{N}_0} \cup \{0\},$
5. $\phi(x) = 1 + 2x + 3x^3 + (x-1)\int_0^8 y\phi(y)\Delta y, \quad \mathscr{T} = 2^{\mathscr{N}_0} \cup \{0\},$
6. $\phi(x) = 1 + \int_0^{16} y^2\phi(y)\Delta y, \quad \mathscr{T} = \mathscr{N}_0^2,$
7. $\phi(x) = 1 + x + \int_{-4}^4 (x+y)\phi(y)\Delta y, \quad \mathscr{T} = \mathscr{Z},$
8. $\phi(x) = -2x + \int_0^4 y\phi(y)\Delta y, \quad \mathscr{T} = \mathscr{Z},$
9. $\phi(x) = -1 + 2x + x^2 + \int_0^9 \phi(y)\Delta y, \quad \mathscr{T} = 3^{\mathscr{N}_0} \cup \{0\}.$

Problem 6 Find a solution to the following equation

$$\phi(x) = \lambda x^2 \int_0^2 \phi(y)\Delta y, \quad \mathscr{T} = 2^{\mathscr{N}_0} \cup \{0\}.$$

Answer $\lambda = \frac{7}{8}, \quad \phi(x) = \frac{7}{8}\alpha x^2$, $\alpha = \text{const}$.

Chapter 6
Hilbert-Schmidt Theory of Generalized Integral Equations with Symmetric Kernels

Assume that $K(x, y)$ is continuous and Hermitian symmetric on $[a, b] \times [a, b]$. By Theorem 10 in Chap. 5, we have that either $||K||^{-1}$ or $-||K||^{-1}$ is an eigenvalue of the operator K. By Theorem 12 in Chap. 5, all eigenvalues of the operator K are real and two eigenfunctions corresponding to different eigenvalues are orthogonal.

6.1 Schmidt's Orthogonalization Process

Theorem 1 *Let $\phi_1(x), \phi_2(x), \ldots, \phi_n(x), \ldots,$ be linearly independent. Let also,*

$$\psi_1(x) = \frac{\phi_1(x)}{||\phi_1||},$$

$$\psi_2(x) = \frac{\phi_2(x) - \langle \phi_2, \psi_1 \rangle \psi_1(x)}{||\phi_2 - \langle \phi_2, \psi_1 \rangle \psi_1||},$$

$$\ldots$$

$$\psi_n(x) = \frac{\phi_n(x) - \sum_{j=1}^{n-1} \langle \phi_n, \psi_j \rangle \psi_j(x)}{\left|\left| \phi_n - \sum_{j=1}^{n-1} \langle \phi_n, \psi_j \rangle \psi_j \right|\right|},$$

$$\ldots .$$

Then $\psi_1(x), \psi_2(x), \ldots, \psi_n(x), \ldots,$ satisfy the orthonormality relations.

Proof Assume that

$$\phi_2(x) - \langle \phi_2, \psi_1 \rangle \psi_1(x) \equiv 0 \quad on \quad [a, b].$$

S.G. Georgiev, *Integral Equations on Time Scales*, Atlantis Studies in Dynamical Systems 5, DOI 10.2991/978-94-6239-228-1_6

Then, using the definition of ψ_1,

$$\phi_2(x) = \langle \phi_2, \psi_1 \rangle \psi_1(x)$$

$$= \langle \phi_2, \psi_1 \rangle \frac{\phi_1(x)}{||\phi_1||},$$

i.e., $\phi_1(x)$ and $\phi_2(x)$ are linearly dependent, which is a contradiction. Therefore

$$\phi_2(x) - \langle \phi_2, \psi_1 \rangle \psi_1(x) \not\equiv 0 \quad on \quad [a, b]$$

and $\psi_2(x)$ is well-defined.
Since ψ_2 is a linear combination of ϕ_2 and ψ_1, then ϕ_3, ψ_1, ψ_2 are linearly independent. Hence,

$$\phi_3(x) - \langle \phi_3, \psi_1 \rangle \psi_1(x) - \langle \phi_3, \psi_2 \rangle \psi_2(x) \not\equiv 0 \quad on \quad [a, b],$$

so we can divide it by its norm and $\psi_3(x)$ is well-defined.
Suppose that $\psi_n(x)$ is well-defined.
Since $\psi_n(x)$ is a linear combination of $\phi_n(x)$, $\psi_1(x)$, $\psi_2(x)$, $\ldots$, $\psi_{n-1}(x)$, then $\phi_{n+1}(x)$, $\psi_1(x)$, $\ldots$, $\psi_n(x)$ are linearly independent. From here,

$$\phi_{n+1}(x) - \sum_{j=1}^{n} \langle \phi_{n+1}, \psi_j \rangle \psi_j(x) \not\equiv 0 \quad on \quad [a, b]$$

and we can divide it by its norm. Consequently $\psi_{n+1}(x)$ is well-defined.
In this way we can define

$$\psi_1(x), \quad \psi_2(x), \quad \ldots, \quad \psi_n(x), \ldots,$$

successively. Clearly, $\psi_n(x)$ is normalized, i.e., $||\psi_n|| = 1$.
Now we will prove that $\{\psi_n\}$ satisfies the orthonormality relations.
By the definition of $\psi_2(x)$ we have

$$\langle \psi_2, \psi_1 \rangle = \langle \phi_2 - \langle \phi_2, \psi_1 \rangle \psi_1, \psi_1 \rangle$$

$$= \langle \phi_2, \psi_1 \rangle - \langle \phi_2, \psi_1 \rangle \langle \psi_1, \psi_1 \rangle$$

$$= \langle \phi_2, \psi_1 \rangle - \langle \phi_2, \psi_1 \rangle$$

$$= 0.$$

Thus, by

$$\begin{aligned}\langle\psi_3,\psi_1\rangle &= \langle\phi_3-\langle\phi_3,\psi_1\rangle\psi_1-\langle\phi_3,\psi_2\rangle\psi_2,\psi_1\rangle\\ &= \langle\phi_3,\psi_1\rangle-\langle\langle\phi_3,\psi_1\rangle\psi_1,\psi_1\rangle-\langle\langle\phi_3,\psi_2\rangle\psi_2,\psi_1\rangle\\ &= \langle\phi_3,\psi_1\rangle-\langle\phi_3,\psi_1\rangle\langle\psi_1,\psi_1\rangle-\langle\phi_3,\psi_2\rangle\langle\psi_2,\psi_1\rangle\\ &= \langle\phi_3,\psi_1\rangle-\langle\phi_3,\psi_1\rangle\\ &= 0.\end{aligned}$$

Similarly, we see, starting with $\langle\psi_3,\psi_2\rangle=0$, that

$$\langle\psi_4,\psi_2\rangle=0,\quad \langle\psi_5,\psi_2\rangle=0,\quad \ldots,\quad \langle\phi_n,\psi_2\rangle=0,\quad \ldots.$$

Repeating this procedure, we finally obtain that $\{\psi_j\}$ satisfies the orthonormality relations.

Corollary 1 *Let $\{\phi_j\}$ and $\{\psi_j\}$ be as in Theorem 1. Then every ψ_n can be written as a linear combination of $\{\phi_j\}$, and moreover, every ϕ_n can be written as a linear combination of $\{\psi_j\}$.*

Theorem 2 *The operator K has at most denumerable many eigenvalues. The set of all eigenvalues has no limiting point except $\pm\infty$.*

Proof By Theorem 12 in Chap. 5, we have that the eigenfunctions corresponding to different eigenvalues are orthogonal.
Assume that there exists a finite limiting point of the eigenvalues of the operator K. Then we have an orthonormal system $\{\phi_j\}$ for which

$$K\phi_j=\lambda_j^{-1}\phi_j,\quad j=1,2,3,\ldots,$$

$$\lambda_j^{-1}\longrightarrow\lambda^{-1}\quad as\quad j\longrightarrow\infty.$$

Since

$$K(\phi_j-\phi_k)=K\phi_j-K\phi_k=\lambda_j^{-1}\phi_j-\lambda_k^{-1}\phi_k,\quad (\phi_j,\phi_k)=\delta_{jk},$$

we obtain that, for sufficiently large j and k, $j\neq k$,

$$||K(\phi_j - \phi_k)||^2 = ||K\phi_j - K\phi_k||^2$$

$$= \langle K\phi_j - K\phi_k, K\phi_j - K\phi_k\rangle$$

$$= \langle \lambda_j^{-1}\phi_j - \lambda_k^{-1}\phi_k, \lambda_j^{-1}\phi_j - \lambda_k^{-1}\phi_k\rangle$$

$$= \langle \lambda_j^{-1}\phi_j, \lambda_j^{-1}\phi_j\rangle - \langle \lambda_j^{-1}\phi_j, \lambda_k^{-1}\phi_k\rangle - \langle \lambda_k^{-1}\phi_k, \lambda_j^{-1}\phi_j\rangle + \langle \lambda_k^{-1}\phi_k, \lambda_k^{-1}\phi_k\rangle$$

$$= \lambda_j^{-2}\langle \phi_j, \phi_j\rangle - \lambda_j^{-1}\lambda_k^{-1}\langle \phi_j, \phi_k\rangle - \lambda_j^{-1}\lambda_k^{-1}\langle \phi_k, \phi_j\rangle + \lambda_k^{-2}\langle \phi_k, \phi_k\rangle$$

$$= \lambda_j^{-2} + \lambda_k^{-2}$$

$$\geq \lambda^{-2},$$

and

$$||\phi_j - \phi_k||^2 = \langle \phi_j - \phi_k, \phi_j - \phi_k\rangle$$

$$= \langle \phi_j, \phi_k\rangle - \langle \phi_k, \phi_j\rangle - \langle \phi_j, \phi_k\rangle + \langle \phi_k, \phi_k\rangle$$

$$= 2,$$

which is a contradiction.
Therefore the number of eigenvalues λ satisfying $n < |\lambda| \leq n+1$ is finite, $n = 0, 1, 2, \ldots$.
Consequently the set of all eigenvalues consists of at most denumerable many points and has no limiting points except for $\pm\infty$, which completes the proof.

Theorem 3 *The multiplicity of every eigenvalue λ of the operator K is finite.*

Proof We note that if

$$K\phi = \lambda^{-1}\phi \quad and \quad K\psi = \lambda^{-1}\psi,$$

then

$$K(\phi + \psi) = \lambda^{-1}(\phi + \psi).$$

Therefore any linear combination of eigenfunctions corresponding to the same eigenvalue is either an eigenfunction corresponding to the same eigenvalue or identically zero. Consequently, for any eigenvalue λ, there exists a set of eigenfunctions, corresponding to the eigenvalue λ, such that they are linearly independent and every eigenfunction corresponding to this eigenvalue is their linear combination. Suppose a set of eigenfunctions contains denumerably many functions ϕ_k. By Schmidt's orthogonalization process, there exists an orthonormal system $\{\psi_k\}$, satisfying the conclusion of Corollary 1. Since

$$K\psi_k = \lambda^{-1}\psi_k,$$

we have

$$K(\psi_j - \psi_k) = K\psi_j - K\psi_k = \lambda^{-1}\psi_j - \lambda^{-1}\psi_k,$$

$$\langle\psi_j, \psi_k\rangle = \delta_{jk}.$$

Hence, if $j \neq k$,

$$||K(\psi_j - \psi_k)||^2 = ||K\psi_j - K\psi_k||^2$$

$$= \langle K\psi_j - K\psi_k, K\psi_j - K\psi_k\rangle$$

$$= \langle\lambda^{-1}\psi_j - \lambda^{-1}\psi_k, \lambda^{-1}\psi_j - \lambda^{-1}\psi_k\rangle$$

$$= \langle\lambda^{-1}\psi_j, \lambda^{-1}\psi_k\rangle - \langle\lambda^{-1}\psi_k, \lambda^{-1}\psi_j\rangle - \langle\lambda^{-1}\psi_j, \lambda^{-1}\psi_k\rangle + \langle\lambda^{-1}\psi_k, \lambda^{-1}\psi_k\rangle$$

$$= \lambda^{-2}\langle\psi_j, \psi_j\rangle - \lambda^{-2}\langle\psi_j, \psi_k\rangle - \lambda^{-2}\langle\psi_k, \psi_j\rangle + \lambda^{-2}\langle\psi_k, \psi_k\rangle$$

$$= 2\lambda^{-2}$$

while

$$||\psi_j - \psi_k||^2 = \langle\psi_j - \psi_k, \psi_j - \psi_k\rangle$$

$$= \langle\psi_j, \psi_j\rangle - \langle\psi_j, \psi_k\rangle - \langle\psi_k, \psi_j\rangle + \langle\psi_k, \psi_k\rangle$$

$$= 2,$$

which is a contradiction.

Remark 1 By Theorems 2 and 3, if the operator K has infinitely many eigenvalues, then we can write its eigenvalues and eigenfunctions as

$$|\lambda_1| \leq |\lambda_2| \leq \ldots, \quad \lim_{j\to\infty} |\lambda_j| = \infty,$$

$$K\phi_j = \lambda_j^{-1}\phi_j, \quad j = 1, 2, \ldots, \quad \langle\phi_j, \phi_k\rangle = \delta_{jk},$$

so that every eigenvalue λ of K is equal to some λ_j and every eigenfunction corresponding to the eigenvalue λ is written as a linear combination of finitely many eigenfunction ϕ_j corresponding to the eigenvalue $\lambda_j = \lambda$.

Definition 1 A Hermitian symmetric kernel $K(x, y)$ which is of the form

$$K(x, y) = \sum_{i=1}^{n} f_i(x)\overline{g_i(y)}$$

is called a degenerated kernel.

Theorem 4 *A degenerated kernel has only a finite number of eigenvalues.*

Proof Let

$$\phi(x) = \lambda \int_a^b K(x, y)\phi(y)\Delta y.$$

Then

$$\phi(x) = \lambda \int_a^b \sum_{i=1}^{n} f_i(x)\overline{g_i(y)}\phi(y)\Delta y$$
$$= \lambda \sum_{i=1}^{n} f_i(x) \int_a^b \overline{g_i(y)}\phi(y)\Delta y.$$

Let

$$c_i = \int_a^b \overline{g_i(y)}\phi(y)\Delta y, \quad i = 1, 2, \ldots, n. \tag{6.1}$$

Then

$$\phi(x) = \lambda \sum_{i=1}^{n} c_i f_i(x). \tag{6.2}$$

We substitute (6.2) in (6.1) and we get

$$c_i = \lambda \int_a^b \overline{g_i(y)} \sum_{j=1}^{n} c_j f_j(y)\Delta y, \quad i = 1, 2, \ldots, n,$$

or

$$\left(1 - \lambda \int_a^b f_i(y)\overline{g_i(y)}\Delta y\right)c_i - \lambda \sum_{j=1, j\neq i}^{n} c_j \int_a^b \overline{g_i(y)} f_j(y)\Delta y = 0, \quad i = 1, 2, \ldots, n.$$

The last system has a solution $c_1, c_2, \ldots, c_n$ if its determinant is equal to zero. Therefore λ must be a root of an algebraic equation of degree n. Consequently the number of all eigenvalues in question is at most n.

6.2 Approximations of Eigenvalues

Let ϕ_n be an orthonormal system and let

$$f(x) = \sum_{n=1}^{\infty} c_n \phi_n(x), \quad g(x) = \sum_{n=1}^{\infty} d_n \phi_n(x)$$

be the Fourier expansions which converge absolutely and uniformly on the interval $[a, b]$. Then

$$\begin{aligned}
\langle f, g \rangle &= \langle \sum_{n=1}^{\infty} c_n \phi_n, \sum_{m=1}^{\infty} d_m \phi_m \rangle \\
&= \sum_{n=1}^{\infty} \langle c_n \phi_n, \sum_{m=1}^{\infty} d_m \phi_m \rangle \\
&= \sum_{n=1}^{\infty} c_n \langle \phi_n, \sum_{m=1}^{\infty} d_m \phi_m \rangle \\
&= \sum_{n=1}^{\infty} c_n \sum_{m=1}^{\infty} \langle \phi_m, d_m \phi_m \rangle \\
&= \sum_{n=1}^{\infty} \sum_{m=1}^{\infty} c_n \overline{d_m} \langle \phi_n, \phi_m \rangle \\
&= \sum_{n=1}^{\infty} c_n \overline{d_n}.
\end{aligned}$$

In particular, if $f(x) \equiv g(x)$ on $[a, b]$, then

$$||f||^2 = \langle f, f \rangle = \sum_{n=1}^{\infty} c_n \overline{c_n} = \sum_{n=1}^{\infty} |c_n|^2. \tag{6.3}$$

Definition 2 The relation (6.3) is called the Parseval completeness relation.

We note that

$$\begin{aligned}
\langle f, \sum_{i=1}^{n} c_i \phi_i \rangle &= \langle \sum_{j=1}^{\infty} c_j \phi_j, \sum_{i=1}^{n} c_i \phi_i \rangle \\
&= \sum_{j=1}^{\infty} c_j \langle \phi_j, \sum_{i=1}^{n} c_i \phi_i \rangle
\end{aligned}$$

$$= \sum_{j=1}^{\infty} \sum_{i=1}^{n} c_j \langle \phi_j, c_i \phi_i \rangle$$

$$= \sum_{j=1}^{\infty} \sum_{i=1}^{n} c_j \overline{c_i} \langle \phi_j, \phi_i \rangle$$

$$= \sum_{i=1}^{n} |c_i|^2$$

for every $n \in \mathcal{N}$.

Theorem 5 *We have*

$$\lim_{n\to\infty} \left|\left| f - \sum_{j=1}^{n} c_j \phi_j \right|\right|^2 = 0,$$

that is,

$$\lim_{n\to\infty} \int_a^b \left| f(x) - \sum_{j=1}^{n} c_j \phi_j(x) \right|^2 \Delta x = 0.$$

Proof Using the Parseval completeness relation, we get

$$\left|\left| f - \sum_{j=1}^{n} c_j \phi_j \right|\right|^2 = \langle f - \sum_{j=1}^{n} c_j \phi_j, f - \sum_{i=1}^{n} c_i \phi_i \rangle$$

$$= \langle f, f - \sum_{j=1}^{n} c_j \phi_j \rangle - \langle \sum_{j=1}^{n} c_j \phi_j, f - \sum_{j=1}^{n} c_j \phi_j \rangle$$

$$= \langle f, f \rangle - \langle f, \sum_{j=1}^{n} c_j \phi_j \rangle - \langle \sum_{j=1}^{n} c_j \phi_j, f \rangle + \langle \sum_{j=1}^{n} c_j \phi_j, \sum_{i=1}^{n} c_i \phi_i \rangle$$

$$= ||f||^2 - \sum_{j=1}^{n} |c_j|^2 - \sum_{j=1}^{n} |c_j|^2 + \sum_{j=1}^{n} |c_j|^2$$

$$= ||f||^2 - \sum_{j=1}^{n} |c_j|^2 \longrightarrow_{n\to\infty} 0,$$

which completes the proof.

Theorem 6 *Suppose that every eigenvalue of the operator K is positive. Let $g_0(x) \not\equiv 0$, $g_1(x) = (Kg_0)(x) \not\equiv 0$,*

$$g_2 = Kg_1 = K^2 g_0, \quad g_3 = Kg_2 = K^2 g_1 = K^3 g_0, \quad \dots,$$

and

$$\beta_n = \frac{||g_n||}{||g_{n+1}||}, \quad \alpha_n = \frac{\langle g_{n+1}, g_n \rangle}{||g_{n+1}||^2}.$$

Then

$$0 < \alpha_n \leq \beta_n$$

and $\{\beta_n\}_{n=1}^{\infty}$ *is monotone decreasing. Further, there exists an eigenvalue* λ *of* K *such that the sequences* $\{\alpha_n\}_{n=1}^{\infty}$ *and* $\{\beta_n\}_{n=1}^{\infty}$ *tend to* λ *as* $n \to \infty$ *and* $\lambda_n \geq \lambda$.

Proof Let $K\phi_j = \lambda_j^{-1}\phi_j$, $j = 1, 2, \ldots$. By the expansion theorem, every g_n can be expanded as follows.

$$\begin{aligned} g_n(x) &= (K^{n-1}g_1)(x) \\ &= \sum_{j=1}^{\infty}\langle K^{n-1}g_1, \phi_j\rangle\phi_j(x) \\ &= \sum_{j=1}^{\infty}\langle g_1, K^{n-1}\phi_j\rangle\phi_j(x) \\ &= \sum_{j=1}^{\infty}\langle g_1, \lambda_j^{-(n-1)}\phi_j\rangle\phi_j(x) \\ &= \sum_{j=1}^{\infty}\langle g_1, \phi_j\rangle\lambda_j^{-(n-1)}\phi_j(x) \\ &= \sum_{j=j_0}^{\infty}\langle g_1, \phi_j\rangle\lambda_j^{-(n-1)}\phi_j(x), \end{aligned}$$

where j_0 is at least j for which $\langle g_1, \phi_j\rangle \neq 0$. If we suppose that such j_0 does not exist, then

$$\langle g_1, \phi_j\rangle = 0 \quad for \quad all \quad j \in \mathscr{N}.$$

Hence and Theorem 14 in Chap. 5, we conclude that

$$g_1(x) = \sum_{j=1}^{\infty}\langle g_1, \phi_j\rangle\phi_j(x) \equiv 0,$$

which is a contradiction with the assumption $g_1(x) \not\equiv 0$. Consequently j_0 exists and

$$g_n(x) = \sum_{j=j_0}^{\infty}\langle g_1, \phi_j\rangle\lambda_j^{-(n-1)}\phi_j(x). \tag{6.4}$$

As in above,

$$g_{n+1}(x) = \sum_{j=j_0}^{\infty}\langle g_1, \phi_j\rangle\lambda_j^{-n}\phi_j(x).$$

Therefore

$$\langle g_{n+1}, g_n\rangle = \sum_{j=j_0}^{\infty} \langle g_1, \phi_j\rangle \overline{\langle g_1, \phi_j\rangle} \lambda_j^{-(n-1)} \lambda_j^{-n}$$
$$= \sum_{j=j_0}^{\infty} \lambda_j^{-2n+1} |\langle g_1, \phi_j\rangle|^2$$

and

$$||g_{n+1}||^2 = \sum_{j=j_0}^{\infty} \langle g_1, \phi_j\rangle \overline{\langle g_1, \phi_j\rangle} \lambda_j^{-n} \lambda_j^{-n}$$
$$= \sum_{j=j_0}^{\infty} \lambda_j^{-2n} |\langle g_1, \phi_j\rangle|^2. \tag{6.5}$$

Hence, by the assumption

$$0 < \lambda_1 \le \lambda_2 \le \lambda_3 \le \ldots$$

we obtain that

$$\alpha_n = \frac{\sum_{j=j_0}^{\infty} \lambda_j^{-2n+1} |\langle g_1, \phi_j\rangle|^2}{\sum_{j=j_0}^{\infty} \lambda_j^{-2n} |\langle g_1, \phi_j\rangle|^2}$$
$$= \frac{\lambda_{j_0}^{-2n+1} |\langle g_1, \phi_{j_0}\rangle|^2 + \sum_{j=j_0+1}^{\infty} \lambda_j^{-2n+1} |\langle g_1, \phi_j\rangle|^2}{\lambda_{j_0}^{-2n} |\langle g_1, \phi_{j_0}\rangle|^2 + \sum_{j=j_0+1}^{\infty} \lambda_j^{-2n} |\langle g_1, \phi_j\rangle|^2}$$
$$= \lambda_{j_0} \frac{|\langle g_1, \phi_{j_0}\rangle|^2 + \sum_{j=j_0+1}^{\infty} (\lambda_j \lambda_{j_0}^{-1})^{-2n+1} |\langle g_1, \phi_j\rangle|^2}{|\langle g_1, \phi_{j_0}\rangle|^2 + \sum_{j=j_0+1}^{\infty} (\lambda_j \lambda_{j_0}^{-1})^{-2n} |\langle g_1, \phi_j\rangle|^2}$$

and

$$\lambda_{j_0} \le \alpha_n,$$

i.e.,

$$\lambda_{j_0} \le \alpha_n = \lambda_{j_0} \frac{|\langle g_1, \phi_{j_0}\rangle|^2 + \sum_{j=j_0+1}^{\infty} (\lambda_j \lambda_{j_0}^{-1})^{-2n+1} |\langle g_1, \phi_j\rangle|^2}{|\langle g_1, \phi_{j_0}\rangle|^2 + \sum_{j=j_0+1}^{\infty} (\lambda_j \lambda_{j_0}^{-1})^{-2n} |\langle g_1, \phi_j\rangle|^2}.$$

By the Cauchy-Schwartz inequality we have

$$|\langle g_{n+1}, g_n\rangle| \le ||g_{n+1}|| ||g_n||.$$

Therefore

$$\begin{aligned}\alpha_n &= \frac{|\langle g_{n+1}, g_n\rangle|}{||g_{n+1}||^2} \\ &\leq \frac{||g_{n+1}||||g_n||}{||g_{n+1}||^2} \\ &= \frac{||g_n||}{||g_{n+1}||} \\ &= \beta_n.\end{aligned}$$

Using the Cauchy-Schwartz inequality, we see that

$$|\langle g_{n+1}, g_{n-1}\rangle| \leq ||g_{n+1}||||g_{n-1}||$$

and since

$$\begin{aligned}|\langle g_{n+1}, g_{n-1}\rangle| &= |\langle K g_n, g_{n-1}\rangle| \\ &= |\langle g_n, K g_{n-1}\rangle| \\ &= |\langle g_n, g_n\rangle| \\ &= ||g_n||^2,\end{aligned}$$

we obtain that

$$||g_n||^2 \leq ||g_{n+1}||||g_{n-1}||,$$

$$\begin{aligned}\frac{||g_n||}{||g_{n-1}||} &= \frac{||g_n||^2}{||g_n||||g_{n-1}||} \\ &\leq \frac{||g_{n+1}||||g_{n-1}||}{||g_n||||g_{n-1}||} \\ &= \frac{||g_{n+1}||}{||g_n||},\end{aligned}$$

i.e.,

$$\beta_n \leq \beta_{n-1}.$$

Therefore the sequence $\{\beta_n\}_{n=1}^{\infty}$ is monotone decreasing. Hence $\lim_{n\to\infty} \beta_n$ does exist. For the sequence $\{||g_n||\}_{n=1}^{\infty}$ of positive numbers, the following inequalities hold

$$\begin{aligned}\underline{\lim}_{n\to\infty}\frac{||g_n||}{||g_{n+1}||} &\leq \underline{\lim}_{n\to\infty}||g_n||^{-\frac{1}{n}}\\ &\leq \overline{\lim}_{n\to\infty}||g_n||^{-\frac{1}{n}} \leq \overline{\lim}_{n\to\infty}\frac{||g_n||}{||g_{n+1}||}.\end{aligned} \tag{6.6}$$

By (6.5) we have

$$\begin{aligned}\sqrt[2n]{||g_{n+1}||^2} &= \sqrt[2n]{\sum_{j=j_0}^{\infty} \lambda_j^{-2n}|\langle g_1, \phi_j\rangle|^2}\\ &= \sqrt[2n]{\sum_{j=j_0}^{\infty} \frac{1}{\lambda_j^{2n}}|\langle g_1, \phi_j\rangle|^2}\\ &= \lambda_{j_0}^{-1}\sqrt[2n]{\sum_{j=j_0}^{\infty}\left(\frac{\lambda_{j_0}}{\lambda_j}\right)^{2n}|\langle g_1, \phi_j\rangle|^2}.\end{aligned} \tag{6.7}$$

Hence, using that $\lambda_j \geq \lambda_{j_0}$ for $j \geq j_0$,

$$\begin{aligned}\overline{\lim}_{n\to\infty}\sqrt[2n]{||g_{n+1}||^2} &= \overline{\lim}_{n\to\infty}\lambda_{j_0}^{-1}\sqrt[2n]{\sum_{j=j_0}^{\infty}\left(\frac{\lambda_{j_0}}{\lambda_j}\right)^{2n}|\langle g_1, \phi_j\rangle|^2}\\ &\leq \overline{\lim}_{n\to\infty}\lambda_{j_0}^{-1}\sqrt[2n]{\sum_{j=j_0}^{\infty}|\langle g_1, \phi_j\rangle|^2}\\ &\leq \lambda_{j_0}^{-1}\overline{\lim}_{n\to\infty}\sqrt[2n]{\sum_{j=1}^{\infty}|\langle g_1, \phi_j\rangle|^2}\\ &= \lambda_{j_0}^{-1}\overline{\lim}_{n\to\infty}\sqrt[2n]{||g_1||^2}\\ &= \lambda_{j_0}^{-1}.\end{aligned}$$

Also, from (6.7), we obtain

$$\begin{aligned}\underline{\lim}_{n\to\infty}\sqrt[2n]{||g_{n+1}||^2} &\geq \underline{\lim}_{n\to\infty}\lambda_{j_0}^{-1}\sqrt[2n]{|\langle g_1, \phi_{j_0}\rangle|^2}\\ &\geq \lambda_{j_0}^{-1}.\end{aligned}$$

Consequently

$$\lim_{n\to\infty}\sqrt[2n]{||g_{n+1}||^2} = \lambda_{j_0}^{-1},$$

whereupon

$$\lim_{n\to\infty} ||g_{n+1}||^{\frac{1}{n+1}} = \lambda_{j_0}^{-1}.$$

Hence and (6.6), we obtain

$$\lim_{n\to\infty} \alpha_n = \lim_{n\to\infty} \beta_n = \lambda_{j_0}^{-1}.$$

Theorem 7 *Suppose that every eigenvalue of the operator K is positive. Let α_n and β_n be as in Theorem 6. Then, for each n, there exists an eigenvalue λ_{j_n} of K such that*

$$\sqrt{\beta_n^2 - \alpha_n^2} \geq \alpha_n - \lambda_{j_n} \geq 0,$$

and for sufficiently large n, $\lambda_{j_n} = \lambda_{j_0}$.

Proof Note that

$$||g_n - \alpha_n g_{n+1}||^2 = \langle g_n - \alpha_n g_{n+1}, g_n - \alpha_n g_{n+1}\rangle$$

$$= \langle g_n, g_n - \alpha_n g_{n+1}\rangle - \langle \alpha_n g_{n+1}, g_n - \alpha_n g_{n+1}\rangle$$

$$= \langle g_n, g_n\rangle - \langle g_n, \alpha_n g_{n+1}\rangle - \langle \alpha_n g_{n+1}, g_n\rangle + \langle \alpha_n g_{n+1}, \alpha_n g_{n+1}\rangle$$

$$= ||g_n||^2 - \alpha_n\langle g_n, g_{n+1}\rangle - \alpha_n\langle g_{n+1}, g_n\rangle + \alpha_n^2\langle g_{n+1}, g_{n+1}\rangle$$

$$= ||g_n||^2 - 2\alpha_n\langle g_n, g_{n+1}\rangle + \alpha_n^2||g_{n+1}||^2$$

$$= ||g_{n+1}||^2 \left(\frac{||g_n||^2}{||g_{n+1}||^2} - 2\alpha_n \frac{\langle g_n, g_{n+1}\rangle}{||g_{n+1}||^2} + \alpha_n^2\right)$$

$$= ||g_{n+1}||^2 \left(\beta_n^2 - 2\alpha_n^2 + \alpha_n^2\right)$$

$$= ||g_{n+1}||^2(\beta_n^2 - \alpha_n^2).$$

Hence and (6.4), we get

$$\beta_n^2 - \alpha_n^2 = \frac{1}{||g_{n+1}||^2}||g_n - \alpha_n g_{n+1}||^2$$

$$= \frac{1}{||g_{n+1}||^2}\left|\left|\sum_{j=j_0}^{\infty}\langle g_1, \phi_j\rangle\lambda_j^{-(n-1)}\phi_j(x) - \alpha_n\sum_{j=j_0}^{\infty}\langle g_1, \phi_j\rangle\lambda_j^{-n}\phi_j(x)\right|\right|^2$$

$$= \frac{1}{||g_{n+1}||^2}\left|\left|\sum_{j=j_0}^{\infty}\left(\lambda_j^{-(n-1)} - \alpha_n\lambda_j^{-n}\right)\langle g_1, \phi_j\rangle\phi_j(x)\right|\right|^2.$$

Now using the Parseval relation, we obtain

$$\beta_n^2 - \alpha_n^2 = \frac{1}{||g_{n+1}||^2} \sum_{j=j_0}^{\infty} \left(\lambda_j^{-(n-1)} - \alpha_n \lambda_j^{-n}\right)^2 |\langle g_1, \phi_j \rangle|^2$$

$$= \frac{1}{||g_{n+1}||^2} \sum_{j=j_0}^{\infty} \lambda_j^{-2n} (\lambda_j - \alpha_n)^2 |\langle g_1, \phi_j \rangle|^2.$$

We set

$$\min_{j \geq j_0} (\lambda_j - \alpha_n)^2 = (\lambda_{j_n} - \alpha_n)^2.$$

Then

$$\beta_n^2 - \alpha_n^2 \geq \frac{(\lambda_{j_n} - \alpha_n)^2}{||g_{n+1}||^2} \sum_{j=j_0}^{\infty} \lambda_j^{-2n} |\langle g_1, \phi_j \rangle|^2.$$

Hence and (6.5), we obtain

$$\beta_n^2 - \alpha_n^2 \geq (\lambda_{j_n} - \alpha_n)^2.$$

Since

$$\lim_{n\to\infty} \beta_n = \lim_{n\to\infty} \alpha_n = \lambda_{j_0},$$

we obtain that $\lambda_{j_n} = \lambda_{j_0}$ for sufficiently large n, which completes the proof.

6.3 Inhomogeneous Generalized Integral Equations

Here we will consider the equation

$$u(x) = \phi(x) - \lambda \int_a^b K(x, y)\phi(y)\Delta y, \tag{6.8}$$

where $K(x, y)$ is continuous and Hermitian symmetric on $[a, b] \times [a, b]$, $u(x)$ is continuous on $[a, b]$.
We denote by $\{\lambda_j\}$ the eigenvalues of the associated homogeneous equation

$$0 = \phi(x) - \lambda \int_a^b K(x, y)\phi(y)\Delta y, \tag{6.9}$$

and by $\{\phi_j(x)\}$ the corresponding complete orthonormal system of eigenfunctions. We have that

$$|\lambda_1| \leq |\lambda_2| \leq \cdots \leq |\lambda_n| \leq \cdots,$$

$$\lim_{n\to\infty} |\lambda_n| = \infty,$$

and

$$K\phi_j = \lambda_j^{-1}\phi_j, \quad \langle\phi_j, \phi_k\rangle = \delta_{jk}.$$

First case. λ is different from any of the eigenvalues λ_j.
Suppose that $\phi(x)$ is a solution to the Eq. (6.8). Then

$$\phi(x) - u(x) = \lambda \int_a^b K(x, y)\phi(y)\Delta y.$$

Hence, by the expansion theorem, $\phi(x) - u(x)$ can be expanded in a Fourier series, which converges absolutely and uniformly on the interval $[a, b]$, with respect to the orthonormal system $\{\phi_j(x)\}$. Let

$$c_j = \langle\phi - u, \phi_j\rangle.$$

Then

$$\begin{aligned}
c_j &= \langle\lambda K\phi, \phi_j\rangle \\
&= \lambda\langle K\phi, \phi_j\rangle \\
&= \lambda\langle\phi, K\phi_j\rangle \\
&= \lambda\langle\phi, \lambda_j^{-1}\phi_j\rangle \\
&= \lambda\lambda_j^{-1}\langle\phi, \phi_j\rangle \\
&= \lambda\lambda_j^{-1}\langle\phi - u + u, \phi_j\rangle \\
&= \lambda\lambda_j^{-1}\left(\langle\phi - u, \phi_j\rangle + \langle u, \phi_j\rangle\right) \\
&= \lambda\lambda_j^{-1}\left(c_j + \langle u, \phi_j\rangle\right) \\
&= \lambda\lambda_j^{-1}(c_j + u_j),
\end{aligned}$$

where $u_j = \langle u, \phi_j\rangle$.
Hence, we obtain that

$$c_j(1 - \lambda\lambda_j^{-1}) = \lambda\lambda_j^{-1}u_j$$

or

$$c_j = \frac{\lambda\lambda_j^{-1}u_j}{1-\lambda\lambda_j^{-1}}$$
$$= \frac{\lambda\lambda_j^{-1}u_j}{\lambda_j^{-1}(\lambda_j-\lambda)}$$
$$= \frac{\lambda u_j}{\lambda_j-\lambda}.$$

Consequently

$$\phi(x) - u(x) = \sum_{j=1}^{\infty} c_j\phi_j(x)$$
$$= \sum_{j=1}^{\infty} \frac{\lambda u_j}{\lambda_j-\lambda}\phi_j(x)$$
$$= \lambda\sum_{j=1}^{\infty} \frac{u_j}{\lambda_j-\lambda}\phi_j(x),$$

or

$$\phi(x) = u(x) + \lambda\sum_{j=1}^{\infty} \frac{u_j}{\lambda_j-\lambda}\phi_j(x). \tag{6.10}$$

Theorem 8 *The series (6.10) converges absolutely and uniformly on the interval $[a, b]$ and satisfies the Eq. (6.8).*

Proof By Bessel inequality we have

$$\sum_{j=1}^{\infty} |u_j|^2 \le \int_a^b |u(x)|^2\Delta x < \infty,$$

$$\sum_{j=1}^{\infty} \lambda_j^{-2}|\phi_j(x)|^2 \le \int_a^b |K(x,y)|^2\Delta y \le \sup_{x\in[a,b]}\int_a^b |K(x,y)|^2\Delta y < \infty.$$

Hence,

$$\left(\sum_{j=k}^{\infty}\left|\frac{\lambda}{\lambda_j-\lambda}u_j\phi_j(x)\right|\right)^2 \le \sum_{j=k}^{\infty}|u_j|^2\sum_{j=k}^{\infty}\left|\frac{\lambda}{\lambda_j-\lambda}\right|^2|\phi_j(x)|^2$$

$$= \sum_{j=k}^{\infty}|u_j|^2\sum_{j=k}^{\infty}\left|\frac{\lambda\lambda_j}{\lambda_j-\lambda}\right|^2|\lambda_j^{-1}\phi_j(x)|^2.$$

Let

$$C = \sup_{j\geq 1} \left| \frac{\lambda\lambda_j}{\lambda_j - \lambda} \right|.$$

We note that C exists and $C < \infty$ because $\lim_{j\to\infty} |\lambda_j| = \infty$ and $\lambda_j \neq \lambda$. Consequently

$$\left(\sum_{j=k}^{\infty} \left|\frac{\lambda}{\lambda_j-\lambda} u_j\phi_j(x)\right|\right)^2 \leq C^2 \sum_{j=k}^{\infty} |u_j|^2 \sum_{j=k}^{\infty} |\lambda_j^{-1}\phi_j(x)|^2$$
$$\leq C^2 \int_a^b |u(x)|^2 \Delta x \sup_{x\in[a,b]} \int_a^b |K(x, y)|^2 \Delta y.$$

Therefore the series (6.10) converges absolutely and uniformly on the interval $[a, b]$. Now we substitute (6.10) in (6.8) and integrate term by term. We get that the right side is equal to

$$F(x) = u(x) + \sum_{j=1}^{\infty} \frac{\lambda}{\lambda_j-\lambda} u_j\phi_j(x) - \left[\lambda \int_a^b K(x, y)u(y)\Delta y \right.$$
$$\left. + \sum_{j=1}^{\infty} \frac{\lambda}{\lambda_j-\lambda} u_j\lambda \int_a^b K(x, y)\phi_j(y)\Delta y\right]. \tag{6.11}$$

On the other hand, by the expansion theorem, we have

$$\begin{aligned}\int_a^b K(x, y)u(y)\Delta y &= \sum_{j=1}^{\infty} \langle Ku, \phi_j\rangle\phi_j(x)\\ &= \sum_{j=1}^{\infty} \langle u, K\phi_j\rangle\phi_j(x)\\ &= \sum_{j=1}^{\infty} \langle u, \lambda_j^{-1}\phi_j\rangle\phi_j(x)\\ &= \sum_{j=1}^{\infty} \lambda_j^{-1}\langle u, \phi_j\rangle\phi_j(x)\\ &= \sum_{j=1}^{\infty} \lambda_j^{-1} u_j\phi_j(x).\end{aligned}$$

Hence,

$$F(x) = u(x) + \sum_{j=1}^{\infty} \frac{\lambda}{\lambda_j - \lambda} u_j \phi_j(x)$$

$$-\left[\sum_{j=1}^{\infty} \lambda \lambda_j^{-1} u_j \phi_j(x) + \sum_{j=1}^{\infty} \frac{\lambda^2}{\lambda_j - \lambda} \lambda_j^{-1} u_j \phi_j(x)\right]$$

$$= u(x) + \sum_{j=1}^{\infty} \frac{\lambda}{\lambda_j - \lambda} u_j \phi_j(x) - \sum_{j=1}^{\infty} \frac{\lambda}{\lambda_j - 1} u_j \phi_j(x) = u(x),$$

which completes the proof.

Second case. $\lambda = \lambda_{j_0}$ for some j_0. Suppose that $\phi(x)$ is a solution of the equation

$$u(x) = \phi(x) - \lambda_{j_0} \int_a^b K(x, y) \phi(y) \Delta y. \tag{6.12}$$

We multiply both sides of (6.12) by $\overline{\phi_j(x)}$ and then we integrate from a to b, we obtain

$$\begin{aligned} u_j &= \langle u, \phi_j \rangle \\ &= \langle \phi - \lambda_{j_0} K\phi, \phi_j \rangle \\ &= \langle \phi, \phi_j \rangle - \langle \lambda_{j_0} K\phi, \phi_j \rangle \\ &= \langle \phi, \phi_j \rangle - \lambda_{j_0} \langle \phi, K\phi_j \rangle \\ &= \langle \phi, \phi_j \rangle - \lambda_{j_0} \langle \phi, \lambda_j^{-1} \phi_j \rangle \\ &= \langle u, \phi_j \rangle - \lambda_{j_0} \lambda_j^{-1} \langle \phi, \phi_j \rangle. \end{aligned}$$

Hence,

$$u_j = 0 \quad if \quad \lambda_j = \lambda_{j_0}^{-1}. \tag{6.13}$$

If $u(x)$ satisfies the condition (6.13), then as in the first case, we have that

$$\phi(x) = u(x) + \sum_{\lambda_j \neq \lambda_{j_0}} \frac{\lambda}{\lambda_j - \lambda} u_j \phi_j(x)$$

converges absolutely and uniformly on $[a, b]$ and satisfies the Eq. (6.12).

Let now $\phi(x)$ and $\xi(x)$ be two solutions of the Eq. (6.12). We set

$$\psi(x) = \phi(x) - \xi(x).$$

We have that $\psi(x)$ satisfies the equation

$$0 = \psi(x) - \lambda_{j_0} \int_a^b K(x, y)\psi(y)\Delta y.$$

Hence, $\psi(x)$ can be written as a linear combination

$$\sum_{\lambda_j = \lambda_{j0}} c_j \phi_j(x)$$

of the eigenfunctions $\phi_j(x)$ corresponding to the eigenvalue $\lambda_j = \lambda_{j_0}$. Here c_j are arbitrary constants.

Then

$$u(x) + \sum_{\lambda_j \neq \lambda_{j0}} \frac{\lambda}{\lambda_j - \lambda} u_j \phi_j(x) + \sum_{\lambda_j = \lambda_{j0}} c_j \phi_j(x)$$

is the general solution of the Eq. (6.12).

Chapter 7
The Laplace Transform Method

This chapter is devoted on applications of the Laplace transform on time scales to dynamic equations, generalized Volterra integral equations and generalized Volterra integro-differential equations.

7.1 The Laplace Transform

7.1.1 Definition and Examples

Throughout this chapter we assume that the time scale $\mathscr{T}_0$ is such that

$$0 \in \mathscr{T}_0 \quad \textit{and} \quad \sup \mathscr{T}_0 = \infty.$$

Definition 1 Let $f : \mathscr{T}_0 \longmapsto \mathscr{C}$ is regulated. We define the set

$$\mathscr{D}\{f\} = \Big\{ z \in \mathscr{C} : 1 + z\mu(t) \neq 0 \quad \textit{for all} \quad t \in \mathscr{T}_0 \quad \textit{and}$$
$$\textit{the improper integral} \quad \int_0^\infty f(y) e^\sigma_{\ominus z}(y, 0) \Delta y \quad \textit{exists} \Big\}.$$

Remark 1 For $z \in \mathscr{C}$ such that $1 + z\mu(t) \neq 0$ for all $t \in \mathscr{T}_0$ we have

$$\ominus z = -\frac{z}{1 + z\mu(t)} \quad \textit{and} \quad 1 + \ominus z\mu(t) = 1 - \frac{z\mu(t)}{1 + z\mu(t)} = \frac{1}{1 + z\mu(t)} \neq 0$$

S.G. Georgiev, *Integral Equations on Time Scales*, Atlantis Studies in Dynamical Systems 5, DOI 10.2991/978-94-6239-228-1_7

for all $t \in \mathscr{T}_0$. Therefore $e_{\ominus z}(y, 0)$ is well defined on $\mathscr{T}_0$. In this case we have

$$\begin{aligned} e_{\ominus z}(y, 0) &= e^{\int_0^y \frac{1}{\mu(\tau)} \mathrm{Log}(1 + \ominus z \mu(\tau)) \Delta \tau} \\ &= e^{\int_0^y \frac{1}{\mu(\tau)} \mathrm{Log} \frac{1}{1+z\mu(\tau)} \Delta \tau} \\ &= e^{-\int_0^y \frac{1}{\mu(\tau)} \mathrm{Log}(1+z\mu(\tau)) \Delta \tau} \end{aligned}$$

and

$$e^{\sigma}_{\ominus z}(y, 0) = e^{-\int_0^{\sigma(y)} \frac{1}{\mu(\tau)} \mathrm{Log}(1+z\mu(\tau)) \Delta \tau}.$$

Definition 2 Assume that $f : \mathscr{T}_0 \longmapsto \mathscr{C}$ is a regulated function. Then the Laplace transform of f is defined by

$$\mathscr{L}(f)(z) = \int_0^\infty f(y) e^{\sigma}_{\ominus z}(y, 0) \Delta y$$

for $z \in \mathscr{D}\{f\}$.

Example 1 Let $\mathscr{T}_0 = 2^{\mathscr{N}_0} \cup \{0\}$. Let also, $\chi_{[1,2^2]}$ be the characteristic function on $\mathscr{T}_0 \cap [1, 2^2]$. Then $\sigma(t) = 2t$ and $\mu(t) = t$, $t \in \mathscr{T}_0$. Also,

$$\begin{aligned} \mathscr{L}(\chi_{[1,2^2]})(z) &= \int_0^\infty \chi_{[0,2^2]} e_{\ominus z}(\sigma(y), 0) \Delta y \\ &= \int_1^{2^2} e_{\ominus z}(\sigma(y), 0) \Delta y \\ &= \int_1^{2^2} e^{-\int_0^{\sigma(y)} \frac{1}{\mu(\tau)} \mathrm{Log}(1+z\mu(\tau)) \Delta \tau} \Delta y \\ &= \int_1^{2^2} e^{-\sum_{s \in [0,\sigma(y))} \frac{1}{\mu(s)} \mathrm{Log}(1+z\mu(s))\mu(s)} \Delta y \\ &= \int_1^{2^2} e^{-\sum}_{s \in [0,\sigma(y))} \mathrm{Log}(1 + z\mu(s)) \Delta y \\ &= \int_1^{2^2} e^{-\sum_{s \in [0,\sigma(y))} \mathrm{Log}(1+sz)} \Delta y \\ &= \int_1^{2^2} \prod_{s \in [0,\sigma(y))} \frac{1}{1+sz} \Delta y \\ &= \prod_{s \in [0,\sigma(1))} \frac{1}{1+sz} \mu(1) + \prod_{s \in [0,\sigma(2))} \frac{1}{1+sz} \mu(2) \\ &= \prod_{s \in [0,2)} \frac{1}{1+sz} \mu(1) + \prod_{s \in [0,2^2)} \frac{1}{1+sz} \mu(2) \\ &= \prod_{s \in [0,2)} \frac{1}{1+sz} + 2 \prod_{s \in [0,2^2)} \frac{1}{1+sz}. \end{aligned}$$

Example 2 Let $\mathscr{T}_0 = h\mathscr{Z}, \quad h > 0$. Then $\sigma(t) = t + h, \mu(t) = h, \quad t \in \mathscr{T}_0$. Hence,

$$
\begin{aligned}
\mathscr{L}(1)(z) &= \int_0^\infty e_{\ominus z}(\sigma(y), 0)\Delta y \\
&= \int_0^\infty e^{-\int_0^{\sigma(y)} \frac{1}{\mu(s)} \mathrm{Log}(1+z\mu(s))\Delta s} \\
&= \int_0^\infty e^{-\sum_{s\in[0,\sigma(y))} \frac{1}{\mu(s)} \mathrm{Log}(1+z\mu(s))\mu(s)} \Delta y
\end{aligned}
$$

$$
\begin{aligned}
&= \int_0^\infty e^{-\sum_{s\in[0,\sigma(y))} \mathrm{Log}(1+hz)} \Delta y \\
&= \int_0^\infty \prod_{s\in[0,\sigma(y))} \frac{1}{1+hz} \Delta y \\
&= \sum_{t=0}^\infty \prod_{s\in[0,\sigma(t))} \frac{1}{1+hz} \mu(t) \\
&= h \sum_{t=0}^\infty \left(\frac{1}{1+hz}\right)^{\frac{t}{h}+1} \\
&= \frac{h}{1+hz} \sum_{t=0}^\infty \left(\frac{1}{1+hz}\right)^{\frac{t}{h}}.
\end{aligned}
$$

Note that $\sum_{t=0}^\infty \left(\frac{1}{1+hz}\right)^{\frac{t}{h}}$ is convergent if and only if

$$
\begin{aligned}
&\left|\tfrac{1}{1+zh}\right| < 1 \iff \\
&\tfrac{1}{|1+zh|} < 1 \iff \\
&1 > |1+hz| \iff \\
&\tfrac{1}{h} > \left|\tfrac{1}{h} + z\right| \iff \\
&\tfrac{1}{h} > \left|z - \left(-\tfrac{1}{h}\right)\right|.
\end{aligned}
$$

If $D(a, r) \subset \mathscr{C}$ denotes the closed ball of radius r about the point a, then the region of convergence is $\mathscr{C} \backslash D\left(-\frac{1}{h}, \frac{1}{h}\right)$, i.e.,

$$
\mathscr{D}\{1\} = \mathscr{C} \backslash D\left(-\frac{1}{h}, \frac{1}{h}\right).
$$

Example 3 Let $\mathscr{T}_0 = \mathscr{N}_0$. Suppose that $f : \mathscr{T}_0 \longmapsto \mathscr{R}$ is regulated and $|f(t)| \leq M$ for all $t \in \mathscr{T}_0$. Then $\sigma(t) = t + 1, \mu(t) = 1, t \in \mathscr{T}_0$. Hence,

$$
\begin{aligned}
\mathscr{L}(f)(z) &= \int_0^\infty f(y)e_{\ominus z}(\sigma(y), 0)\Delta y \\
&= \int_0^\infty f(y)e^{-\int_0^{\sigma(y)} \frac{1}{\mu(\tau)}\operatorname{Log}(1+z\mu(\tau))\Delta\tau}\Delta y \\
&= \int_0^\infty f(y)e^{-\sum_{s\in[0,\sigma(y))} \frac{1}{\mu(s)}\operatorname{Log}(1+z\mu(s))\mu(s)}\Delta y \\
&= \int_0^\infty f(y)e^{-\sum_{s\in[0,y+1)} \operatorname{Log}(1+z)}\Delta y \\
&= \int_0^\infty f(y) \prod_{s\in[0,y+1)} \frac{1}{1+z}\Delta y \\
&= \int_0^\infty f(y)\left(\frac{1}{1+z}\right)^{y+1}\Delta y \\
&= \frac{1}{1+z}\int_0^\infty \frac{f(y)}{(1+z)^y}\Delta y \\
&= \frac{1}{1+z}\sum_{t=0}^\infty \frac{f(t)}{(1+z)^t}\mu(t) \\
&= \frac{1}{1+z}\sum_{t=0}^\infty \frac{f(t)}{(1+z)^t}.
\end{aligned}
$$

Since $|f(t)| \leq M$ for all $t \in \mathscr{T}_0$, we have that $\mathscr{L}(f)(z)$ is convergent if and only if

$$
\begin{aligned}
&\left|\tfrac{1}{1+z}\right| < 1 \iff \\
&\tfrac{1}{|1+z|} < 1 \iff \\
&|1+z| > 1.
\end{aligned}
$$

Lemma 1 *Suppose that the time scale $\mathscr{T}_0$ is such that $\mu(t) < M$ for all $t \in \mathscr{T}_0$ and for some positive constant M. Let also, $f : \mathscr{T}_0 \longmapsto \mathscr{C}$, $z \in \mathscr{C}$ be such that $1 + z\mu(t) \neq 0$ for all $t \in \mathscr{T}_0$, and*

$$\lim_{t\longrightarrow\infty} f(t)e_{\ominus z}(\sigma(t), 0) = 0.$$

Then

$$\lim_{t\longrightarrow\infty} f(t)e_{\ominus z}(t, 0) = 0.$$

Proof Using the properties of the exponential function, we have

$$\begin{aligned} e_{\ominus z}(\sigma(t), 0) &= (1 + \ominus z(t)\mu(t))e_{\ominus z}(t, 0) \\ &= \left(1 - \frac{z\mu(t)}{1 + z\mu(t)}\right) e_{\ominus z}(t, 0) \\ &= \frac{1}{1 + z\mu(t)} e_{\ominus z}(t, 0). \end{aligned}$$

Hence,

$$0 = \lim_{t \longrightarrow \infty} f(t)e_{\ominus z}(\sigma(t), 0) = \lim_{t \longrightarrow \infty} \frac{f(t)}{1 + z\mu(t)} e_{\ominus z}(t, 0).$$

Because $\mu(t) < M$ for all $t \in \mathscr{T}_0$, then

$$\lim_{t \longrightarrow \infty} f(t)e_{\ominus z}(t, 0) = 0.$$

Remark 2 Note that the Laplace transform maps functions defined on time scales to functions defined on some subsets of complex numbers. The region of convergence of the transform, $\mathscr{D}\{f\}$, varies not only with the function f but also with the time scale.

7.1.2 Properties of the Laplace Transform

Theorem 1 *Let f and g be regulated functions on $\mathscr{T}_0$ and let $\alpha, \beta \in \mathscr{C}$. Then*

$$\mathscr{L}(\alpha f + \beta g)(z) = \alpha\mathscr{L}(f)(z) + \beta\mathscr{L}(g)(z)$$

for all $z \in \mathscr{D}\{f\} \cap \mathscr{D}\{g\}$.

Proof Let $z \in \mathscr{D}\{f\} \cap \mathscr{D}\{g\}$ be arbitrarily chosen. Then $\mathscr{L}(f)(z)$ and $\mathscr{L}(g)(z)$ exist. Hence $\alpha\mathscr{L}(f)(z)$ and $\beta\mathscr{L}(g)(z)$ exist, from where $\alpha\mathscr{L}(f)(z) + \beta\mathscr{L}(g)(z)$ exists. Also, using the properties of the improper integral, we have

$$\begin{aligned} \alpha\mathscr{L}(f)(z) + \beta\mathscr{L}(g)(z) &= \alpha \int_0^\infty f(y)e_{\ominus z}(\sigma(y), 0)\Delta y + \beta \int_0^\infty g(y)e_{\ominus z}(\sigma(y), 0)\Delta y \\ &= \int_0^\infty \alpha f(y)e_{\ominus z}(\sigma(y), 0)\Delta y + \int_0^\infty \beta g(y)e_{\ominus z}(\sigma(y), 0)\Delta y \\ &= \int_0^\infty (\alpha f(y)e_{\ominus z}(\sigma(y), 0) + \beta g(y)e_{\ominus z}(\sigma(y), 0))\, \Delta y \\ &= \int_0^\infty (\alpha f(y) + \beta g(y))e_{\ominus z}(\sigma(y), 0)\Delta y \end{aligned}$$

$$= \int_0^\infty (\alpha f + \beta g)(y) e_{\ominus z}(\sigma(y), 0) \Delta y$$
$$= \mathscr{L}(\alpha f + \beta g)(z).$$

Theorem 2 *Let $f : \mathscr{T}_0 \longmapsto \mathscr{C}$ be such that f^Δ is regulated. Then*

$$\mathscr{L}(f^\Delta)(z) = z\mathscr{L}(f)(z) - f(0) + \lim_{y\to\infty} f(y) e_{\ominus z}(y, 0)$$

for all $\mathscr{D}\{f\}$ such that the limit exists.

Proof Using integration by parts, we obtain

$$f(y) e_{\ominus z}(y, 0)\Big|_{y=0}^{y=\infty} = \int_0^\infty f(y) e_{\ominus z}^\Delta(y, 0) \Delta y + \int_0^\infty f^\Delta(y) e_{\ominus z}(\sigma(y), 0) \Delta y$$

or

$$\lim_{y\to\infty} f(y) e_{\ominus z}(y, 0) - f(0) e_{\ominus z}(0, 0) = \int_0^\infty f^\Delta(y) e_{\ominus z}(\sigma(y), 0) \Delta y$$
$$+ \int_0^\infty (f(y))(\ominus z)(y) e_{\ominus z}(y, 0) \Delta y,$$

or

$$\int_0^\infty f^\Delta(y) e_{\ominus z}(\sigma(y), 0) \Delta y = \lim_{y\to\infty} f(y) e_{\ominus z}(y, 0) - f(0)$$
$$- \int_0^\infty f(y)(\ominus z)(y) e_{\ominus z}(y, 0) \Delta y,$$

or

$$\mathscr{L}(f^\Delta)(z) = \lim_{y\to\infty} f(y) e_{\ominus z}(y, 0) - f(0) - \int_0^\infty f(y)(\ominus z) e_{\ominus z}(y, 0) \Delta y. \tag{7.1}$$

We note that

$$-(\ominus z) e_{\ominus z}(y, 0) = \frac{z}{1 + z\mu(y)} e_{\ominus z}(y, 0)$$
$$= z e_{\ominus z}(\sigma(y), 0).$$

Hence and (7.1), we get

$$\mathscr{L}(f^\Delta)(z) = \lim_{y\to\infty} f(y) e_{\ominus z}(y, 0) - f(0) - \int_0^\infty f(y) z e_{\ominus z}(\sigma(y), 0) \Delta y$$

$$= z\int_0^\infty f(y)e_{\ominus z}(\sigma(y),0)\Delta y - f(0) + \lim_{y\to\infty} f(y)e_{\ominus z}(y,0)$$
$$= z\mathscr{L}(f)(z) - f(0) + \lim_{y\to\infty} f(y)e_{\ominus z}(y,0),$$

which completes the proof.

Corollary 1 *Let $f : \mathscr{T}_0 \longmapsto \mathscr{C}$ be such that f^{Δ^n} is regulated for some $n \in \mathscr{N}$. Then*

$$\mathscr{L}(f^{\Delta^n})(z) = z^n\mathscr{L}(f)(z) - \sum_{l=0}^{n-1} z^l f^{\Delta^{n-1-l}}(0) \tag{7.2}$$

for those $z \in \mathscr{D}\{f\}$ such that

$$\lim_{y\to\infty} f^{\Delta^l}(y)e_{\ominus z}(y,0) = 0, \quad 0 \le l \le n-1.$$

Proof We will use mathematical induction.

1. $n = 1$. The assertion follows from Theorem 2.
2. Assume that (7.2) is true for some $n \in \mathscr{N}$.
3. We will prove that

$$\mathscr{L}(f^{\Delta^{n+1}})(z) = z^{n+1}\mathscr{L}(f)(z) - \sum_{l=0}^{n} z^l f^{\Delta^{n-l}}(0)$$

for those $z \in \mathscr{D}\{f\}$ for which

$$\lim_{y\to\infty} f^{\Delta^l} e_{\ominus z}(y,0) = 0, \quad 0 \le l \le n,$$

and $f^{\Delta^{n+1}}$ is regulated.
Really, applying (7.2), we obtain

$$\begin{aligned}
\mathscr{L}(f^{\Delta^{n+1}})(z) &= \mathscr{L}((f^{\Delta^n})^{\Delta})(z)\\
&= z\mathscr{L}(f^{\Delta^n})(z) - f^{\Delta^n}(0)\\
&= z\left(z^n\mathscr{L}(f)(z) - \sum_{l=0}^{n-1} z^l f^{\Delta^{n-1-l}}(0)\right) - f^{\Delta^n}(0)\\
&= z^{n+1}\mathscr{L}(f)(z) - \sum_{l=0}^{n-1} z^{l+1} f^{\Delta^{n-1-l}}(0) - f^{\Delta^n}(0)
\end{aligned}$$

$$= z^{n+1}\mathscr{L}(f)(z) - \sum_{l=1}^{n} z^l f^{\Delta^{n-l}}(0) - f^{\Delta^n}(0)$$

$$= z^{n+1}\mathscr{L}(f)(z) - \sum_{l=0}^{n} z^l f^{\Delta^{n-l}}(0),$$

which completes the proof.

Theorem 3 *Assume $f : \mathscr{T}_0 \longmapsto \mathscr{C}$ is regulated. If*

$$F(x) = \int_0^x f(y)\Delta y$$

for $x \in \mathscr{T}_0$, then

$$\mathscr{L}(F)(z) = \frac{1}{z}\mathscr{L}(f)(z)$$

for those $z \in \mathscr{D}\{f\}\backslash\{0\}$ satisfying

$$\lim_{x\to\infty}\left(e_{\ominus z}(x,0)\int_0^x f(y)\Delta y\right) = 0.$$

Proof Using integration by parts, we get

$$\begin{aligned}
\mathscr{L}(F)(z) &= \int_0^\infty F(y)e_{\ominus z}(\sigma(y),0)\Delta y \\
&= \int_0^\infty F(y)(1+\ominus z\mu(y))e_{\ominus z}(y,0)\Delta y \\
&= \int_0^\infty F(y)\frac{1}{1+z\mu(y)}e_{\ominus z}(y,0)\Delta y \\
&= -\frac{1}{z}\int_0^\infty F(y)\frac{-z}{1+z\mu(y)}e_{\ominus z}(y,0)\Delta y \\
&= -\frac{1}{z}\int_0^\infty F(y)\ominus z(y)e_{\ominus z}(y,0)\Delta y \\
&= -\frac{1}{z}\int_0^\infty F(y)e^{\Delta}_{\ominus z}(y,0)\Delta y \\
&= -\frac{1}{z}\left(\lim_{y\to\infty} F(y)e_{\ominus z}(y,0) - F(0)e_{\ominus z}(0,0)\right) \\
&\quad + \frac{1}{z}\int_0^\infty F^{\Delta}(y)e_{\ominus z}(\sigma(y),0)\Delta y
\end{aligned}$$

$$= \frac{1}{z} \int_0^\infty f(y) e_{\ominus z}(\sigma(y), 0) \Delta y$$
$$= \frac{1}{z} \mathscr{L}(f)(z),$$

which completes the proof.

Theorem 4 *For all $n \in \mathscr{N}_0$ we have*

$$\mathscr{L}(h_n(x, 0))(z) = \frac{1}{z^{n+1}}, \quad x \in \mathscr{T}_0, \tag{7.3}$$

for those $z \in \mathscr{C} \backslash \{0\}$ such that $1 + z\mu(x) \neq 0$, $x \in \mathscr{T}_0$, and

$$\lim_{x \to \infty} (h_n(x, 0) e_{\ominus z}(x, 0)) = 0. \tag{7.4}$$

Proof We note that (7.4) implies

$$\lim_{x \to \infty} (h_l(x, 0) e_{\ominus z}(x, 0)) = 0 \quad for \quad all \quad 0 \leq l \leq n.$$

To prove our assertion we will use mathematical induction.

1. $n = 0$. We have that $h_0(x, 0) = 1$ and
$$\mathscr{L}(1)(z) = \frac{1}{z}.$$
2. Assume that (7.3) holds for some $n \in \mathscr{N}_0$.
3. We will prove that
$$\mathscr{L}(h_{n+1}(x, 0))(z) = \frac{1}{z^{n+2}}$$
for those $z \in \mathscr{C} \backslash \{0\}$ such that $1 + z\mu(x) \neq 0$, $x \in \mathscr{T}_0$, and
$$\lim_{x \to \infty} (h_{n+1}(x, 0) e_{\ominus z}(x, 0)) = 0.$$
Really,
$$\mathscr{L}(h_{n+1}(x, 0))(z) = \int_0^\infty h_{n+1}(y, 0) e_{\ominus z}(\sigma(y), 0) \Delta y$$
$$= \int_0^\infty \left(\int_0^y h_n(t, 0) \Delta t \right) e_{\ominus z}(\sigma(y), 0) \Delta y.$$
From here, using Theorem 3 and (7.3), we obtain

$$\begin{aligned}\mathscr{L}(h_{n+1}(x,0))(z) &= \frac{1}{z}\mathscr{L}(h_n(x,0))(z)\\ &= \frac{1}{z}\frac{1}{z^{n+1}}\\ &= \frac{1}{z^{n+2}},\end{aligned}$$

which completes the proof.

Theorem 5 *Let $\alpha \in \mathscr{C}$ and $1+\alpha\mu(x) \neq 0$ for $x \in \mathscr{T}_0$. Then*

$$\mathscr{L}(e_\alpha(x,0))(z) = \frac{1}{z-\alpha}, \quad x \in \mathscr{T}_0,$$

provided

$$\lim_{x\to\infty} e_{\alpha\ominus z}(x,0) = 0.$$

Proof We have

$$\begin{aligned}\mathscr{L}(e_\alpha(x,0))(z) &= \int_0^\infty e_\alpha(y,0)e_{\ominus z}(\sigma(y),0)\Delta y\\ &= \int_0^\infty e_\alpha(y,0)\left(1+(\ominus z)(y)\mu(y)\right)e_{\ominus z}(y,0)\Delta y\\ &= \int_0^\infty \frac{1}{1+z\mu(y)}e_\alpha(y,0)e_{\ominus z}(y,0)\Delta y\\ &= \int_0^\infty \frac{1}{1+z\mu(y)}e_{\alpha\ominus z}(y,0)\Delta y\\ &= \frac{1}{\alpha-z}\int_0^\infty \frac{\alpha-z}{1+z\mu(y)}e_{\alpha\ominus z}(y,0)\Delta y,\end{aligned}$$

i.e.,

$$\mathscr{L}(e_\alpha(x,0))(z) = \frac{1}{\alpha-z}\int_0^\infty \frac{\alpha-z}{1+z\mu(y)}e_{\alpha\ominus z}(y,0)\Delta y. \tag{7.5}$$

we note that

$$\begin{aligned}\alpha\ominus z &= \alpha\oplus(\ominus z)\\ &= \alpha+(\ominus z)+\alpha(\ominus z)\mu(y)\\ &= \alpha-\frac{z}{1+z\mu(y)}-\frac{\alpha z\mu(y)}{1+z\mu(y)}\\ &= \frac{\alpha-z}{1+z\mu(y)}.\end{aligned}$$

Hence and (7.5), we obtain

$$\begin{aligned}
\mathscr{L}(e_\alpha(x,0))(z) &= \frac{1}{\alpha - z}\int_0^\infty \frac{\alpha - z}{1 + z\mu(y)} e_{\alpha\ominus z}(y,0)\Delta y \\
&= \frac{1}{\alpha - z}\int_0^\infty \alpha \ominus z(y) e_{\alpha\ominus z}(y,0)\Delta y \\
&= \frac{1}{\alpha - z}\int_0^\infty e^{\Delta}_{\alpha\ominus z}(y,0)\Delta y \\
&= \frac{1}{\alpha - z}\left(\lim_{y\to\infty} e_{\alpha\ominus z}(y,0) - e_{\alpha\ominus z}(0,0)\right) \\
&= \frac{1}{z - \alpha}
\end{aligned}$$

Corollary 2 *We have*

1. $\mathscr{L}(\cos_\alpha(x,0))(x) = \dfrac{z}{z^2 + \alpha^2}$,
2. $\mathscr{L}(\sin_\alpha(x,0))(z) = \dfrac{\alpha}{z^2 + \alpha^2}$

provided that

$$\lim_{x\to\infty} e_{i\alpha\ominus z}(x,0) = \lim_{x\to\infty} e_{-i\alpha\ominus z}(x,0) = 0.$$

Proof 1. From the definition for $\cos_\alpha(x,0)$ we have

$$\cos_\alpha(x,0) = \frac{e_{i\alpha}(x,0) + e_{-i\alpha}(x,0)}{2}.$$

Hence,

$$\begin{aligned}
\mathscr{L}(\cos_\alpha(x,0))(z) &= \mathscr{L}\left(\frac{e_{i\alpha}(x,0) + e_{-i\alpha}(x,0)}{2}\right) \\
&= \frac{1}{2}\mathscr{L}(e_{i\alpha}(x,0) + e_{-i\alpha}(x,0))(z) \\
&= \frac{1}{2}\left(\mathscr{L}(e_{i\alpha}(x,0))(z) + \mathscr{L}(e_{-i\alpha}(x,0))(z)\right) \\
&= \frac{1}{2}\left(\frac{1}{z - i\alpha} + \frac{1}{z + i\alpha}\right) \\
&= \frac{1}{2}\frac{z - i\alpha + z + i\alpha}{(z - i\alpha)(z + i\alpha)} \\
&= \frac{z}{z^2 + \alpha^2}.
\end{aligned}$$

2. From the definition for $\sin_\alpha(x,0)$ we have

$$\sin_\alpha(x,0) = \frac{e_{i\alpha}(x,0) - e_{-i\alpha}(x,0)}{2i}.$$

Hence,

$$\begin{aligned}
\mathscr{L}(\sin_\alpha(x,0))(z) &= \mathscr{L}\left(\frac{e_{i\alpha}(x,0) - e_{-i\alpha}(x,0)}{2i}\right)\\
&= \frac{1}{2i}\mathscr{L}(e_{i\alpha}(x,0) - e_{-i\alpha}(x,0))(z)\\
&= \frac{1}{2i}\left(\mathscr{L}(e_{i\alpha}(x,0))(z) - \mathscr{L}(e_{-i\alpha}(x,0))(z)\right)\\
&= \frac{1}{2i}\left(\frac{1}{z-i\alpha} - \frac{1}{z+i\alpha}\right)\\
&= \frac{1}{2i}\frac{z+i\alpha - z + i\alpha}{(z-i\alpha)(z+i\alpha)}\\
&= \frac{\alpha}{z^2+\alpha^2}.
\end{aligned}$$

Definition 3 Let $f : \mathscr{N}_0 \longmapsto \mathscr{R}$ and let $z \in \mathscr{R}$. Then the $\mathscr{Z}$–transform is defined by

$$\mathscr{Z}(f)(z) = \sum_{t=0}^{\infty} \frac{f(t)}{(z+1)^{t+1}}$$

provided the series converges.

Theorem 6 *Let $\mathscr{T}_0 = \mathscr{N}_0$. Then*

$$\mathscr{L}(f)(z) = \mathscr{Z}(f)(z)$$

for every $f : \mathscr{T}_0 \longmapsto \mathscr{R}$ and every $z \subset \mathscr{D}\{f\}$.

Proof For $y \in \mathscr{T}_0 \cap [0,\infty)$ we have

$$\begin{aligned}
e_{\ominus z}(\sigma(y),0) &= (1 + (\ominus z)\mu(y))e_{\ominus z}(y,0)\\
&= \left(1 - \frac{z\mu(y)}{1+z\mu(y)}\right)e_{\ominus z}(y,0)\\
&= \frac{1}{1+z\mu(y)}e_{\ominus z}(y,0)\\
&= \frac{1}{1+z}e_{\ominus z}(y,0)
\end{aligned}$$

$$
\begin{aligned}
&= \frac{1}{1+z} e^{\int_0^y \frac{1}{\mu(y)} \mathrm{Log}(1+(\ominus z)\mu(y))\Delta y} \\
&= \frac{1}{1+z} e^{\int_0^y \mathrm{Log}\frac{1}{1+z}\Delta y} \\
&= \frac{1}{1+z}\left(\frac{1}{1+z}\right)^y \\
&= \frac{1}{(1+z)^{y+1}}.
\end{aligned}
$$

Hence,

$$
\begin{aligned}
\mathscr{L}(f)(z) &= \int_0^\infty \frac{f(y)}{(1+z)^{y+1}}\Delta y \\
&= \sum_{t=0}^{\infty} \frac{f(t)}{(1+z)^{t+1}} \\
&= \mathscr{Z}(f)(z).
\end{aligned}
$$

Exercise 1 Let $\alpha > 0$. Prove that

$$
\mathscr{Z}\left(\alpha^t\right)(z) = \frac{1}{z+1-\alpha}
$$

for every $z \in \mathscr{D}\{\alpha^t\}$ such that $|z+1| > \alpha$.

Exercise 2 Let $f : \mathscr{N}_0 \longmapsto \mathscr{R}$. Prove that

1. $\mathscr{Z}(f^\sigma)(z) = (z+1)\mathscr{Z}(f)(z) - f(0)$,
2. $\mathscr{Z}(f^{\sigma\sigma}) = (z+1)^2\mathscr{Z}(f)(z) - (z+1)f(0) - f(1)$,
3. $\mathscr{Z}\left(f^{\sigma^l}\right)(z) = (z+1)^l\mathscr{Z}(f)(z) - \sum_{k=0}^{l-1}(z+1)^{l-1-k}f(k),\ l \in \mathscr{N}$

for every $z \in \mathscr{D}\{f\}$.

7.1.3 Convolution and Shifting Properties of Special Functions

The usual convolution of two functions f and g on the real interval $[0, \infty)$ is defined by

$$
(f \star g)(x) = \int_0^x f(x-y)g(y)ds \quad for \quad x \geq 0.
$$

However, this definition does not work for general time scales because $x, y \in \mathscr{T}_0$ does not imply that $x - y \in \mathscr{T}_0$.

Definition 4 Assume that f is one of the functions $e_\alpha(x,0)$, $\sinh_\alpha(x,0)$, $\cosh_\alpha(x,0)$, $\cos_\alpha(x,0)$, $\sin_\alpha(x,0)$, or $h_k(x,0)$, $k \in \mathscr{N}_0$. If g is a regulated function on $\mathscr{T}_0$, then we define the convolution of f with g by

$$(f \star g)(x) = \int_0^x f(x, \sigma(y)) g(y) \Delta y \quad \textit{for} \quad x \in \mathscr{T}_0.$$

Theorem 7 (Convolution Theorem) *Assume that* $\alpha \in \mathscr{R}$ *and* f *is one of the functions* $e_\alpha(x,0)$, $\sinh_\alpha(x,0)$, $\cosh_\alpha(x,0)$, $\cos_\alpha(x,0)$, $\sin_\alpha(x,0)$, *or* $h_k(x,0)$, $k \in \mathscr{N}_0$. *If* g *is a regulated function on* $\mathscr{T}_0$ *such that*

$$\lim_{x\to\infty} e_{\ominus z}(x,0)(f \star g)(x) = 0,$$

then

$$\mathscr{L}(f \star g)(z) = \mathscr{L}(f)(z)\mathscr{L}(g)(z). \tag{7.6}$$

Proof 1. $f(x,0) = e_\alpha(x,0)$. Consider the initial value problem for the dynamic equation

$$l^\Delta - \alpha l = g(x), \quad l(0) = 0. \tag{7.7}$$

The solution of this problem is given by

$$l(x) = \int_0^x e_\alpha(x, \sigma(y)) g(y) \Delta y,$$

which can be rewritten in the form

$$l(x) = (e_\alpha(x, \sigma(y)) \star g)(x).$$

Now we apply the Laplace transform to both sides of the Eq. (7.7) to get

$$\mathscr{L}\left(l^\Delta - \alpha l\right)(z) = \mathscr{L}(g)(z)$$

or

$$\mathscr{L}(l^\Delta)(z) - \alpha \mathscr{L}(l)(z) = \mathscr{L}(g)(z),$$

or

$$z\mathscr{L}(l)(z) - \alpha \mathscr{L}(l)(z) = \mathscr{L}(g)(z),$$

whereupon

$$\mathscr{L}(l)(z) = \frac{1}{z-\alpha}\mathscr{L}(g)(z).$$

Since

$$\mathscr{L}(e_\alpha(x,0))(z) = \frac{1}{z-\alpha},$$

we conclude that

$$\mathscr{L}(l)(z) = \mathscr{L}(f)(z)\mathscr{L}(g)(z)$$

or we get (7.6).

2. Let $f(x,0) = \cosh_\alpha(x,0)$. We have

$$\cosh_\alpha(x,0) = \frac{e_\alpha(x,0) + e_{-\alpha}(x,0)}{2}.$$

Then

$$\begin{aligned}
(f \star g)(x) &= \int_0^x \cosh_\alpha(x,\sigma(y))g(y)\Delta y \\
&= \frac{1}{2}\int_0^x \left(e_\alpha(x,\sigma(y)) + e_{-\alpha}(x,\sigma(y))\right) g(y)\Delta y \\
&= \frac{1}{2}\int_0^x e_\alpha(x,\sigma(y))g(y)\Delta y + \frac{1}{2}\int_0^x e_{-\alpha}(x,\sigma(y))g(y)\Delta y \\
&= \frac{1}{2}(e_\alpha(x,\sigma(y)) \star g)(x) + \frac{1}{2}(e_{-\alpha}(x,\sigma(y)) \star g)(x).
\end{aligned}$$

From here,

$$\begin{aligned}
\mathscr{L}(f \star g)(z) &= \mathscr{L}\left(\frac{1}{2}(e_\alpha(x,\sigma(y)) \star g)(x) + \frac{1}{2}(e_{-\alpha}(x,\sigma(y)) \star g)(x)\right)(z) \\
&= \frac{1}{2}\mathscr{L}(e_\alpha(x,\sigma(y)) \star g)(z) + \frac{1}{2}\mathscr{L}(e_{-\alpha}(x,\sigma(y)) \star g)(z) \\
&= \frac{1}{2}\mathscr{L}(e_\alpha(x,\sigma(y)))(z)\mathscr{L}(g)(z) + \frac{1}{2}\mathscr{L}(e_{-\alpha}(x,\sigma(y)))(z)\mathscr{L}(g)(z) \\
&= \frac{1}{2}\frac{1}{z-\alpha}\mathscr{L}(g)(z) + \frac{1}{2}\frac{1}{z+\alpha}\mathscr{L}(g)(z) \\
&= \frac{1}{2}\left(\frac{1}{z-\alpha} + \frac{1}{z+\alpha}\right)\mathscr{L}(g)(z) \\
&= \frac{z}{z^2-\alpha^2}\mathscr{L}(g)(z) \\
&= \mathscr{L}(\cosh_\alpha(x,0))(z)\mathscr{L}(g)(z) \\
&= \mathscr{L}(f)(z)\mathscr{L}(g)(z).
\end{aligned}$$

3. Let $f(x,0) = \sinh_\alpha(x,0)$. Then

$$\sinh_\alpha(x,0) = \frac{e_\alpha(x,0) - e_{-\alpha}(x,0)}{2},$$

and

$$\begin{aligned}(f \star g)(x) &= \int_0^x \sinh_\alpha(x, \sigma(y))g(y)\Delta y \\ &= \frac{1}{2}\int_0^x \left(e_\alpha(x, \sigma(y)) - e_{-\alpha}(x, \sigma(y))\right) g(y)\Delta y \\ &= \frac{1}{2}\int_0^x e_\alpha(x, \sigma(y))g(y)\Delta y - \frac{1}{2}\int_0^x e_{-\alpha}(x, \sigma(y))g(y)\Delta y \\ &= \frac{1}{2}(e_\alpha(x, \sigma(y)) \star g)(x) - \frac{1}{2}(e_{-\alpha}(x, \sigma(y)) \star g)(x).\end{aligned}$$

Hence,

$$\begin{aligned}\mathscr{L}(f \star g)(z) &= \mathscr{L}\left(\frac{1}{2}(e_\alpha(x, \sigma(y)) \star g)(x) - \frac{1}{2}(e_{-\alpha}(x, \sigma(y)) \star g)(x)\right)(z) \\ &= \frac{1}{2}\mathscr{L}(e_\alpha(x, \sigma(y)) \star g)(z) - \frac{1}{2}\mathscr{L}(e_{-\alpha}(x, \sigma(y)) \star g)(z) \\ &= \frac{1}{2}\mathscr{L}(e_\alpha(x, \sigma(y)))(z)\mathscr{L}(g)(z) - \frac{1}{2}\mathscr{L}(e_{-\alpha}(x, \sigma(y)))(z)\mathscr{L}(g)(z) \\ &= \frac{1}{2}\frac{1}{z-\alpha}\mathscr{L}(g)(z) - \frac{1}{2}\frac{1}{z+\alpha}\mathscr{L}(g)(z) \\ &= \frac{1}{2}\left(\frac{1}{z-\alpha} - \frac{1}{z+\alpha}\right)\mathscr{L}(g)(z) \\ &= \frac{\alpha}{z^2-\alpha^2}\mathscr{L}(g)(z) \\ &= \mathscr{L}(\sinh_\alpha(x, 0))(z)\mathscr{L}(g)(z) \\ &= \mathscr{L}(f)(z)\mathscr{L}(g)(z).\end{aligned}$$

4. Let $f(x, 0) = \cos_\alpha(x, 0)$. We have

$$\cos_\alpha(x, 0) = \frac{e_{i\alpha}(x, 0) + e_{-i\alpha}(x, 0)}{2}.$$

Then

$$\begin{aligned}(f \star g)(x) &= \int_0^x \cos_\alpha(x, \sigma(y))g(y)\Delta y \\ &= \frac{1}{2}\int_0^x \left(e_{i\alpha}(x, \sigma(y)) + e_{-i\alpha}(x, \sigma(y))\right) g(y)\Delta y \\ &= \frac{1}{2}\int_0^x e_{i\alpha}(x, \sigma(y))g(y)\Delta y + \frac{1}{2}\int_0^x e_{-i\alpha}(x, \sigma(y))g(y)\Delta y \\ &= \frac{1}{2}(e_{i\alpha}(x, \sigma(y)) \star g)(x) + \frac{1}{2}(e_{-i\alpha}(x, \sigma(y)) \star g)(x).\end{aligned}$$

From here,

$$\begin{aligned}
\mathscr{L}(f \star g)(z) &= \mathscr{L}\left(\frac{1}{2}(e_{i\alpha}(x, \sigma(y)) \star g)(x) + \frac{1}{2}(e_{-i\alpha}(x, \sigma(y)) \star g)(x)\right)(z) \\
&= \frac{1}{2}\mathscr{L}(e_{i\alpha}(x, \sigma(y)) \star g)(z) + \frac{1}{2}\mathscr{L}(e_{-i\alpha}(x, \sigma(y)) \star g)(z) \\
&= \frac{1}{2}\mathscr{L}(e_{i\alpha}(x, \sigma(y)))(z)\mathscr{L}(g)(z) + \frac{1}{2}\mathscr{L}(e_{-i\alpha}(x, \sigma(y)))(z)\mathscr{L}(g)(z) \\
&= \frac{1}{2}\frac{1}{z - i\alpha}\mathscr{L}(g)(z) + \frac{1}{2}\frac{1}{z + i\alpha}\mathscr{L}(g)(z) \\
&= \frac{1}{2}\left(\frac{1}{z - i\alpha} + \frac{1}{z + i\alpha}\right)\mathscr{L}(g)(z) \\
&= \frac{z}{z^2 + \alpha^2}\mathscr{L}(g)(z) \\
&= \mathscr{L}(\cos_\alpha(x, 0))(z)\mathscr{L}(g)(z) \\
&= \mathscr{L}(f)(z)\mathscr{L}(g)(z).
\end{aligned}$$

5. Let $f(x, 0) = \sin_\alpha(x, 0)$. Then

$$\sin_\alpha(x, 0) = \frac{e_{i\alpha}(x, 0) - e_{-i\alpha}(x, 0)}{2i},$$

and

$$\begin{aligned}
(f \star g)(x) &= \int_0^x \sin_\alpha(x, \sigma(y)) g(y) \Delta y \\
&= \frac{1}{2i}\int_0^x (e_{i\alpha}(x, \sigma(y)) - e_{-i\alpha}(x, \sigma(y)))\, g(y)\Delta y \\
&= \frac{1}{2i}\int_0^x e_{i\alpha}(x, \sigma(y)) g(y)\Delta y - \frac{1}{2}\int_0^x e_{-i\alpha}(x, \sigma(y)) g(y)\Delta y \\
&= \frac{1}{2i}(e_{i\alpha}(x, \sigma(y)) \star g)(x) - \frac{1}{2}(e_{-i\alpha}(x, \sigma(y)) \star g)(x).
\end{aligned}$$

Hence,

$$\begin{aligned}
\mathscr{L}(f \star g)(z) &= \mathscr{L}\left(\frac{1}{2i}(e_{i\alpha}(x, \sigma(y)) \star g)(x) - \frac{1}{2i}(e_{-i\alpha}(x, \sigma(y)) \star g)(x)\right)(z) \\
&= \frac{1}{2i}\mathscr{L}(e_{i\alpha}(x, \sigma(y)) \star g)(z) - \frac{1}{2i}\mathscr{L}(e_{-i\alpha}(x, \sigma(y)) \star g)(z) \\
&= \frac{1}{2i}\mathscr{L}(e_{i\alpha}(x, \sigma(y)))(z)\mathscr{L}(g)(z) - \frac{1}{2i}\mathscr{L}(e_{-i\alpha}(x, \sigma(y)))(z)\mathscr{L}(g)(z) \\
&= \frac{1}{2i}\frac{1}{z - i\alpha}\mathscr{L}(g)(z) - \frac{1}{2i}\frac{1}{z + i\alpha}\mathscr{L}(g)(z) \\
&= \frac{1}{2i}\left(\frac{1}{z - i\alpha} - \frac{1}{z + i\alpha}\right)\mathscr{L}(g)(z)
\end{aligned}$$

$$= \frac{\alpha}{z^2+\alpha^2}\mathscr{L}(g)(z)$$
$$= \mathscr{L}(\sin_\alpha(x,0))(z)\mathscr{L}(g)(z)$$
$$= \mathscr{L}(f)(z)\mathscr{L}(g)(z).$$

6. $f(x,0) = h_k(x,0), \quad k \in \mathscr{N}_0$. Consider the initial value problem for the dynamic equation

$$l^{\Delta^{k+1}}(x) = g(x), \quad l^{\Delta^j}(0) = 0, \quad j = 0, 1, \ldots, k. \tag{7.8}$$

The solution of the equation (7.8) is given by

$$l(x) = \int_0^x h_k(x, \sigma(y))g(y)\Delta y$$

or

$$l(x) = (h_k(x, \sigma(y)) \star g)(x).$$

Taking the Laplace transform of both sides of (7.8), gives

$$\mathscr{L}\left(l^{\Delta^{k+1}}\right)(z) = \mathscr{L}(g)(z),$$

whereupon

$$z^{k+1}\mathscr{L}(l)(z) = \mathscr{L}(g)(z)$$

or

$$\mathscr{L}(l)(z) = \frac{1}{z^{k+1}}\mathscr{L}(g)(z)$$
$$= \mathscr{L}(h_k(x,0))(z)\mathscr{L}(g)(z)$$
$$= \mathscr{L}(f)(z)\mathscr{L}(g)(z).$$

Theorem 8 *Assume f and g are each one of the functions $e_\alpha(x,0)$, $\cosh_\alpha(x,0)$, $\sinh_\alpha(x,0)$, $\cos_\alpha(x,0)$, $\sin_\alpha(x,0)$, or $h_k(x,0)$, not both $h_k(x,0)$. Then*

$$f \star g = g \star f.$$

Proof 1. Let $f(x,0) = e_\alpha(x,0)$ and $g(x,0) = e_\beta(x,0)$. Let also,

$$p(x) = e_\alpha(x,0) \star e_\beta(x,0) \quad \textit{and} \quad q(x) = e_\beta(x,0) \star e_\alpha(x,0).$$

Note that

$$p(0) = q(0) = 0.$$

Also, $p(x)$ and $q(x)$ are solutions to the initial value problems

$$p^{\Delta} - \alpha p = e_{\beta}(x, 0), \quad p(0) = 0,$$

and

$$q^{\Delta} - \beta q = e_{\alpha}(x, 0), \quad q(0) = 0,$$

respectively.
Then

$$\begin{aligned} p^{\Delta}(0) &= \alpha p(0) + e_{\beta}(0, 0) = 1, \\ q^{\Delta}(0) &= \beta q(0) + e_{\alpha}(0, 0) = 1. \end{aligned}$$

We claim that $p(x)$ and $q(x)$ are solutions to the initial value problem

$$m^{\Delta^2} - (\alpha + \beta)m^{\Delta} + \alpha\beta m = 0, \quad m(0) = 0, \quad m^{\Delta}(0) = 1. \tag{7.9}$$

We have

$$p^{\Delta} = \alpha p + e_{\beta}(x, 0),$$

from where, after differentiating, we obtain

$$p^{\Delta^2} = \alpha p^{\Delta} + \beta e_{\beta}(x, 0).$$

Hence,

$$\begin{aligned} &p^{\Delta^2} - (\alpha + \beta)p^{\Delta} + \alpha\beta p = \alpha p^{\Delta} + \beta e_{\beta}(x, 0) - (\alpha + \beta)p^{\Delta} + \alpha\beta p \\ &= \beta e_{\beta}(x, 0) - \beta p^{\Delta} + \alpha\beta p \\ &= \beta e_{\beta}(x, 0) - \beta(\alpha p + e_{\beta}(x, 0)) + \alpha\beta p \\ &= \beta e_{\beta}(x, 0) - \alpha\beta p - \beta e_{\beta}(x, 0) + \alpha\beta p \\ &= 0. \end{aligned}$$

Also,

$$q^{\Delta} = \beta q + e_{\alpha}(x, 0),$$

whereupon, differentiating, we obtain

$$q^{\Delta^2} = \beta q^{\Delta} + \alpha e_{\alpha}(x, 0).$$

Hence,

$$\begin{aligned} &q^{\Delta^2} - (\alpha + \beta)q^{\Delta} + \alpha\beta q = \beta q^{\Delta} + \alpha e_{\alpha}(x, 0) - (\alpha + \beta)q^{\Delta} + \alpha\beta q \\ &= \alpha e_{\alpha}(x, 0) - \alpha q^{\Delta} + \alpha\beta q \\ &= \alpha e_{\alpha}(x, 0) - \alpha(\beta q + e_{\alpha}(x, 0)) + \alpha\beta q \\ &= \alpha e_{\alpha}(x, 0) - \alpha\beta q - \alpha e_{\alpha}(x, 0) + \alpha\beta q \\ &= 0. \end{aligned}$$

Since the problem (7.9) has unique solution, we conclude that

$$p(x) = q(x).$$

2. Next we consider

$$\begin{aligned}
e_\alpha(x,0) \star \cosh_\beta(x,0) &= e_\alpha(x,0) \star \frac{e_\beta(x,0)+e_{-\beta}(x,0)}{2} \\
&= \frac{1}{2}\left(e_\alpha(x,0) \star (e_\beta(x,0)+e_{-\beta}(x,0))\right) \\
&= \frac{1}{2}\left(e_\alpha(x,0) \star e_\beta(x,0)+e_\alpha(x,0) \star e_{-\beta}(x,0)\right) \\
&= \frac{1}{2}\left(e_\beta(x,0) \star e_\alpha(x,0)+e_{-\beta}(x,0) \star e_\alpha(x,0)\right) \\
&= \frac{1}{2}\left(e_\beta(x,0)+e_{-\beta}(x,0)\right) \star e_\alpha(x,0) \\
&= \cosh_\beta(x,0) \star e_\alpha(x,0).
\end{aligned}$$

3. Let

$$z(x) = e_\alpha(x,0) \star h_k(x,0) \quad and \quad q(x) = h_k(x,0) \star e_\alpha(x,0).$$

We have that $z(x)$ is the solution to the initial value problem

$$z^\Delta(x) - \alpha z(x) = h_k(x,0), \quad z(0) = 0.$$

Differentiating this equation i times gives

$$z^{\Delta^{i+1}}(x) - \alpha z^\Delta(x) = h_{k-i}(x,0), \quad i = 1, 2, \ldots, k.$$

Also,

$$z^{\Delta^i}(0) = 0, \quad 0 \le i \le k, \quad z^{\Delta^{k+1}}(0) = 1.$$

Thus we get that $z(x)$ is the solution to the initial value problem

$$\begin{aligned}
&z^{\Delta^{k+2}}(x) - \alpha z^{\Delta^{k+1}}(x) = 0, \\
&z^{\Delta^i}(0) = 0, \quad 0 \le i \le k, \quad z^{\Delta^{k+1}}(0) = 1.
\end{aligned}$$

Since $h_k(0,0) = 0, \quad k > 0$, we obtain that

$$q^\Delta(0) = 0, \quad 0 \le i \le k.$$

Also, q is the solution of the initial value problem

$$q^{\Delta^{k+1}}(x) = e_\alpha(x,0), \quad q^{\Delta^i}(0) = 0, \quad 0 \le i \le k.$$

Hence, differentiating the last equation, we obtain

$$\begin{aligned} q^{\Delta^{k+2}}(x) &= e_\alpha^\Delta(x,0) \\ &= \alpha e_\alpha(x,0) \\ &= \alpha q^{\Delta^{k+1}}(x) \end{aligned}$$

and

$$q^{\Delta^{k+1}}(0) = e_\alpha(0,0) = 1.$$

Consequently $q(x)$ is the solution of the initial value problem

$$q^{\Delta^{k+2}}(x) - \alpha q^{\Delta^{k+1}}(x) = 0, \quad q^{\Delta^i}(0) = 0, \quad q^{\Delta^{k+1}}(0) = 1, \quad 0 \le i \le k.$$

Therefore $z(x)$ and $q(x)$ are solutions to the same initial value problem. Hence, they must be equal.

Exercise 3 Prove that

1. $e_\alpha(x,0) \star \sinh_\beta(x,0) = \sinh_\beta(x,0) \star e_\alpha(x,0)$,
2. $e_\alpha(x,0) \star \cos_\beta(x,0) = \cos_\beta(x,0) \star e_\alpha(x,0)$,
3. $e_\alpha(x,0) \star \sin_\beta(x,0) = \sin_\beta(x,0) \star e_\alpha(x,0)$,
4. $\cosh_\alpha(x,0) \star \cosh_\beta(x,0) = \cosh_\beta(x,0) \star \cosh_\alpha(x,0)$,
5. $\cosh_\alpha(x,0) \star \sinh_\beta(x,0) = \sinh_\beta(x,0) \star \cosh_\alpha(x,0)$,
6. $\cosh_\alpha(x,0) \star \cos_\beta(x,0) = \cos_\beta(x,0) \star \cosh_\alpha(x,0)$,
7. $\cosh_\alpha(x,0) \star \sin_\beta(x,0) = \sin_\beta(x,0) \star \cosh_\alpha(x,0)$,
8. $\cosh_\alpha(x,0) \star h_k(x,0) = h_k(x,0) \star \cosh_\alpha(x,0)$,
9. $\sinh_\alpha(x,0) \star \sinh_\beta(x,0) = \sinh_\beta(x,0) \star \sinh_\alpha(x,0)$,
10. $\sinh_\alpha(x,0) \star \cos_\beta(x,0) = \cos_\beta(x,0) \star \sinh_\alpha(x,0)$,
11. $\sinh_\alpha(x,0) \star \sin_\beta(x,0) = \sin_\beta(x,0) \star \sinh_\alpha(x,0)$,
12. $\sinh_\alpha(x,0) \star h_k(x,0) = h_k(x,0) \star \sinh_\alpha(x,0)$,
13. $\cos_\alpha(x,0) \star \cos_\beta(x,0) = \cos_\beta(x,0) \star \cos_\alpha(x,0)$,
14. $\cos_\alpha(x,0) \star \sin_\beta(x,0) = \sin_\beta(x,0) \star \cos_\alpha(x,0)$,
15. $\cos_\alpha(x,0) \star h_k(x,0) = h_k(x,0) \star \cos_\alpha(x,0)$,
16. $\sin_\alpha(x,0) \star \sin_\beta(x,0) = \sin_\beta(x,0) \star \sin_\alpha(x,0)$,
17. $\sin_\alpha(x,0) \star h_k(x,0) = h_k(x,0) \star \sin_\alpha(x,0)$.

Theorem 9 *Let $\alpha, \beta \in \mathscr{R}$ and $1+\alpha\mu(x) \neq 0$, $1+\beta\mu(x) \neq 0$ for all $x \in \mathscr{T}$. Then*

$$\mathscr{L}\left(e_\alpha(x,0)\sin_{\frac{\beta}{1+\alpha\mu}}(x,0)\right)(z) = \frac{\beta}{(z-\alpha)^2+\beta^2}$$

provided

$$\lim_{x\to\infty} e_\alpha(x,0)\sin_{\frac{\beta}{1+\mu\alpha}}(x,0) = 0 \quad \textit{and} \quad \lim_{x\to\infty} e_\alpha(x,0)\left(\sin_{\frac{\beta}{1+\mu\alpha}}(x,0)\right)^\Delta = 0.$$

Proof Let

$$p(x) = e_\alpha(x,0)\sin_{\frac{\beta}{1+\mu\alpha}}(x,0).$$

Then

$$\begin{aligned}
p^\Delta(x) &= e_\alpha^\Delta(x,0)\sin_{\frac{\beta}{1+\mu\alpha}}(x,0) + e_\alpha(\sigma(x),0)\sin^\Delta_{\frac{\beta}{1+\mu\alpha}}(x,0)\\
&= \alpha e_\alpha(x,0)\sin_{\frac{\beta}{1+\mu\alpha}}(x,0) + (1+\alpha\mu(x))e_\alpha(x,0)\frac{\beta}{1+\alpha\mu(x)}\cos_{\frac{\beta}{1+\mu\alpha}}(x,0)\\
&= \alpha e_\alpha(x,0)\sin_{\frac{\beta}{1+\mu\alpha}}(x,0) + \beta e_\alpha(x,0)\cos_{\frac{\beta}{1+\mu\alpha}}(x,0).
\end{aligned}$$

We differentiate the last equation and we get

$$\begin{aligned}
p^{\Delta^2}(x) &= \alpha e_\alpha^\Delta(x,0)\sin_{\frac{\beta}{1+\mu\alpha}}(x,0) + \alpha e_\alpha(\sigma(x),0)\sin^\Delta_{\frac{\beta}{1+\mu\alpha}}(x,0)\\
&+ \beta e_\alpha^\Delta(x,0)\cos_{\frac{\beta}{1+\mu\alpha}}(x,0) + \beta e_\alpha(\sigma(x),0)\cos^\Delta_{\frac{\beta}{1+\mu\alpha}}(x,0)\\
&= \alpha^2 e_\alpha(x,0)\sin_{\frac{\beta}{1+\mu\alpha}}(x,0) + \alpha(1+\alpha\mu(x))e_\alpha(x,0)\frac{\beta}{1+\alpha\mu(x)}\cos_{\frac{\beta}{1+\mu\alpha}}(x,0)\\
&+ \alpha\beta e_\alpha(x,0)\cos_{\frac{\beta}{1+\mu\alpha}}(x,0) - \beta(1+\alpha\mu(x))e_\alpha(x,0)\frac{\beta}{1+\alpha\mu(x)}\sin_{\frac{\beta}{1+\mu\alpha}}(x,0)\\
&= \alpha^2 e_\alpha(x,0)\sin_{\frac{\beta}{1+\mu\alpha}}(x,0) + \alpha\beta e_\alpha(x,0)\cos_{\frac{\beta}{1+\mu\alpha}}(x,0)\\
&+ \alpha\beta e_\alpha(x,0)\cos_{\frac{\beta}{1+\mu\alpha}}(x,0) - \beta^2 e_\alpha(x,0)\sin_{\frac{\beta}{1+\mu\alpha}}(x,0)\\
&= (\alpha^2-\beta^2)p(x) + 2\alpha\beta e_\alpha(x,0)\cos_{\frac{\beta}{1+\mu\alpha}}(x,0).
\end{aligned}$$

In this way we obtain the system

$$\begin{cases}
p^\Delta(x) = \alpha p(x) + \beta e_\alpha(x,0)\cos_{\frac{\beta}{1+\mu\alpha}}(x,0)\\[2ex]
p^{\Delta^2}(x) = (\alpha^2-\beta^2)p(x) + 2\alpha\beta e_\alpha(x,0)\cos_{\frac{\beta}{1+\mu\alpha}}(x,0).
\end{cases}$$

Hence,

$$p^{\Delta^2}(x) - 2\alpha p^\Delta(x) = -(\alpha^2+\beta^2)p(x)$$

or

$$p^{\Delta^2}(x) + 2\alpha p^\Delta(x) + (\alpha^2+\beta^2)p(x) = 0.$$

Also,

$$\begin{aligned}
&p(0) = e_\alpha(0,0)\sin_{\frac{\beta}{1+\mu\alpha}}(0,0) = 0,\\
&p^\Delta(0) = \alpha e_\alpha(0,0)\sin_{\frac{\beta}{1+\mu\alpha}}(0,0) + \beta e_\alpha(0,0)\cos_{\frac{\beta}{1+\mu\alpha}}(0,0) = \beta.
\end{aligned}$$

Consequently we obtain the following initial value problem

$$\begin{cases}
p^{\Delta^2}(x) - 2\alpha p^\Delta(x) + (\alpha^2+\beta^2)p(x) = 0\\
p(0) = 0, \quad p^\Delta(0) = \beta.
\end{cases} \tag{7.10}$$

Now we apply the Laplace transform to both sides of the dynamic equation (7.10) and we obtain

$$\mathscr{L}\left(p^{\Delta^2}(x) - 2\alpha p^{\Delta}(x) + (\alpha^2+\beta^2)p(x)\right)(z) = 0$$

or

$$\mathscr{L}\left(p^{\Delta^2}(x)\right)(z) - 2\alpha\mathscr{L}\left(p^{\Delta}(x)\right)(z) + (\alpha^2+\beta^2)\mathscr{L}(p(x))(z) = 0,$$

or

$$z^2\mathscr{L}(p)(z) - p^{\Delta}(0) - zp(0) - 2\alpha z\mathscr{L}(p)(z) + 2\alpha p(0) + (\alpha^2+\beta^2)\mathscr{L}(p)(z) = 0,$$

or

$$z^2\mathscr{L}(p)(z) - \beta - 2\alpha z\mathscr{L}(p)(z) + (\alpha^2+\beta^2)\mathscr{L}(p)(z) = 0$$

or

$$\left(z^2 - 2\alpha z + \alpha^2 + \beta^2\right)\mathscr{L}(p)(z) = \beta,$$

whereupon

$$\mathscr{L}(p)(z) = \frac{\beta}{(z-\alpha)^2+\beta^2},$$

which completes the proof.

Exercise 4 Let $\alpha, \beta \in \mathscr{R}$ and $1+\alpha\mu(x) \neq 0$, $1+\beta\mu(x) \neq 0$ for all $x \in \mathscr{T}$. Prove that

$$\mathscr{L}\left(e_{\alpha}(x,0)\cos_{\frac{\beta}{1+\alpha\mu}}(x,0)\right)(z) = \frac{z-\alpha}{(z-\alpha)^2+\beta^2},$$

provided that

$$\lim_{x\to\infty} e_{\alpha}(x,0)\cos_{\frac{\beta}{1+\alpha\mu}}(x,0) = \lim_{x\to\infty} e_{\alpha}(x,0)\cos^{\Delta}_{\frac{\beta}{1+\alpha\mu}}(x,0) = 0.$$

Definition 5 Let $a \in \mathscr{T}_0,\quad a > 0$. Define the step function u_a by

$$u_a = \begin{cases} 0 & \textit{if} \quad x \in \mathscr{T}_0 \cap (-\infty, a) \\ 1 & \textit{if} \quad x \in \mathscr{T}_0 \cap [a, \infty). \end{cases}$$

Theorem 10 *Let* $a \in \mathscr{T}_0,\quad a > 0$. *Then*

$$\mathscr{L}(u_a(x))(z) = \frac{e_{\ominus z}(a,0)}{z}$$

for those $z \in \mathscr{D}\{u_a\}$ *such that*

$$\lim_{x\to\infty} e_{\ominus z}(x,0) = 0.$$

Proof We have

$$\begin{aligned}
\mathscr{L}(u_a(x))(z) &= \int_0^\infty u_a(y)e_{\ominus z}(\sigma(y),0)\Delta y\\
&= \int_a^\infty e_{\ominus z}(\sigma(y),0)\Delta y\\
&= \int_a^\infty (1+\mu\ominus z)e_{\ominus z}(y,0)\Delta y\\
&= \int_a^\infty \frac{1}{1+\mu z}e_{\ominus z}(y,0)\Delta y\\
&= -\frac{1}{z}\int_a^\infty \frac{-z}{1+\mu z}e_{\ominus z}(y,0)\Delta y\\
&= -\frac{1}{z}\int_a^\infty \ominus z e_{\ominus z}(y,0)\Delta y\\
&= -\frac{1}{z}\int_a^\infty e^{\Delta}_{\ominus z}(y,0)\Delta y\\
&= -\frac{1}{z}e_{\ominus z}(y,0)\Big|_{y=a}^{y=\infty}\\
&= -\frac{1}{z}\left(\lim_{y\to\infty} e_{\ominus z}(y,0)-e_{\ominus z}(a,0)\right)\\
&= \frac{1}{z}e_{\ominus z}(a,0).
\end{aligned}$$

Theorem 11 *Let* $a\in\mathscr{T}_0,\quad a>0$. *Assume that* f *is one of the functions* $e_\alpha(x,0)$, $\cos_\alpha(x,0)$, $\sin_\alpha(x,a)$, $\sinh_\alpha(x,0)$, $\cosh_\alpha(x,0)$. *If* $1+z\mu(x)\neq 0,\ 1+\alpha\mu(x)\neq 0$ *for all* $x\in\mathscr{T}_0$, *and*

$$\lim_{x\to\infty} e_{\alpha\ominus z}(x,a)=\lim_{x\to\infty} e_{i\alpha\ominus z}(x,a)=\lim_{x\to\infty} e_{-i\alpha\ominus z}(x,a)=0,$$

then

$$\mathscr{L}(u_a(x)f(x,a))=e_{\ominus z}(a,0)\mathscr{L}(f(x,a))(z).$$

Proof 1. Let $f(x,a)=e_\alpha(x,a)$. Then

$$\begin{aligned}
e_\alpha(x,a)e_{\ominus z}(\sigma(x),0) &= (1+\mu(x)\ominus z)e_\alpha(x,a)e_{\ominus z}(x,0)\\
&= \frac{1}{1+z\mu(x)}e_\alpha(x,a)e_{\ominus z}(x,0)\frac{e_{\ominus z}(0,a)}{e_{\ominus z}(0,a)}\\
&= \frac{1}{1+z\mu(x)}e_\alpha(x,a)e_{\ominus z}(x,a)\frac{1}{e_{\ominus z}(0,a)}\\
&= \frac{1}{1+z\mu(x)}e_{\alpha\ominus z}(x,a)e_{\ominus z}(a,0)
\end{aligned}$$

$$= \frac{1}{\alpha - z} \frac{\alpha - z}{1 + z\mu(x)} e_{\alpha\ominus z}(x, a) e_{\ominus z}(a, 0)$$
$$= \frac{1}{\alpha - z} (\alpha \ominus z) e_{\alpha\ominus z}(x, a) e_{\ominus z}(a, 0)$$
$$= \frac{1}{\alpha - z} e^{\Delta}_{\alpha\ominus z}(x, a) e_{\ominus z}(a, 0).$$

Thus

$$\begin{aligned}
\mathscr{L}(u_a(x)f(x)) &= \int_0^{\infty} u_a(x)f(x, a)e_{\ominus z}(\sigma(x), 0)\Delta x \\
&= \int_a^{\infty} e_\alpha(x, a)e_{\ominus z}(\sigma(x), 0)\Delta x \\
&= \frac{1}{\alpha - z} e_{\ominus z}(a, 0) \int_a^{\infty} e^{\Delta}_{\alpha\ominus z}(x, a)\Delta x \\
&= \frac{1}{\alpha - z} e_{\ominus z}(a, 0) e_{\alpha\ominus z}(x, a)\Big|_{x=a}^{x=\infty} \\
&= \frac{1}{\alpha - z} e_{\ominus z}(a, 0) \left(\lim_{x\to\infty} e_{\alpha\ominus z}(x, a) - e_{\alpha\ominus z}(a, a)\right) \\
&= \frac{1}{z - \alpha} e_{\ominus z}(a, 0) \\
&= e_{\ominus z}(a, 0)\mathscr{L}(e_\alpha(x, 0))(z).
\end{aligned}$$

2. Let $f(x, a) = \cos_\alpha(x, a)$. Then

$$f(x, a) = \frac{1}{2}\left(e_{i\alpha}(x, a) + e_{-\imath\alpha}(x, a)\right).$$

Hence,

$$\begin{aligned}
&\mathscr{L}(u_a(x)f(x, a)) = \mathscr{L}\left(u_a(x, a)\tfrac{1}{2}(e_{i\alpha}(x, a) + e_{-\imath\alpha}(x, a))\right)(z) \\
&= \mathscr{L}\left(\tfrac{1}{2}u_a(x)e_{i\alpha}(x, a) + \tfrac{1}{2}u_a(x)e_{-i\alpha}(x, a)\right)(z) \\
&= \tfrac{1}{2}\mathscr{L}(u_a(x)e_{i\alpha}(x, a))(z) + \tfrac{1}{2}\mathscr{L}(u_a(x)e_{-i\alpha}(x, a))(z) \\
&= \tfrac{1}{2}e_{\ominus z}(a, 0)\mathscr{L}(e_{i\alpha}(x, 0)) + \tfrac{1}{2}e_{\ominus z}(a, 0)\mathscr{L}(e_{-i\alpha}(x, 0))(z) \\
&= e_{\ominus z}(a, 0)\mathscr{L}\left(\tfrac{1}{2}e_{i\alpha}(x, 0)\right) + e_{\ominus z}(a, 0)\mathscr{L}\left(\tfrac{1}{2}e_{-i\alpha}(x, 0)\right)(z) \\
&= e_{\ominus z}(a, 0)\mathscr{L}\left(\tfrac{1}{2}(e_{i\alpha}(x, 0) + e_{-i\alpha}(x, 0))\right)(z) \\
&= e_{\ominus z}(a, 0)\mathscr{L}(\cos_\alpha(x, 0))(z).
\end{aligned}$$

Definition 6 Let $a, b, \alpha \in \mathscr{T}$ and $f : \mathscr{T} \longmapsto \mathscr{R}$ be continuous. If $\delta_\alpha(x), \quad x \in \mathscr{T}$, satisfies the following conditions

$$\int_a^b f(x)\delta_\alpha(x)\Delta x = \begin{cases} f(\alpha) & \text{if} \quad \alpha \in [a, b) \\ 0 & \text{otherwise}, \end{cases}$$

then $\delta_\alpha(x)$ will be called Dirac delta function.

Theorem 12 *Let* $\alpha \in \mathscr{T}_0, \quad \alpha \geq 0$. *Then*

$$\mathscr{L}(\delta_\alpha(x))(z) = e^\sigma_{\ominus z}(\alpha, 0).$$

Proof We have

$$\begin{aligned}\mathscr{L}(\delta_\alpha(x))(z) &= \int_0^\infty \delta_\alpha(y) e_{\ominus z}(\sigma(y), 0)\Delta y\\ &= e_{\ominus z}(\sigma(\alpha), 0).\end{aligned}$$

Exercise 5 Prove the following relations.

1. If $\alpha \neq \beta$, then
$$e_\alpha(x,0) \star e_\beta(x,0) = \frac{1}{\beta - \alpha}(e_\beta(x,0) - e_\alpha(x,0)).$$
2.
$$e_\alpha(x,0) \star e_\alpha(x,0) = e_\alpha(x,0)\int_0^x \frac{1}{1+\alpha\mu(y)}\Delta y.$$
3. If $\alpha^2 + \beta^2 \neq 0$, then
$$e_\alpha(x,0) \star \sin_\beta(x,0) = \frac{\beta e_\alpha(x,0) - \alpha \sin_\beta(x,0) - \beta\cos_\beta(x,0)}{\alpha^2+\beta^2.}$$
4. If $\alpha \neq 0$, $\alpha \neq \beta$, then
$$\cos_\alpha(x,0) \star \cos_\beta(x,0) = \frac{-\beta \sin_\beta(x,0) + \alpha\sin_\alpha(x,0)}{\alpha^2 - \beta^2}.$$
5. If $\alpha \neq 0$, then
$$\cos_\alpha(x,0) \star \cos_\alpha(x,0) = \frac{1}{\alpha}\sin_\alpha(x,0) + \frac{1}{2}x\cos_\alpha(x,0).$$
6. If $k \geq 0$, then
$$\sin_\alpha(x,0) \star h_k(x,0)$$
$$= \begin{cases} (-1)^{\frac{(k+1)(k+2)}{2}} \frac{1}{\alpha^{k+1}} \cos_\alpha(x,0) + \sum_{j=0}^{\frac{k}{2}} (-1)^j \frac{h_{k-2j}(x,0)}{\alpha^{2j+1}} & \text{if } k \text{ is even}\\ (-1)^{\frac{(k+1)(k+2)}{2}} \frac{1}{\alpha^{k+1}} \sin_\alpha(x,0) + \sum_{j=0}^{\frac{k-1}{2}} (-1)^j \frac{h_{k-2j}(x,0)}{\alpha^{2j+1}} & \text{if } k \text{ is odd}.\end{cases}$$

7.2 Applications to Dynamic Equations

The Laplace transform method for solving dynamic equations can be summarized by the following steps.

1. Take the Laplace transform of both sides of the equation. This result in what is called the transformed equation.
2. Obtain the equation
$$\mathscr{L}(\phi)(z) = F(z),$$
where $F(z)$ is an algebraic expression in the variable z.
3. Finding the solution ϕ of the considered equation using the algebraic expression $F(z)$ and the properties of the Laplace transform.

Note that the initial conditions of the problem are absorbed into the method, unlike other approaches to problems of this type, i.e., the method of variation of parameters or undetermined coefficients.

Example 4 We consider the initial value problem

$$\begin{cases} \phi^{\Delta^2}(x) + 5\phi^{\Delta}(x) + 6\phi(x) = 0, \\ \phi(0) = 1, \quad \phi^{\Delta}(0) = -5. \end{cases}$$

The, using the properties of the Laplace transform, we have

$$\begin{aligned} \mathscr{L}(\phi^{\Delta^2})(z) &= z^2\mathscr{L}(\phi)(z) - \phi^{\Delta}(0) - z\phi(0) \\ &= z^2\mathscr{L}(\phi)(z) + 5 - z, \\ \mathscr{L}(\phi^{\Delta})(z) &= z\mathscr{L}(\phi)(z) - \phi(0) \\ &= z\mathscr{L}(\phi)(z) - 1. \end{aligned}$$

Hence, taking tha Laplace transform of both sides of the given equation, we get

$$\mathscr{L}(\phi^{\Delta^2} + 5\phi^{\Delta} + 6\phi)(z) = \mathscr{L}(0)(z)$$

or

$$\begin{aligned} 0 &= \mathscr{L}(\phi^{\Delta^2}(z) + 5\mathscr{L}(\phi^{\Delta})(z) + 6\mathscr{L}(\phi)(z) \\ &= z^2\mathscr{L}(\phi)(z) - z + 5 + 5\,(z\mathscr{L}(\phi)(z) - 1) + 6\mathscr{L}(\phi)(z) \\ &= (z^2 + 5z + 6)\mathscr{L}(\phi)(z) - z, \end{aligned}$$

whereupon

$$\begin{aligned}\mathscr{L}(\phi)(z) &= \frac{z}{z^2+5z+6}\\ &= \frac{z}{(z+2)(z+3)}\\ &= \frac{3}{z+3} - \frac{2}{z+2}\\ &= 3\mathscr{L}(e_3(x,0))(z) - 2\mathscr{L}(e_{-2}(x,0))(z)\\ &= \mathscr{L}(3e_{-3}(x,0) - 2e_{-2}(x,0))(z).\end{aligned}$$

Therefore

$$\phi(x) = 3e_{-3}(x,0) - 2e_{-2}(x,0), \quad x \in \mathscr{T}_0.$$

Example 5 Consider the initial value problem

$$\begin{cases}\phi^{\Delta^2} - 4\phi^{\Delta} + 13\phi = 0\\ \phi(0) = 1, \quad \phi^{\Delta}(0) = 1.\end{cases}$$

Using the properties of the Laplace transform, we have

$$\begin{aligned}\mathscr{L}(\phi^{\Delta^2})(z) &= z^2\mathscr{L}(\phi)(z) - \phi^{\Delta}(0) - z\phi(0)\\ &= z^2\mathscr{L}(\phi)(z) - 1 - z,\\ \mathscr{L}(\phi^{\Delta})(z) &= z\mathscr{L}(\phi)(z) - \phi(0)\\ &= z\mathscr{L}(\phi)(z) - 1\end{aligned}$$

Taking the Laplace transform of both sides of the given equation, we obtain

$$\mathscr{L}(\phi^{\Delta^2} - 4\phi^{\Delta} + 13\phi)(z) = \mathscr{L}(0)(z)$$

or

$$\mathscr{L}(\phi^{\Delta^2})(z) - 4\mathscr{L}(\phi^{\Delta})(z) + 13\mathscr{L}(\phi)(z) = 0,$$

or

$$z^2\mathscr{L}(\phi)(z) - 1 - z - 4\left(z\mathscr{L}(\phi)(z) - 1\right) + 13\mathscr{L}(\phi)(z) = 0,$$

or

$$z^2\mathscr{L}(\phi)(z) - 1 - z - 4z\mathscr{L}(\phi)(z) + 4 + 13\mathscr{L}(\phi)(z) = 0,$$

or

$$(z^2 - 4z + 13)\mathscr{L}(\phi)(z) = z - 3,$$

or

$$\begin{aligned}
\mathscr{L}(\phi)(z) &= \frac{z-3}{z^2-4z+13}\\
&= \frac{z-3}{(z-2)^2+9}\\
&= \frac{z-2-1}{(z-2)^2+9}\\
&= \frac{z-2}{(z-2)^2+9} - \frac{1}{(z-2)^2+9}\\
&= \frac{z-2}{(z-2)^2+9} - \frac{1}{3}\frac{3}{(z-2)^2+9}\\
&= \mathscr{L}\left(e_2(x,0)\cos_{\frac{3}{1+2\mu(x)}}(x,0)\right)(z) - \frac{1}{3}\mathscr{L}\left(e_2(x,0)\sin_{\frac{3}{1+2\mu(x)}}(x,0)\right)(z)\\
&= \mathscr{L}\left(e_2(x,0)\cos_{\frac{3}{1+2\mu(x)}}(x,0) - \frac{1}{3}e_2(x,0)\sin_{\frac{3}{1+2\mu(x)}}(x,0)\right)(z).
\end{aligned}$$

Therefore

$$\phi(x) = e_2(x,0)\cos_{\frac{3}{1+2\mu(x)}}(x,0) - \frac{1}{3}e_2(x,0)\sin_{\frac{3}{1+2\mu(x)}}(x,0).$$

Example 6 Consider the initial value problem

$$\begin{aligned}
&\phi^{\Delta^2} + 2\phi^{\Delta} - 3\phi = 0\\
&\phi(0) = 5, \quad \phi^{\Delta}(0) = 1.
\end{aligned}$$

Using the properties of the Laplace transform, we have

$$\begin{aligned}
\mathscr{L}(\phi^{\Delta^2})(z) &= z^2\mathscr{L}(\phi)(z) - \phi^{\Delta}(0) - z\phi(0)\\
&= z^2\mathscr{L}(\phi)(z) - 1 - 5z,\\
\mathscr{L}(\phi^{\Delta})(z) &= z\mathscr{L}(\phi)(z) - \phi(0)\\
&= z\mathscr{L}(\phi)(z) - 5.
\end{aligned}$$

Taking the Laplace transform of both sides of the given equation, we get

$$\mathscr{L}\left(\phi^{\Delta^2} + 2\phi^{\Delta} - 3\phi\right)(z) = \mathscr{L}(0)(z)$$

or

$$\mathscr{L}(\phi^{\Delta^2}) + 2\mathscr{L}(\phi^{\Delta})(z) - 3\mathscr{L}(\phi)(z) = 0,$$

or

$$z^2\mathscr{L}(\phi)(z) - 1 - 5z + 2\left(z\mathscr{L}(\phi)(z) - 5\right) - 3\mathscr{L}(\phi)(z) = 0,$$

or

$$z^2\mathscr{L}(\phi)(z) - 1 - 5z + 2z\mathscr{L}(\phi)(z) - 10 - 3\mathscr{L}(\phi)(z) = 0,$$

or

$$(z^2 + 2z - 3)\mathscr{L}(\phi)(z) = 5z + 11,$$

or

$$\begin{aligned}
\mathscr{L}(\phi)(z) &= \frac{5z+11}{z^2+2z-3} \\
&= \frac{5z+11}{(z+1)^2-2} \\
&= \frac{5(z+1)+6}{(z+1)^2-2} \\
&= 5\frac{z+1}{(z+1)^2-2} + \frac{6}{(z+1)^2-2} \\
&= 5\frac{z+1}{(z+1)^2-(\sqrt{2})^2} + \frac{6}{\sqrt{2}}\frac{\sqrt{2}}{(z+1)^2-(\sqrt{2})^2} \\
&= 5\frac{z+1}{(z+1)^2-(\sqrt{2})^2} + 3\sqrt{2}\frac{\sqrt{2}}{(z+1)^2-(\sqrt{2})^2} \\
&= 5\mathscr{L}\left(e_{-1}(x,0)\cosh_{\frac{\sqrt{2}}{1-\mu(x)}}(x,0)\right)(z) + 3\sqrt{2}\mathscr{L}\left(e_{-1}(x,0)\sinh_{\frac{\sqrt{2}}{1-\mu(x)}}(x,0)\right)(z) \\
&= \mathscr{L}\left(5e_{-1}(x,0)\cosh_{\frac{\sqrt{2}}{1-\mu(x)}}(x,0) + 3\sqrt{2}e_{-1}(x,0)\sinh_{\frac{\sqrt{2}}{1-\mu(x)}}(x,0)\right)(z).
\end{aligned}$$

Therefore

$$\phi(x) = 5e_{-1}(x,0)\cosh_{\frac{\sqrt{2}}{1-\mu(x)}}(x,0) + 3\sqrt{2}e_{-1}(x,0)\sinh_{\frac{\sqrt{2}}{1-\mu(x)}}(x,0).$$

Example 7 Consider the initial value problem

$$\begin{cases} \phi^{\Delta^3} + \phi^{\Delta} = e_1(x,0) \\ \phi(0) = \phi^{\Delta}(0) = \phi^{\Delta^2}(0) = 0. \end{cases}$$

Using the properties of the Laplace transform, we have

$$\begin{aligned}
\mathscr{L}(\phi^{\Delta^3})(z) &= z^3\mathscr{L}(\phi)(z) - \phi^{\Delta^2}(0) - z\phi^{\Delta}(0) - z^2\phi(0) \\
&= z^3\mathscr{L}(\phi)(z), \\
\mathscr{L}(\phi^{\Delta})(z) &= z\mathscr{L}(\phi)(z) - \phi(0) \\
&= z\mathscr{L}(\phi)(z), \\
\mathscr{L}(e_1(x,0))(z) &= \frac{1}{z-1}.
\end{aligned}$$

Taking the Laplace transform of both sides of the given equation, we get

$$\mathscr{L}\left(\phi^{\Delta^3}+\phi^{\Delta}\right)(z)=\mathscr{L}(e_1(x,0))(z)$$

or

$$\mathscr{L}(\phi^{\Delta^3})(z)+\mathscr{L}(\phi^{\Delta})(z)=\frac{1}{z-1},$$

or

$$z^3\mathscr{L}(\phi)(z)+z\mathscr{L}(\phi)(z)=\frac{1}{z-1},$$

or

$$(z^3+z)\mathscr{L}(\phi)(z)=\frac{1}{z-1},$$

or

$$\begin{aligned}
\mathscr{L}(\phi)(z) &= \frac{1}{(z-1)(z^3+z)}\\
&= \frac{1}{(z^2-z)(z^2+1)}\\
&= -\frac{1}{2}\frac{z}{z^2-z}+\frac{1}{z^2-z}+\frac{1}{2}\frac{z}{z^2+1}-\frac{1}{2}\frac{1}{z^2+1}\\
&= -\frac{1}{2}\frac{1}{z-1}+\frac{1}{z(z-1)}+\frac{1}{2}\frac{z}{z^2+1}-\frac{1}{2}\frac{1}{z^2+1}\\
&= -\frac{1}{2}\frac{1}{z-1}+\frac{1}{z-1}-\frac{1}{z}+\frac{1}{2}\frac{z}{z^2+1}-\frac{1}{2}\frac{1}{z^2+1}\\
&= \frac{1}{2}\frac{1}{z-1}-\frac{1}{z}+\frac{1}{2}\frac{z}{z^2+1}-\frac{1}{2}\frac{1}{z^2+1}\\
&= \frac{1}{2}\mathscr{L}(e_1(x,0))(z)-\mathscr{L}(h_0(x,0))(z)+\frac{1}{2}\mathscr{L}(\cos_1(x,0))(z)\\
&\quad -\frac{1}{2}\mathscr{L}(\sin_1(x,0))(z)\\
&= \mathscr{L}\left(\frac{1}{2}e_1(x,0)-h_0(x,0)+\frac{1}{2}\cos_1(x,0)-\frac{1}{2}\sin_1(x,0)\right)(z).
\end{aligned}$$

Therefore

$$\phi(x)=\frac{1}{2}e_1(x,0)-1+\frac{1}{2}\cos_1(x,0)-\frac{1}{2}\sin_1(x,0).$$

Exercise 6 Use the Laplace transform to solve the following initial value problems.

1.

$$\begin{cases}\phi^{\Delta^2}-6\phi^{\Delta}+13\phi=0\\ \phi(0)=\phi^{\Delta}(0)=1.\end{cases}$$

2.

$$\begin{cases} \phi^{\Delta^2} + 4\phi = \cos_2(x, 0) \\ \phi(0) = \phi^{\Delta}(0) = 1. \end{cases}$$

3.

$$\begin{cases} \phi^{\Delta^2} - 4\phi^{\Delta} + 20\phi = 0 \\ \phi(0) = 1, \quad \phi^{\Delta}(0) = 2. \end{cases}$$

Answer

1.

$$\phi(x) = e_3(x, 0)\cos_{\frac{2}{1+3\mu(x)}}(x, 0) - e_3(x, 0)\sin_{\frac{2}{1+3\mu(x)}}(x, 0).$$

2.

$$\phi(x) = \frac{1}{2}\cos_2(x, 0) \star \sin_2(x, 0) + \cos_2(x, 0) + \frac{1}{2}\sin_2(x, 0).$$

3.

$$\phi(x) = e_2(x, 0)\cos_{\frac{4}{1+2\mu(x)}}(x, 0).$$

7.3 Generalized Volterra Integral Equations of the Second Kind

Consider the generalized Volterra integral equation of the second kind

$$\phi(x) = u(x) + \lambda \int_0^x K(x, y)\phi(y)\Delta y, \tag{7.11}$$

where $K(x, y)$ is $e_\alpha(x, \sigma(y))$, $\cosh_\alpha(x, \sigma(y))$, $\sinh_\alpha(x, \sigma(y))$, $\cos_\alpha(x, \sigma(y))$, $\sin_\alpha(x, \sigma(y))$, or $h_k(x, \sigma(y))$, $u : \mathscr{T}_0 \longmapsto \mathscr{R}$ is a given continuous function, λ is a parameter.

The Eq. (7.11) we can rewrite in the following form

$$\phi(x) = u(x) + \lambda(K \star \phi)(x).$$

Taking the Laplace transform of both sides of the last equation, we get

$$\mathscr{L}(\phi)(z) = \mathscr{L}(u)(z) + \lambda\mathscr{L}(K \star \phi)(z),$$

whereupon, applying the convolution theorem for the Laplace transform, we obtain

$$\mathscr{L}(\phi)(z) = \mathscr{L}(u)(z) + \lambda\mathscr{L}(K)(z)\mathscr{L}(\phi)(z).$$

Hence,

$$(1-\lambda\mathscr{L}(K)(z))\mathscr{L}(\phi)(z)=\mathscr{L}(u)(z)$$

or

$$\mathscr{L}(\phi)(z)=\frac{\mathscr{L}(u)(z)}{1-\lambda\mathscr{L}(K)(z)}.$$

Example 8 Consider the equation

$$\phi(x)=1+2\int_0^x e_2(x,y+1)\phi(y)\Delta y,\quad \mathscr{T}_0=\mathscr{Z}.$$

Here

$$\sigma(x)=x+1,\quad \mu(x)=1,\quad x\in\mathscr{T}_0,$$
$$K(x,\sigma(y))=e_2(x,y+1),\quad u(x)=1,\quad \lambda=2.$$

The given equation we can rewrite in the following form

$$\phi(x)=1+2\left(e_2(x,0)\star\phi(x)\right).$$

We take the Laplace transform of both sides of the given equation and we find

$$\begin{aligned}\mathscr{L}(\phi)(z)&=\mathscr{L}(1)(z)+2\mathscr{L}\left(e_2(x,0)\star\phi(x)\right)(z)\\&=\frac{1}{z}+2\mathscr{L}(e_2(x,0))(z)\mathscr{L}(\phi)(z)\\&=\frac{1}{z}+\frac{2}{z-2}\mathscr{L}(\phi)(z),\end{aligned}$$

whereupon

$$\left(1-\frac{2}{z-2}\right)\mathscr{L}(\phi)(z)=\frac{1}{z},$$

or

$$\frac{z-2-2}{z-2}\mathscr{L}(\phi)(z)=\frac{1}{z},$$

or

$$\frac{z-4}{z-2}\mathscr{L}(\phi)(z)=\frac{1}{z},$$

or

$$\begin{aligned}\mathscr{L}(\phi)(z)&=\frac{z-2}{z(z-4)}\\&=\frac{1}{2}\frac{1}{z}+\frac{1}{2}\frac{1}{z-4}\end{aligned}$$

$$= \frac{1}{2}\mathscr{L}(h_0(x,0))(z) + \frac{1}{2}\mathscr{L}(e_4(x,0))(z)$$
$$= \mathscr{L}\left(\frac{1}{2}h_0(x,0) + \frac{1}{2}e_4(x,0)\right)(z).$$

Consequently

$$\phi(x) = \frac{1}{2}h_0(x,0) + \frac{1}{2}e_4(x,0).$$

Example 9 Consider the equation

$$\phi(x) = e_3(x,0) - \int_0^x \cosh_2(x,2y)\phi(y)\Delta y, \quad \mathscr{T}_0 = 2^{\mathscr{N}_0} \cup \{0\}.$$

Here

$$\sigma(x) = 2x, \quad \mu(x) = x, \quad x \in \mathscr{T}_0,$$
$$u(x) = e_3(x,0), \quad K(x,\sigma(y)) = \cosh_2(x,2y), \quad \lambda = -1.$$

Then the given equation can be rewritten in the following form

$$\phi(x) = e_3(x,0) - (\cosh_2(x,0) \star \phi(x)).$$

We take the Laplace transform of both sides of the last equation and we get

$$\mathscr{L}(\phi)(z) = \mathscr{L}(e_3(x,0))(z) - \mathscr{L}(\cosh_2(x,0) \star \phi(x))(z)$$
$$= \frac{1}{z-3} - \mathscr{L}(\cosh_2(x,0))(z)\mathscr{L}(\phi)(z)$$
$$= \frac{1}{z-3} - \frac{z}{z^2-4}\mathscr{L}(\phi)(z),$$

whereupon

$$\left(1 + \frac{z}{z^2-4}\right)\mathscr{L}(\phi)(z) = \frac{1}{z-3},$$

or

$$\frac{z^2+z-4}{z^2-4}\mathscr{L}(\phi)(z) = \frac{1}{z-3},$$

or

$$\mathscr{L}(\phi)(z) = \frac{z^2-4}{(z-3)(z^2+z-4)}$$
$$= \frac{5}{8}\frac{1}{z-3} + \frac{\frac{3}{8}z + \frac{1}{2}}{z^2+z-4}$$

$$= \tfrac{5}{8}\tfrac{1}{z-3} + \tfrac{3}{8}\tfrac{z+\frac{1}{2}}{\left(z+\frac{1}{2}\right)^2-\frac{17}{4}} + \tfrac{5}{16}\tfrac{1}{\left(z+\frac{1}{2}\right)^2-\frac{17}{4}}$$

$$= \tfrac{5}{8}\tfrac{1}{z-3} + \tfrac{3}{8}\tfrac{z+\frac{1}{2}}{\left(z+\frac{1}{2}\right)^2-\frac{17}{4}} + \tfrac{5\sqrt{17}}{136}\tfrac{\frac{\sqrt{17}}{2}}{\left(z+\frac{1}{2}\right)^2-\frac{17}{4}}.$$

Consequently

$$\phi(x) = \frac{5}{8}e_3(x,0) + \frac{3}{8}e_{-\frac{1}{2}}(x,0)\cosh_{\frac{\sqrt{17}}{2-x}}(x,0) + \frac{5\sqrt{17}}{136}e_{-\frac{1}{2}}(x,0)\sinh_{\frac{\sqrt{17}}{2-x}}.$$

Example 10 Consider the equation

$$\phi(x) = \sin_2(x,0) + \int_0^x \cos_2(x,3y)\phi(y)\Delta y, \quad \mathscr{T}_0 = 3^{\mathscr{N}_0} \cup \{0\}.$$

Here

$$\sigma(x) = 3x, \quad \mu(x) = 2x, \quad x \in \mathscr{T}_0,$$
$$u(x) = \sin_2(x,0), \quad K(x,\sigma(y)) = \cos_2(x,3y), \quad \lambda = 1.$$

The given equation we can rewrite in the following form.

$$\phi(x) = \sin_2(x,0) + \cos_2(x,0) \star \phi(x).$$

We take the Laplace transform of both sides of the last equation and we find

$$\begin{aligned}\mathscr{L}(\phi)(z) &= \mathscr{L}(\sin_2(x,0) + \cos_2(x,0) \star \phi(x))(z)\\ &= \mathscr{L}(\sin_2(x,0))(z) + \mathscr{L}(\cos_2(x,0))(z)\mathscr{L}(\phi)(z)\\ &= \frac{2}{z^2+4} + \frac{z}{z^2+4}\mathscr{L}(\phi)(z),\end{aligned}$$

or

$$\left(1 - \frac{z}{z^2+4}\right)\mathscr{L}(\phi)(z) = \frac{2}{z^2+4},$$

or

$$\frac{z^2-z+4}{z^2+4}\mathscr{L}(\phi)(z) = \frac{2}{z^2+4},$$

or

$$\begin{aligned}\mathscr{L}(\phi)(z) &= \frac{2}{z^2 - z + 4}\\ &= \frac{2}{\left(z - \frac{1}{2}\right)^2 + \frac{15}{4}}\\ &= \frac{2}{\frac{\sqrt{15}}{2}} \frac{\frac{\sqrt{15}}{2}}{\left(z - \frac{1}{2}\right)^2 + \left(\frac{\sqrt{15}}{2}\right)^2}\\ &= \frac{4\sqrt{15}}{15}\mathscr{L}\left(e_{\frac{1}{2}}(x, 0)\sin_{\frac{\sqrt{15}}{2+2x}}(x, 0)\right)(z).\end{aligned}$$

Consequently

$$\phi(x) = \frac{4\sqrt{15}}{15} e_{\frac{1}{2}}(x, 0)\sin_{\frac{\sqrt{15}}{2+2x}}(x, 0).$$

Exercise 7 Solve each of the following equations.

1. $\phi(x) = e_2(x, 0) + 4\int_0^x \phi(s)\Delta s,$
2. $\phi(x) = x + 4\int_0^x (x - \sigma(y))\phi(y)\Delta y.$

Answer

1. $\phi(x) = -e_2(x, 0) + 2e_4(x, 0),$
2. $\phi(x) = \frac{1}{2}\sinh_2(x, 0).$

Remark 3 Consider the generalized Volterra integral equation of the second kind

$$\phi(x) = u(x) + \lambda\int_0^x \phi(y)\Delta y. \tag{7.12}$$

We can rewrite this equation in the form

$$\phi(x) = u(x) + \lambda(1 \star \phi)(x).$$

Taking the Laplace transform of both sides of the last equation, we get

$$\begin{aligned}\mathscr{L}(\phi)(z) &= \mathscr{L}(u)(z) + \lambda\mathscr{L}(1 \star \phi)(z)\\ &= \mathscr{L}(u)(z) + \lambda\mathscr{L}(1)(z)\mathscr{L}(\phi)(z)\\ &= \mathscr{L}(u)(z) + \lambda\frac{1}{z}\mathscr{L}(\phi)(z)\end{aligned}$$

or

$$\left(1 - \lambda\frac{1}{z}\right)\mathscr{L}(\phi)(z) = \mathscr{L}(u)(z). \tag{7.13}$$

If we use that

$$\mathscr{L}\left(\int_0^x \phi(y)\Delta y\right)(z) = \frac{1}{z}\mathscr{L}(\phi)(z),$$

taking the Laplace transform of both sides of (7.12) we obtain (7.13).

7.4 Generalized Volterra Integral Equations of the First Kind

Here we consider the equation

$$u(x) = \lambda \int_0^x K(x, \sigma(y))\phi(y)\Delta y, \tag{7.14}$$

where $K(x, 0)$ is $e_\alpha(x, 0)$, $\cosh_\alpha(x, 0)$, $\sinh_\alpha(x, 0)$, $\cos_\alpha(x, 0)$, $\sin_\alpha(x, 0)$, or $h_k(x, 0)$, $u : \mathscr{T}_0 \longmapsto \mathscr{R}$ is a given continuous function, λ is a parameter.
We can rewrite the Eq. (7.14) in the following form

$$u(x) = \lambda K(x, 0) \star \phi(x). \tag{7.15}$$

We take the Laplace transform of both sides of (7.15) and we obtain

$$\mathscr{L}(u)(z) = \lambda \mathscr{L}(K(x, 0) \star \phi(x))(z).$$

Now applying the convolution theorem for the Laplace transform, we obtain

$$\mathscr{L}(u)(z) = \lambda \mathscr{L}(K)(z)\mathscr{L}(\phi)(z).$$

Example 11 Consider the equation

$$\sin_2(x, 0) = \int_0^x e_{-3}(x, y+1)\phi(y)\Delta y, \quad \mathscr{T}_0 = \mathscr{Z}.$$

Here

$$\sigma(x) = x + 1, \quad \mu(x) = 1, \quad x \in \mathscr{T}_0.$$

The given equation we can rewrite in the form

$$\sin_2(x, 0) = \int_0^x e_{-3}(x, \sigma(y))\phi(y)\Delta y$$

or

$$\sin_2(x, 0) = e_{-3}(x, 0) \star \phi(x).$$

We take the Laplace transform of both sides of the last equation and we get

$$\begin{aligned}\mathscr{L}(\sin_2(x,0))(z) &= \mathscr{L}(e_{-3}(x,0) \star \phi(x))(z)\\ &= \mathscr{L}(e_{-3}(x,0))(z)\mathscr{L}(\phi)(z)\\ &= \frac{1}{z+3}\mathscr{L}(\phi)(z),\end{aligned}$$

whereupon

$$\frac{2}{z^2+4} = \frac{1}{z+3}\mathscr{L}(\phi)(z).$$

Hence,

$$\begin{aligned}\mathscr{L}(\phi)(z) &= 2\frac{z+3}{z^2+4}\\ &= 2\frac{z}{z^2+4} + 3\frac{2}{z^2+4}\\ &= 2\mathscr{L}(\cos_2(x,0))(z) + 3\mathscr{L}(\sin_2(x,0))(z)\\ &= \mathscr{L}(2\cos_2(x,0) + 3\sin_2(x,0))(z).\end{aligned}$$

Consequently

$$\phi(x) = 2\cos_2(x,0) + 3\sin_2(x,0).$$

Example 12 Consider the equation

$$\sinh_3(x,0) = 2\int_0^x e_3(x,2y)\phi(y)\Delta y, \quad \mathscr{T}_0 = 2^{\mathscr{N}_0} \cup \{0\}.$$

Here

$$\sigma(x) = 2x, \quad \mu(x) = x, \quad x \in \mathscr{T}_0.$$

Then the given equation we can rewrite in the following form

$$\sinh_3(x,0) = 2\int_0^x e_3(x,\sigma(y))\phi(y)\Delta y$$

or

$$\sinh_3(x,0) = 2(e_3(x,0) \star \phi(x)).$$

We take the Laplace transform of both sides of the last equation and we obtain

$$\begin{aligned}\mathscr{L}(\sinh_3(x,0))(z) &= \mathscr{L}\left(2(e_3(x,0)\star\phi(x))\right)(z)\\&= 2\mathscr{L}(e_3(x,0)\star\phi(x))(z)\\&= 2\mathscr{L}(e_3(x,0))(z)\mathscr{L}(\phi)(z)\\&= 2\frac{1}{z-3}\mathscr{L}(\phi)(z),\end{aligned}$$

whereupon

$$\frac{3}{z^2-9} = 2\frac{1}{z-3}\mathscr{L}(\phi)(z).$$

Hence,

$$\begin{aligned}\mathscr{L}(\phi)(z) &= \frac{3}{2}\frac{1}{z+3}\\&= \frac{3}{2}\mathscr{L}(e_{-3}(x,0))(z)\\&= \mathscr{L}\left(\frac{3}{2}e_{(-3)}(x,0)\right)(z).\end{aligned}$$

Consequently

$$\phi(x) = \frac{3}{2}e_{-3}(x,0).$$

Example 13 Consider the equation

$$\cos_2(x,0)\star\sinh_3(x,0) = \int_0^x e_3(x,\sigma(y))\phi(y)\Delta y, \quad \mathscr{T}_0 = \mathscr{Z}.$$

Here

$$\sigma(x) = x+1, \quad \mu(x) = 1, \quad x\in\mathscr{T}_0.$$

The given equation we can rewrite in the following form

$$\cos_2(x,0)\star\sinh_3(x,0) = \int_0^x e_3(x,\sigma(y))\phi(y)\Delta y$$

or

$$\cos_2(x,0)\star\sinh_3(x,0) = e_3(x,0)\star\phi(x).$$

We take the Laplace transform of both sides of the last equation and we obtain

$$\mathscr{L}(\cos_2(x,0)\star\sinh_3(x,0))(z) = \mathscr{L}(e_3(x,0)\star\phi(x))(z)$$

or

$$\mathscr{L}(\cos_2(x,0))(z)\mathscr{L}(\sinh_3(x,0))(z) = \mathscr{L}(e_3(x,0))(z)\mathscr{L}(\phi)(z),$$

whereupon

$$\frac{z}{z^2+4}\frac{3}{z^2-9} = \frac{1}{z-3}\mathcal{L}(\phi)(z).$$

Hence,

$$\begin{aligned}
\mathcal{L}(\phi)(z) &= \frac{3z}{(z+3)(z^2+4)} \\
&= -\frac{9}{13}\frac{1}{z+3} + \frac{\frac{9}{13}z+\frac{12}{13}}{z^2+4} \\
&= -\frac{9}{13}\frac{1}{z+3} + \frac{9}{13}\frac{z}{z^2+4} + \frac{6}{13}\frac{2}{z^2+4} \\
&= -\frac{9}{13}\mathcal{L}(e_{(-3)}(x,0))(z) + \frac{9}{13}\mathcal{L}(\cos_2(x,0))(z) + \frac{6}{13}\mathcal{L}(\sin_2(x,0))(z) \\
&= \mathcal{L}\left(-\frac{9}{13}e_{(-3)}(x,0) + \frac{9}{13}\cos_2(x,0) + \frac{6}{13}\sin_2(x,0)\right).
\end{aligned}$$

Consequently

$$\phi(x) = -\frac{9}{13}e_{(-3)}(x,0) + \frac{9}{13}\cos_2(x,0) + \frac{6}{13}\sin_2(x,0).$$

Exercise 8 Use the Laplace transform to solve the following equations.

1. $e_3(x,0) = \int_0^x e_{-2}^{\sigma}(x,y)\phi(y)\Delta y, \quad \mathcal{T}_0 = \mathcal{Z},$
2. $\cosh_4(x,0) = \int_0^x e_2^{\sigma}(x,y)\phi(y)\Delta y, \quad \mathcal{T}_0 = 2^{\mathcal{N}_0} \cup \{0\},$
3. $\sinh_2(x,0) = \int_0^x \cosh_4^{\sigma}(x,y)\phi(y)\Delta y, \quad \mathcal{T}_0 = \mathcal{N}_0,$
4. $e_{-3}(x,0) \star e_2(x,0) = \int_0^x e_5(x,\sigma(y))\phi(y)\Delta y, \quad \mathcal{T}_0 = \mathcal{Z},$
5. $e_1(x,0) \star e_{-7}(x,0) = 4\int_0^x \cosh_2(x,\sigma(y))\phi(y)\Delta y, \quad \mathcal{T}_0 = \mathcal{Z},$
6. $\sinh_2(x,0) \star \cos_{-3}(x,0) = 2\int_0^x \sinh_4(x,\sigma(y))\phi(y)\Delta y, \quad \mathcal{T}_0 = \mathcal{Z}.$

7.5 Generalized Volterra Integro-Differential Equations of the Second Kind

Here we consider the generalized Volterra integro-differential equation of the second kind

$$\phi^{\Delta^n}(x) = u(x) + \int_0^x K(x,\sigma(y))\phi(y)\Delta y, \quad x \in \mathcal{T}_0, \tag{7.16}$$

$$\phi(0) = \phi_0, \quad \phi^{\Delta}(0) = \phi_1, \quad \ldots, \quad \phi^{\Delta^{n-1}}(0) = \phi_{n-1}, \tag{7.17}$$

where $K(x, 0)$ is $e_\alpha(x, 0)$, $\cosh_\alpha(x, 0)$, $\sinh_\alpha(x, 0)$, $\cos_\alpha(x, 0)$, $\sin_\alpha(x, 0)$, or $h_k(x, 0)$, $u : \mathscr{T}_0 \longmapsto \mathscr{R}$ be a given continuous function, ϕ_m, $0 \le m \le n-1$, are given constants.

The Eq. (7.16) we can rewrite in the following form

$$\phi^{\Delta^n}(x) = -u(x) + K(x, 0) \star \phi(x).$$

We take the Laplace transform of both sides of the last equation and we get

$$\begin{aligned} \mathscr{L}(\phi^{\Delta^n})(z) &= \mathscr{L}(u(x) + K(x, 0) \star \phi(x))(z) \\ &= \mathscr{L}(u)(z) + \mathscr{L}(K(x, 0) \star \phi(x))(z) \\ &= \mathscr{L}(u)(z) + \mathscr{L}(K)(z)\mathscr{L}(\phi)(z), \end{aligned}$$

whereupon

$$z^n \mathscr{L}(\phi)(z) - \sum_{l=0}^{n-1} z^l \phi^{\Delta^{n-1-l}}(0) = \mathscr{L}(u)(z) + \mathscr{L}(K)(z)\mathscr{L}(\phi)(z).$$

Hence, using (7.17), we obtain

$$\begin{aligned} (z^n - \mathscr{L}(K)(z))\mathscr{L}(\phi)(z) &= \mathscr{L}(u)(z) + \sum_{l=0}^{n-1} z^l \phi^{\Delta^{n-1-l}}(0) \\ &= \mathscr{L}(u)(z) + \sum_{l=0}^{n-1} z^l \phi_{n-1-l}. \end{aligned}$$

Example 14 Consider the equation

$$\begin{aligned} &\phi^{\Delta^2}(x) = -1 - \textstyle\int_0^x e_2(x, y+1)\phi(y)\Delta y, \quad \mathscr{T}_0 = \mathscr{Z}, \\ &\phi(0) = 1, \quad \phi^{\Delta}(0) = -1. \end{aligned}$$

Here

$$\sigma(x) = x + 1, \quad \mu(x) = 1, \quad x \in \mathscr{T}_0.$$

Then the given equation we can rewrite in the form

$$\phi^{\Delta^2}(x) = -1 - \int_0^x e_2(x, \sigma(y))\phi(y)\Delta y$$

or

$$\phi^{\Delta^2}(x) = -1 - e_2(x, 0) \star \phi(x).$$

We take the Laplace transform of both sides of the last equation and we get

$$\begin{aligned}\mathscr{L}(\phi^{\Delta^2}(z)) &= \mathscr{L}(-1 - e_2(x,0) \star \phi(x))(z)\\ &= \mathscr{L}(-1)(z) - \mathscr{L}(e_2(x,0) \star \phi(x))(z)\\ &= -\frac{1}{z} - \mathscr{L}(e_2(x,0))(z)\mathscr{L}(\phi)(z)\\ &= -\frac{1}{z} - \frac{1}{z-2}\mathscr{L}(\phi)(z).\end{aligned}$$

Hence,

$$z^2\mathscr{L}(\phi)(z) - \phi^{\Delta}(0) - z\phi(0) = -\frac{1}{z} - \frac{1}{z-2}\mathscr{L}(\phi)(z),$$

or

$$z^2\mathscr{L}(\phi)(z) + 1 - z = -\frac{1}{z} - \frac{1}{z-2}\mathscr{L}(\phi)(z),$$

or

$$\left(z^2 + \frac{1}{z-2}\right)\mathscr{L}(\phi)(z) = z - 1 - \frac{1}{z},$$

or

$$\frac{z^3 - 2z^2 + 1}{z-2}\mathscr{L}(\phi)(z) = \frac{z^2 - z - 1}{z},$$

or

$$\frac{(z-1)(z^2 - z - 1)}{z-2}\mathscr{L}(\phi)(z) = \frac{z^2 - z - 1}{z},$$

or

$$\frac{z-1}{z-2}\mathscr{L}(\phi)(z) = \frac{1}{z},$$

or

$$\begin{aligned}\mathscr{L}(\phi)(z) &= \frac{z-2}{z(z-1)}\\ &= \frac{2}{z} - \frac{1}{z-1}\\ &= 2\mathscr{L}(h_0(x,0))(z) - \mathscr{L}(e_1(x,0))(z)\\ &= \mathscr{L}(2 - e_1(x,0))(z).\end{aligned}$$

Consequently

$$\phi(x) = 2 - e_1(x,0).$$

Example 15 Consider the equation

$$\phi^{\Delta}(x) = h_1(x,0) + \int_0^x h_1(x,\sigma(y))\phi(y)\Delta y, \quad \phi(0) = 0.$$

The given equation we can rewrite in the form

$$\phi^{\Delta}(x) = h_1(x,0) + h_1(x,0) \star \phi(x).$$

We take the Laplace transform of both sides of the last equation and we obtain

$$\begin{aligned}
\mathscr{L}(\phi^{\Delta})(z) &= \mathscr{L}(h_1(x,0) + h_1(x,0) \star \phi(x))(z) \\
&= \mathscr{L}(h_1)(z) + \mathscr{L}(h_1(x,0) \star \phi(x))(z) \\
&= \mathscr{L}(h_1)(z) + \mathscr{L}(h_1)(z)\mathscr{L}(\phi)(z) \\
&= \frac{1}{z^2} + \frac{1}{z^2}\mathscr{L}(\phi)(z).
\end{aligned}$$

Hence,

$$z\mathscr{L}(\phi)(z) - \phi(0) = \frac{1}{z^2} + \frac{1}{z^2}\mathscr{L}(\phi)(z),$$

or

$$z\mathscr{L}(\phi)(z) = \frac{1}{z^2} + \frac{1}{z^2}\mathscr{L}(\phi)(z),$$

or

$$\left(z - \frac{1}{z^2}\right)\mathscr{L}(\phi)(z) = \frac{1}{z^2},$$

or

$$\frac{z^3-1}{z^2}\mathscr{L}(\phi)(z) = \frac{1}{z^2},$$

or

$$\begin{aligned}
\mathscr{L}(\phi)(z) &= \frac{1}{z^3-1} \\
&= \frac{1}{(z-1)(z^2+z+1)} \\
&= \frac{1}{3}\frac{1}{z-1} - \frac{1}{3}\frac{z+2}{z^2+z+1} \\
&= \frac{1}{3}\mathscr{L}(e_1(x,0))(z) - \frac{1}{3}\frac{z+2}{\left(z+\frac{1}{2}\right)^2 + \left(\frac{\sqrt{3}}{2}\right)^2} \\
&= \frac{1}{3}\mathscr{L}(e_1(x,0))(z) - \frac{1}{3}\frac{z+\frac{1}{2}+\frac{3}{2}}{\left(z+\frac{1}{2}\right)^2 + \left(\frac{\sqrt{3}}{2}\right)^2} \\
&= \frac{1}{3}\mathscr{L}(e_1(x,0))(z) - \frac{1}{3}\frac{z+\frac{1}{2}}{\left(z+\frac{1}{2}\right)^2 + \left(\frac{\sqrt{3}}{2}\right)^2} - \frac{1}{2}\frac{1}{\left(z+\frac{1}{2}\right)^2 + \left(\frac{\sqrt{3}}{2}\right)^2}
\end{aligned}$$

$$= \frac{1}{3}\mathscr{L}(e_1(x,0))(z) - \frac{1}{3}\frac{z+\frac{1}{2}}{\left(z+\frac{1}{2}\right)^2+\left(\frac{\sqrt{3}}{2}\right)^2} - \frac{1}{2}\frac{1}{\frac{\sqrt{3}}{2}}\frac{\frac{\sqrt{3}}{2}}{\left(z+\frac{1}{2}\right)^2+\left(\frac{\sqrt{3}}{2}\right)^2}$$

$$= \frac{1}{3}\mathscr{L}(e_1(x,0))(z) - \frac{1}{3}\frac{z+\frac{1}{2}}{\left(z+\frac{1}{2}\right)^2+\left(\frac{\sqrt{3}}{2}\right)^2} - \frac{\sqrt{3}}{3}\frac{\left(\frac{\sqrt{3}}{2}\right)^2}{\left(z+\frac{1}{2}\right)^2+\left(\frac{\sqrt{3}}{2}\right)^2}$$

$$= \frac{1}{3}\mathscr{L}(e_1(x,0)) - \frac{1}{3}\mathscr{L}\left(e_{-\frac{1}{2}}(x,0)\cos_{\frac{\sqrt{3}}{2-\mu(x)}}\right)$$

$$-\frac{\sqrt{3}}{3}\mathscr{L}\left(e_{-\frac{1}{2}}(x,0)\sin_{\frac{\sqrt{3}}{2-\mu(x)}}(x,0)\right)$$

$$= \mathscr{L}\left(\frac{1}{3}e_1(x,0) - \frac{1}{3}e_{-\frac{1}{2}}(x,0)\cos_{\frac{\sqrt{3}}{2-\mu(x)}}(x,0) - \frac{\sqrt{3}}{3}e_{-\frac{1}{2}}(x,0)\sin_{\frac{\sqrt{3}}{2-\mu(x)}}(x,0)\right).$$

Consequently

$$\phi(x) = \frac{1}{3}e_1(x,0) - \frac{1}{3}e_{-\frac{1}{2}}(x,0)\cos_{\frac{\sqrt{3}}{2-\mu(x)}}(x,0) - \frac{\sqrt{3}}{3}e_{-\frac{1}{2}}(x,0)\sin_{\frac{\sqrt{3}}{2-\mu(x)}}(x,0).$$

Example 16 Consider the equation

$$\phi^{\Delta}(x) = e_2(x,0) + 4\int_0^x e_3(x,\sigma(y))\phi(y)\Delta y, \quad \phi(0) = 0,$$

The given equation we can rewrite in the form

$$\phi^{\Delta}(x) = e_2(x,0) + 4e_3(x,0) \star \phi(x).$$

We take the Laplace transform of both sides of the last equation and we get

$$\begin{aligned}
\mathscr{L}(\phi^{\Delta})(z) &= \mathscr{L}(e_2(x,0) + 4e_3(x,0)\star\phi(x))(z)\\
&= \mathscr{L}(e_2(x,0))(z) + 4\mathscr{L}(e_3(x,0)\star\phi(x))(z)\\
&= \frac{1}{z-2} + 4\mathscr{L}(e_3(x,0))(z)\mathscr{L}(\phi)(z)\\
&= \frac{1}{z-2} + \frac{4}{z-3}\mathscr{L}(\phi)(z).
\end{aligned}$$

Hence,

$$z\mathscr{L}(\phi)(z) = \frac{1}{z-2} + \frac{4}{z-3}\mathscr{L}(\phi)(z),$$

or

$$\left(z - \frac{4}{z-3}\right)\mathscr{L}(\phi)(z) = \frac{1}{z-2},$$

or

$$\frac{z^2-3z-4}{z-3}\mathscr{L}(\phi)(z) = \frac{1}{z-2},$$

or

$$\frac{(z-4)(z+1)}{z-3}\mathscr{L}(\phi)(z) = \frac{1}{z-2}$$

or

$$\begin{aligned}
\mathscr{L}(\phi)(z) &= \frac{z-3}{(z-2)(z-4)(z+1)}\\
&= \frac{z-2-1}{(z-2)(z-4)(z+1)}\\
&= \frac{1}{(z-4)(z+1)} - \frac{1}{(z-2)(z-4)(z+1)}\\
&= \frac{1}{5}\left(\frac{1}{z-4} - \frac{1}{z+1}\right)\left(1 - \frac{1}{z-2}\right)\\
&= \frac{1}{5}\frac{1}{z-4} - \frac{1}{5}\frac{1}{z+1} - \frac{1}{5}\frac{1}{(z-4)(z-2)}\\
&\quad + \frac{1}{5}\frac{1}{z+1}\frac{1}{z-2}
\end{aligned}$$

$$\begin{aligned}
&= \frac{1}{5}\frac{1}{z-4} - \frac{1}{5}\frac{1}{z+1} - \frac{1}{10}\left(\frac{1}{z-4} - \frac{1}{z-2}\right)\\
&\quad + \frac{1}{15}\left(\frac{1}{z-2} - \frac{1}{z+1}\right)\\
&= \frac{1}{5}\frac{1}{z-4} - \frac{1}{5}\frac{1}{z+1} - \frac{1}{10}\frac{1}{z-4} + \frac{1}{10}\frac{1}{z-2}\\
&\quad + \frac{1}{15}\frac{1}{z-2} - \frac{1}{15}\frac{1}{z+1}\\
&= \frac{1}{10}\frac{1}{z-4} + \frac{1}{6}\frac{1}{z-2} - \frac{4}{15}\frac{1}{z+1}\\
&= \frac{1}{10}\mathscr{L}(e_4(x,0))(z) + \frac{1}{6}\mathscr{L}(e_2(x,0))(z) - \frac{4}{15}\mathscr{L}(e_{-1}(x,0))(z)\\
&= \mathscr{L}\left(\frac{1}{10}e_4(x,0) + \frac{1}{6}e_2(x,0) - \frac{4}{15}e_{-1}(x,0)\right)(z).
\end{aligned}$$

Consequently

$$\phi(x) = \frac{1}{10}e_4(x,0) + \frac{1}{6}e_2(x,0) - \frac{4}{15}e_{-1}(x,0).$$

Exercise 9 Use the Laplace transform to solve the following equations.

1.
$$\begin{cases} \phi^{\Delta^2}(x) = e_1(x,0) + \int_0^x h_1(x,\sigma(y))\phi(y)\Delta y \\ \phi(0) = \phi^{\Delta}(0) = 1, \end{cases}$$

2.
$$\begin{cases} \phi^{\Delta^2}(x) = h_1(x,0) + \int_0^x e_7(x,\sigma(y))\phi(y)\Delta y \\ \phi(0) = 0, \quad \phi^{\Delta}(0) = 1, \end{cases}$$

3.
$$\begin{cases} \phi^{\Delta^2}(x) = -1 + 2\int_0^x h_3(x,\sigma(y))\phi(y)\Delta y \\ \phi(0) = \phi^{\Delta}(0) = 0. \end{cases}$$

7.6 Generalized Volterra Integro-Differential Equations of the First Kind

Here we consider

$$u(x) = \lambda_1 \int_0^x K_1(x,\sigma(y))\phi(y)\Delta y + \lambda_2 \int_0^x K_2(x,\sigma(y))\phi^{\Delta^n}(y)\Delta y, \tag{7.18}$$

$$\phi(0) = \phi_0, \quad \phi^{\Delta}(0) = \phi_1, \quad \phi^{\Delta^{n-1}}(0) = \phi_{n-1}, \tag{7.19}$$

where $K_1(x,0)$ and $K_2(x,0)$ are $e_\alpha(x,0)$, $\cosh_\alpha(x,0)$, $\sinh_\alpha(x,0)$, $\cos_\alpha(x,0)$, $\sin_\alpha(x,0)$ or $h_k(x,0)$, $u : \mathscr{T}_0 \longmapsto \mathscr{R}$ be a given continuous function, λ_1 and λ_2 are parameters.
The Eq. (7.18) we can rewrite in the following form.

$$u(x) = \lambda_1 K_1(x,0) \star \phi(x) + \lambda_2 K_2(x,0) \star \phi^{\Delta^n}(x),$$

We take the Laplace transform of both sides of the last equation and we find

$$\begin{aligned} \mathscr{L}(u)(z) &= \mathscr{L}(\lambda_1 K_1(x,0) \star \phi(x) + \lambda_2 K_2(x,0) \star \phi^{\Delta^n}(x))(z) \\ &= \lambda_1 \mathscr{L}(K_1(x,0) \star \phi(x))(z) + \lambda_2 \mathscr{L}(K_2(x,0) \star \phi^{\Delta^n}(x))(z) \\ &= \lambda_1 \mathscr{L}(K_1)(z)\mathscr{L}(\phi)(z) + \lambda_2 \mathscr{L}(K_2)(z)\mathscr{L}(\phi^{\Delta^n})(z) \\ &= \lambda_1 \mathscr{L}(K_1)\mathscr{L}(\phi)(z) + \lambda_2 \mathscr{L}(K_2)(z)\left(z^n \mathscr{L}(\phi)(z) - \sum_{l=0}^{n-1} z^l \phi^{\Delta^{n-1-l}}(0)\right) \\ &= \lambda_1 \mathscr{L}(K_1)(z)\mathscr{L}(\phi)(z) + z^n \lambda_2 \mathscr{L}(K_2)(z)\mathscr{L}(\phi)(z) \\ &\quad - \lambda_2 \mathscr{L}(K_2)(z) \sum_{i=0}^{n-1} z^l \phi_{n-1-l}, \end{aligned}$$

whereupon

$$\left(\lambda_1 \mathscr{L}(K_1)(z) + z^n \lambda_2 \mathscr{L}(K_2)(z)\right) \mathscr{L}(\phi)(z) = \mathscr{L}(u)(z) + \lambda_2 \mathscr{L}(K_2)(z) \sum_{l=0}^{n-1} z^l \phi_{n-1-l}.$$

Example 17 Consider the equation

$$\cosh_2(x, 0) = -3 \int_0^x e_3(x, \sigma(y))\phi(y)\Delta y + \int_0^x e_3(x, \sigma(y))\phi^{\Delta}(y)\Delta y, \quad \phi(0) = 0.$$

The given equation we can rewrite in the following form

$$\cosh_2(x, 0) = -3e_3(x, 0) \star \phi(x) + e_3(x, 0) \star \phi^{\Delta}(x).$$

We take the Laplace transform of both sides of the last equation and we get

$$\begin{aligned}
\mathscr{L}(\cosh_2(x, 0))(z) &= \mathscr{L}(-3e_3(x, 0) \star \phi(x) + e_3(x, 0) \star \phi^{\Delta}(x))(z) \\
&= \mathscr{L}(-3e_3(x, 0) \star \phi(x))(z) + \mathscr{L}(e_3(x, 0) \star \phi^{\Delta}(x))(z) \\
&= -3\mathscr{L}(e_3(x, 0) \star \phi(x))(z) + \mathscr{L}(e_3(x, 0))\mathscr{L}(\phi^{\Delta}(x))(z) \\
&= -3\mathscr{L}(e_3(x, 0))(z)\mathscr{L}(\phi)(z) + \frac{1}{z-3}(z\mathscr{L}(\phi)(z) - \phi(0)) \\
&= -\frac{3}{z-3}\mathscr{L}(\phi)(z) + \frac{z}{z-3}\mathscr{L}(\phi)(z) \\
&= \frac{z-3}{z-3}\mathscr{L}(\phi)(z) \\
&= \mathscr{L}(\phi)(z),
\end{aligned}$$

i.e.,

$$\mathscr{L}(\phi)(z) = \mathscr{L}(\cosh_2(x, 0))(z).$$

Consequently

$$\phi(x) = \cosh_2(x, 0).$$

Example 18 Consider the equation

$$\begin{cases} h_1(x, 0) = \int_0^x \phi(y)\Delta y + 4\int_0^x h_2(x, \sigma(y))\phi^{\Delta^2}(y)\Delta y \\ \phi(0) = \phi^{\Delta}(0) = 0. \end{cases}$$

The given equation we can rewrite in the following form

$$h_1(x, 0) = h_0(x, 0) \star \phi(x) + 4h_2(x, 0) \star \phi^{\Delta^2}(x).$$

We take the Laplace transform of both sides of the last equation and we find

$$\begin{aligned}\mathscr{L}(h_1(x,0))(z) &= \mathscr{L}(h_0(x,0)\star\phi(x)+4h_2(x,0)\star\phi^{\Delta^2}(x))(z)\\ &= \mathscr{L}(h_0(x,0)\star\phi(x))(z)+4\mathscr{L}(h_2(x,0)\star\phi^{\Delta^2}(x))(z)\\ &= \mathscr{L}(h_0(x,0))(z)\mathscr{L}(\phi)(z)+4\mathscr{L}(h_2(x,0))(z)\mathscr{L}(\phi^{\Delta^2})(z)\\ &= \frac{1}{z}\mathscr{L}(\phi)(z)+4\frac{1}{z^3}\left(z^2\mathscr{L}(\phi)(z)-\phi^{\Delta}(0)-z\phi(0)\right)\\ &= \frac{1}{z}\mathscr{L}(\phi)(z)+\frac{4}{z}\mathscr{L}(\phi)(z)\\ &= \frac{5}{z}\mathscr{L}(\phi)(z).\end{aligned}$$

Hence,

$$\frac{5}{z}\mathscr{L}(\phi)(z)=\frac{1}{z^2},$$

whereupon

$$\begin{aligned}\mathscr{L}(\phi)(z) &= \frac{1}{5}\frac{1}{z}\\ &= \frac{1}{5}\mathscr{L}(1)(z).\end{aligned}$$

Consequently

$$\phi(x)=\frac{1}{5}.$$

Example 19 Consider the equation

$$h_1(x,0)=-\int_0^x\phi(y)\Delta y+2\int_0^x h_1(x,\sigma(y))\phi^{\Delta}(y)\Delta y,\quad \phi(0)=1.$$

The given equation we can rewrite in the following form

$$h_1(x,0)=-h_0(x,0)\star\phi(x)+2h_1(x,0)\star\phi^{\Delta}(x).$$

We take the Laplace transform of both sides of the last equation and we find

$$\begin{aligned}
\mathscr{L}(h_1(x,0))(z) &= \mathscr{L}(-h_0(x,0)\star\phi(x)+2h_1(x,0)\star\phi^{\Delta}(x))(z)\\
&= -\mathscr{L}(h_0(x,0)\star\phi(x))(z)+2\mathscr{L}(h_1(x,0)\star\phi^{\Delta}(x))(z)\\
&= -\mathscr{L}(h_0(x,0))\mathscr{L}(\phi)(z)+2\mathscr{L}(h_1(x,0))(z)\mathscr{L}(\phi^{\Delta})(z)\\
&= -\frac{1}{z}\mathscr{L}(\phi)(z)+\frac{2}{z^2}(z\mathscr{L}(\phi)(z)-\phi(0))\\
&= -\frac{1}{z}\mathscr{L}(\phi)(z)+\frac{2}{z}\mathscr{L}(\phi)(z)-\frac{2}{z^2}\\
&= \frac{1}{z}\mathscr{L}(\phi)(z)-\frac{2}{z^2},
\end{aligned}$$

whereupon

$$\frac{1}{z}\mathscr{L}(\phi)(z)-\frac{2}{z^2}=\frac{1}{z^2},$$

or

$$\frac{1}{z}\mathscr{L}(\phi)(z)=\frac{3}{z^2},$$

or

$$\mathscr{L}(\phi)(z)=\frac{3}{z}.$$

Consequently

$$\phi(x)=3.$$

Because $\phi(0)=1$ the considered problem has no solution.

Exercise 10 Use the Laplace transform to solve the following equations.

1.
$$\begin{cases} \cosh_3(x,0)=\int_0^x e_1(x,\sigma(y))\phi(y)\Delta y-3\int_0^x e_2(x,\sigma(y))\phi^{\Delta^2}(y)\Delta y\\ \phi(0)=\phi^{\Delta}(0)=0,\end{cases}$$

2.
$$\begin{cases} \sinh_3(x,0)=\int_0^x \phi(y)\Delta y+\int_0^x \cos_3(x,\sigma(y))\phi^{\Delta^2}(y)\Delta y\\ \phi(0)=1,\quad \phi^{\Delta}(0)=0,\end{cases}$$

3.
$$\begin{cases} e_4(x,0)=-2\int_0^x e_{-1}(x,\sigma(y))\phi(y)\Delta y+\int_0^x e_2(x,\sigma(y))\phi^{\Delta^2}(y)\Delta y\\ \phi(0)=\phi^{\Delta}(0)=1.\end{cases}$$

7.7 Advanced Practical Exercises

Problem 1 Let $\alpha, \beta \in \mathscr{R}$ and $1 + \alpha\mu(x) \neq 0$, $1 + \beta\mu(x) \neq 0$ for all $x \in \mathscr{T}$. Prove that

$$\mathscr{L}\left(e_\alpha(x,0)\sinh_{\frac{\beta}{1+\alpha\mu}}(x,0)\right)(z) = \frac{\beta}{(z-\alpha)^2 - \beta^2}$$

whenever

$$\lim_{x\to\infty} e_\alpha(x,0)\sinh_{\frac{\beta}{1+\alpha\mu}}(x,0) = \lim_{x\to\infty} e_\alpha(x,0)\sinh^{\Delta}_{\frac{\beta}{1+\alpha\mu}}(x,0) = 0.$$

Problem 2 Let $\alpha, \beta \in \mathscr{R}$ and $1 + \alpha\mu(x) \neq 0$, $1 + \beta\mu(x) \neq 0$ for all $x \in \mathscr{T}$. Prove that

$$\mathscr{L}\left(e_\alpha(x,0)\cosh_{\frac{\beta}{1+\alpha\mu}}(x,0)\right)(z) = \frac{z-\alpha}{(z-\alpha)^2 - \beta^2}$$

whenever

$$\lim_{x\to\infty} e_\alpha(x,0)\cosh_{\frac{\beta}{1+\alpha\mu}}(x,0) = \lim_{x\to\infty} e_\alpha(x,0)\cosh^{\Delta}_{\frac{\beta}{1+\alpha\mu}}(x,0) = 0.$$

Problem 3 Prove the following relations.

1. If $\alpha \neq 0$, then

$$e_\alpha(x,0) \star h_k(x,0) = \frac{1}{\alpha^{k+1}} e_\alpha(x,0) - \sum_{j=0}^{k} \frac{1}{\alpha^{k+1-j}} h_j(x,0).$$

2. If $\alpha^2 + \beta^2 \neq 0$, then

$$e_\alpha(x,0) \star \cos_\beta(x,0) = \frac{\alpha e_\alpha(x,0) + \beta\sin_\beta(x,0) - \alpha\cos_\beta(x,0)}{\alpha^2 + \beta^2}.$$

3. If $\alpha \neq 0$, $\alpha \neq \beta$, then

$$\sin_\alpha(x,0) \star \sin_\beta(x,0) = \frac{\alpha\sin_\beta(x,0) - \beta\sin_\alpha(x,0)}{\alpha^2 - \beta^2}.$$

4. If $\alpha \neq 0$, $\alpha \neq \beta$, then

$$\sin_\alpha(x,0) \star \cos_\beta(x,0) = \frac{\alpha\cos_\beta(x,0) - \alpha\cos_\alpha(x,0)}{\alpha^2 - \beta^2}.$$

5. If $\alpha \neq 0$, then

$$\sin_\alpha(x,0) \star \sin_\alpha(x,0) = \frac{1}{\alpha}\sin_\alpha(x,0) - \frac{1}{2}x\cos_\alpha(x,0).$$

6. If $k \geq 0$, then

$$\cos_\alpha(x,0) \star h_k(x,0)$$
$$= \begin{cases} (-1)^{\frac{k(k+1)}{2}} \frac{1}{\alpha^{k+1}} \sin_\alpha(x,0) + \sum_{j=0}^{\frac{k-2}{2}} (-1)^j \frac{h_{k-2j-1}(x,0)}{\alpha^{2j+2}} & \textit{if } k \textit{ is even} \\ (-1)^{\frac{k(k+1)}{2}} \frac{1}{\alpha^{k+1}} \cos_\alpha(x,0) + \sum_{j=0}^{\frac{k-1}{2}} (-1)^j \frac{h_{k-2j-1}(x,0)}{\alpha^{2j+2}} & \textit{if } k \textit{ is odd.} \end{cases}$$

Problem 4 Use Laplace transform to solve the following initial value problem

$$\phi^{\Delta^2} + 9\phi = \sin_3(x,0)$$
$$\phi(0) = \phi^{\Delta}(0) = 1.$$

Answer

$$\phi(x) = \frac{1}{3}\sin_3(x,0) \star \sin_3(x,0) + \frac{1}{3}\sin_3(x,0) + \cos_3(x,0).$$

Problem 5 Solve the equation

$$\phi(x) = 2e_3(x,0) - 5\int_0^x e_4(x,\sigma(y))\phi(y)\Delta y.$$

Answer

$$\phi(x) = \frac{5}{2}e_{-1}(x,0) - \frac{1}{2}e_3(x,0).$$

Problem 6 Use the Laplace transform to solve the following equations.

1. $e_4(x,0) + e_{-2}(x,0) = \int_0^x e_2^\sigma(x,y)\phi(y)\Delta y, \quad \mathscr{T}_0 = \mathscr{Z},$
2. $\sinh_4(x,0) \star e_{-2}(x,0) = \int_0^x e_4^\sigma(x,y)\phi(y)\Delta y, \quad \mathscr{T}_0 = 2^{\mathscr{N}_0} \cup \{0\},$
3. $\sinh_2(x,0) \star \cosh_2(x,0) = \int_0^x \cosh_2^\sigma(x,y)\phi(y)\Delta y, \quad \mathscr{T}_0 = \mathscr{N}_0,$
4. $e_3(x,0) \star e_{-2}(x,0) + \cosh_2(x,0) = \int_0^x e_1(x,\sigma(y))\phi(y)\Delta y, \quad \mathscr{T}_0 = \mathscr{Z},$
5. $e_3(x,0) + e_{-5}(x,0) = 4\int_0^x \cos_2(x,\sigma(y))\phi(y)\Delta y, \quad \mathscr{T}_0 = \mathscr{Z},$
6. $\sinh_2(x,0) + \cos_2(x,0) = 2\int_0^x \sinh_1(x,\sigma(y))\phi(y)\Delta y, \quad \mathscr{T}_0 = \mathscr{Z}.$

7. $e_{-3}(x,0)+h_2(x,0)\star e_{-2}(x,0)+\cosh_2(x,0)=\int_0^x e_4(x,\sigma(y))\phi(y)\Delta y,$
$\mathcal{T}_0=\mathcal{L}$,
8. $e_{-3}(x,0)+e_5(x,0)\star \sinh_1(x,0)=\int_0^x \cosh_3(x,\sigma(y))\phi(y)\Delta y, \quad \mathcal{T}_0=\mathcal{L},$
9. $\sin_4(x,0)\star e_{-2}(x,0)+\cosh_{-2}(x,0)=\int_0^x \sinh_1(x,\sigma(y))\phi(y)\Delta y, \quad \mathcal{T}_0=\mathcal{L}.$

Problem 7 Use the Laplace transform to solve the following equations.

1.
$$\begin{cases}\phi^{\Delta^2}(x)=e_2(x,0)-3\int_0^x e_1(x,\sigma(y))\phi(y)\Delta y\\ \phi(0)=\phi^{\Delta}(0)=1,\end{cases}$$

2.
$$\begin{cases}\phi^{\Delta^2}(x)=h_1(x,0)-4\int_0^x h_1(x,\sigma(y))\phi(y)\Delta y\\ \phi(0)=1,\quad \phi^{\Delta}(0)=2,\end{cases}$$

3.
$$\begin{cases}\phi^{\Delta^2}(x)=-10e_1(x,0)+2\int_0^x h_1(x,\sigma(y))\phi(y)\Delta y\\ \phi(0)=\phi^{\Delta}(0)=0.\end{cases}$$

4.
$$\begin{cases}\phi^{\Delta^3}(x)=e_2(x,0)-2\int_0^x h_2(x,\sigma(y))\phi(y)\Delta y\\ \phi(0)=\phi^{\Delta}(0)=\phi^{\Delta^2}(0)=0.\end{cases}$$

Problem 8 Use the Laplace transform to solve the following equations.

1.
$$\begin{cases}\cosh_2(x,0)=-4\int_0^x \sinh_1(x,\sigma(y))\phi(y)\Delta y-\int_0^x e_2(x,\sigma(y))\phi^{\Delta^2}(y)\Delta y\\ \phi(0)=0,\quad \phi^{\Delta}(0)=1,\end{cases}$$

2.
$$\begin{cases}e_{-3}(x,0)=-\int_0^x \sin_1(x,\sigma(y))\phi(y)\Delta y+\int_0^x \cosh_3(x,\sigma(y))\phi^{\Delta^2}(y)\Delta y\\ \phi(0)=1,\quad \phi^{\Delta}(0)=2,\end{cases}$$

3.
$$\begin{cases}\sinh_4(x,0)=2\int_0^x e_3(x,\sigma(y))\phi(y)\Delta y-2\int_0^x e_1(x,\sigma(y))\phi^{\Delta^2}(y)\Delta y\\ \phi(0)=\phi^{\Delta}(0)=0,\end{cases}$$

4.
$$\begin{cases}\sinh_1(x,0)=\int_0^x \sin_3(x,\sigma(y))\phi^{\Delta}(y)\Delta y-2\int_0^x e_1(x,\sigma(y))\phi^{\Delta^3}(y)\Delta y\\ \phi(0)=\phi^{\Delta}(0)=\phi^{\Delta^2}(0)=1,\end{cases}$$

5.

$$\begin{cases} e_3(x,0) = 5\int_0^x e_3(x,\sigma(y))\phi(y)\Delta y + \int_0^x e_1(x,\sigma(y))\phi^{\Delta^3}(y)\Delta y \\ \phi(0) = \phi^{\Delta}(0) = 0, \quad \phi^{\Delta^2}(0) = -1, \end{cases}$$

6.

$$\begin{cases} \cosh_4(x,0) = \int_0^x \sinh_3(x,\sigma(y))\phi(y)\Delta y + 2\int_0^x e_1(x,\sigma(y))\phi^{\Delta^3}(y)\Delta y \\ \phi(0) = \phi^{\Delta}(0) = 0, \quad \phi^{\Delta^2}(0) = -3. \end{cases}$$

Chapter 8
The Series Solution Method

In this chapter we describe the series solution method for generalized Volterra integral equations and generalized Volterra integro-differential equations.

8.1 Generalized Volterra Integral Equations of the Second Kind

Definition 1 A real function $u(x)$ is called analytic if it has delta derivative of all orders such that the Taylor series at any point α in its domain

$$g(x) = \sum_{k=0}^{n} f^{\Delta^k}(\alpha) h_k(x, \alpha)$$

converges to $f(x)$ in a neighbourhood of α.

For simplicity, the generic form of Taylor series at $x = 0$ can be written as

$$f(x) = \sum_{n=0}^{\infty} f_n h_n(x, 0). \tag{8.1}$$

In this section we will present a useful method for solving generalized Volterra integral equations of the second kind.
We will assume that the solution $\phi(x)$ of the generalized Volterra integral equation of the second kind

$$\phi(x) = f(x) + \lambda \int_a^x K(x, y)\phi(y)\Delta y \tag{8.2}$$

S.G. Georgiev, *Integral Equations on Time Scales*, Atlantis Studies in Dynamical Systems 5, DOI 10.2991/978-94-6239-228-1_8

is analytic, and therefore possesses a Taylor series of the form given by

$$\phi(x)=\sum_{n=0}^{\infty}a_n h_n(x,0), \tag{8.3}$$

where the coefficients a_n, $n\geq 0$, will be determined recurrently. Also, we assume that $f(x)$ is of the form given by (8.1).
Substituting (8.1) and (8.3) in (8.2) gives

$$\sum_{n=0}^{\infty}a_n h_n(x,0)=\sum_{n=0}^{\infty}f_n h_n(x,0)+\lambda\int_0^x K(x,y)\left(\sum_{n=0}^{\infty}h_n(y,0)\right)\Delta y.$$

Next we equate the coefficients of $h_n(x,0)$ to obtain a recurrence relation in a_n, $n\geq 0$. Solving the recurrence relation will lead to a complete determination of the coefficients a_n, $n\geq 0$. Having determined the coefficients a_n, $n\geq 0$, the series solution follows immediately upon substituting the derived coefficients into (8.3). The exact solution may be obtained if such an exact solution exists. If an exact solution is not obtainable, then the obtained series can be used for numerical purposes.

Example 1 Consider the equation

$$\phi(x)=1+\int_0^x \phi(y)\Delta y.$$

We have

$$\phi(x)=\sum_{n=0}^{\infty}a_n h_n(x,0)$$

and

$$\begin{aligned}\sum_{n=0}^{\infty}a_n h_n(x,0)&=1+\int_0^x\sum_{n=0}^{\infty}a_n h_n(y,0)\Delta y\\&=1+\sum_{n=0}^{\infty}a_n\int_0^x h_n(y,0)\Delta y\\&=1+\sum_{n=0}^{\infty}a_n h_{n+1}(x,0),\end{aligned}$$

whereupon

$$a_0h_0(x,0)+a_1h_1(x,0)+a_2h_2(x,0)+\cdots=h_0(x,0)+a_0h_1(x,0)+a_1h_2(x,0)+\cdots.$$

Hence,

$$\begin{cases} a_0 = 1 \\ a_1 = a_0 \\ a_2 = a_1 \\ \dots \\ a_{n+1} = a_n, \quad n \geq 2. \end{cases}$$

Therefore

$$a_n = 1, \quad n \geq 0.$$

Consequently

$$\begin{aligned} \phi(x) &= h_0(x,0) + h_1(x,0) + \cdots + h_n(x,0) + \cdots \\ &= e_1(x,0). \end{aligned}$$

Example 2 Consider the equation

$$\phi(x) = \sinh_1(x,0) + 2\int_0^x \phi(y)\Delta y.$$

We have

$$\sinh_1(x,0) = \sum_{n=0}^{\infty} h_{2n+1}(x,0) \quad \textit{and} \quad \phi(x) = \sum_{n=0}^{\infty} a_n h_n(x,0).$$

Then

$$\begin{aligned} \sum_{n=0}^{\infty} a_n h_n(x,0) &= \sum_{n=0}^{\infty} h_{2n+1}(x,0) + 2\int_0^x \sum_{n=0}^{\infty} a_n h_n(y,0)\Delta y \\ &= \sum_{n=0}^{\infty} h_{2n+1}(x,0) + 2\sum_{n=0}^{\infty} a_n \int_0^x h_n(y,0)\Delta y \\ &= \sum_{n=0}^{\infty} h_{2n+1}(x,0) + 2\sum_{n=0}^{\infty} a_n h_{n+1}(x,0). \end{aligned}$$

Hence,

$$a_0 h_0(x,0) + a_1 h_1(x,0) + a_2 h_2(x,0) + a_3 h_3(x,0) + a_4 h_4(x,0) + \cdots$$

$$= h_1(x,0) + h_3(x,0) + h_5(x,0) + \cdots$$

$$+ 2a_0 h_1(x,0) + 2a_1 h_2(x,0) + 2a_2 h_3(x,0) + 2a_3 h_4(x,0) + \cdots .$$

Therefore

$$\begin{cases} a_0 = 0 \\ a_1 = 1 + 2a_0 \\ a_2 = 2a_1 \\ a_3 = 1 + 2a_2 \\ a_4 = 2a_3 \\ \cdots \\ a_{2n} = 2a_{2n-1} \\ a_{2n+1} = 1 + 2a_{2n}. \end{cases}$$

From here,

$$a_0 = 0$$

$$a_1 = 1$$

$$a_2 = 2$$

$$a_3 = 1 + 2^2 = 5$$

$$a_4 = 10$$

$$\ldots,$$

and

$$\phi(x) = h_1(x,0) + 2h_2(x,0) + 5h_3(x,0) + 10h_4(x,0) + \cdots .$$

Example 3 Consider the equation

$$\phi(x) = 2 + e_1(x,0) - 3\int_0^x \phi(y)\Delta y.$$

We have

$$e_1(x,0)=\sum_{n=0}^{\infty}h_n(x,0) \quad and \quad \phi(x)=\sum_{n=0}^{\infty}a_nh_n(x,0).$$

Substituting them in the given equation, we get

$$\begin{aligned}\sum_{n=0}^{\infty}a_nh_n(x,0)&=2+\sum_{n=0}^{\infty}h_n(x,0)-3\int_0^x\sum_{n=0}^{\infty}a_nh_n(x,0)\Delta x\\&=2+\sum_{n=0}^{\infty}h_n(x,0)-3\sum_{n=0}^{\infty}a_n\int_0^x h_n(y,0)\Delta y\\&=2h_0(x,0)+\sum_{n=0}^{\infty}h_n(x,0)-3\sum_{n=0}^{\infty}a_nh_{n+1}(x,0).\end{aligned}$$

Hence,

$$\begin{aligned}&a_0h_0(x,0)+a_1h_1(x,0)+a_2h_2(x,0)+a_3h_3(x,0)+a_4h_4(x,0)+\cdots\\&=2h_0(x,0)+h_0(x,0)+h_1(x,0)+h_2(x,0)+h_3(x,0)+h_4(x,0)+\cdots\\&-3a_0h_1(x,0)-3a_1h_2(x,0)-3a_2h_3(x,0)-3a_3h_4(x,0)-\cdots.\end{aligned}$$

Therefore

$$\begin{cases}a_0=3\\a_1=1-3a_0\\a_2=1-3a_1\\a_3=1-3a_2\\\ldots\\a_n=1-3a_{n-1}, \quad n\geq 4,\end{cases}$$

i.e.,

$$\begin{cases}a_0=3\\a_1=-8\\a_2=25\\a_3=-74\\\ldots\end{cases}$$

and

$$\phi(x) = 3 - 8h_1(x,0) + 25h_2(x,0) - 74h_3(x,0) + \cdots .$$

Exercise 1 Find a solution $\phi(x) = \sum_{n=0}^{\infty} a_n h_n(x,0)$ of the following equations.

1. $2\phi(x) = 3 - e_2(x,0) + e_1(x,0) - 4\int_0^x \phi(y)\Delta y,$
2. $\phi(x) = 3\cos_1(x,0) - 2\sin_2(x,0) + \int_0^x \phi(y)\Delta y,$
3. $\phi(x) = e_{-1}(x,0) + 2\cosh_1(x,0) - 4\int_0^x \phi(y)\Delta y,$
4. $\phi(x) = 3\cosh_1(x,0) - 2\sinh_2(x,0) + \int_0^x \phi(y)\Delta y,$
5. $\phi(x) = 4\cosh_2(x,0) - 3e_{-1}(x,0) + e_{-i}(x,0) + 2\int_0^x \phi(y)\Delta y,$
6. $\phi(x) = 1 - 3\cos_{-3}(x,0) + \int_0^x \phi(y)\Delta y,$
7. $\phi(x) = 2e_{-i}(x,0) - 3e_i(x,0) + 2 + \int_0^x \phi(y)\Delta y,$
8. $\phi(x) = 1 - 3e_{-4}(x,0) - \int_0^x \phi(y)\Delta y,$
9. $\phi(x) = 2\cosh_1(x,0) + 3\sinh_2(x,0) + 2\int_0^x \phi(y)\Delta y.$

Lemma 1 *For every $k, l \in \mathcal{N}$ we have*

$$\begin{aligned} h_{k+l}(x,0) &= \int_0^x h_{l-1}(x,\sigma(s))h_k(s,0)\Delta s \\ &= \int_0^x h_{k-1}(x,\sigma(s))h_l(s,0)\Delta s. \end{aligned}$$

Proof Note that

$$\begin{cases} h_{k+l}^{\Delta^l}(x,0) = h_k(x,0), \\ h_{k+l}^{\Delta^i}(0,0) = 0, \quad 0 \le i \le l-1. \end{cases}$$

Therefore $h_{k+l}(x,0)$ is given by

$$h_{k+l}(x,0) = \int_0^x h_{l-1}(x,\sigma(s))h_k(s,0)\Delta s.$$

Also,

$$\begin{cases} h_{k+l}^{\Delta^k}(x, 0) = h_l(x, 0), \\ h_{k+l}^{\Delta^i}(0, 0) = 0, \quad 0 \le i \le k - 1. \end{cases}$$

Consequently

$$h_{k+l}(x, 0) = \int_0^x h_{k-1}(x, \sigma(s))h_l(s, 0)\Delta s,$$

which completes the proof.

Example 4 Let $\mathcal{T} = \mathcal{Z}$. Consider the equation

$$\phi(x) = \cos_1(x, 0) + 2\int_0^x h_2(x, y + 1)\phi(y)\Delta y.$$

Here

$$\sigma(x) = x + 1, \quad x \in \mathcal{T}.$$

Then the given equation we can rewrite in the following form

$$\phi(x) = \cos_1(x, 0) + 2\int_0^x h_2(x, \sigma(y))\phi(y)\Delta y.$$

Let

$$\phi(x) = \sum_{n=0}^{\infty} a_n h_n(x, 0).$$

Then

$$\begin{aligned} \sum_{n=0}^{\infty} a_n h_n(x, 0) &= \sum_{n=0}^{\infty}(-1)^n h_{2n}(x, 0) + 2\int_0^x h_2(x, \sigma(y))\sum_{n=0}^{\infty} a_n h_n(y, 0)\Delta y \\ &= \sum_{n=0}^{\infty}(-1)^n h_{2n}(x, 0) + 2\sum_{n=0}^{\infty} a_n \int_0^x h_2(x, \sigma(y))h_n(y, 0)\Delta y \\ &= \sum_{n=0}^{\infty}(-1)^n h_{2n}(x, 0) + 2\sum_{n=0}^{\infty} a_n h_{n+3}(x, 0). \end{aligned}$$

Hence,

$$a_0h_0(x,0)+a_1h_1(x,0)+a_2h_2(x,0)+a_3h_3(x,0)+a_4h_4(x,0)+\cdots$$

$$=h_0(x,0)-h_2(x,0)+h_4(x,0)-\cdots$$

$$+2a_0h_3(x,0)+2a_1h_4(x,0)+2a_2h_5(x,0)+\cdots.$$

Then

$$\begin{cases} a_0=1\\ a_1=0\\ a_2=-1\\ a_3=2a_0\\ a_4=1+2a_1\\ a_5=2a_2\\ \ldots\\ a_{2n}=(-1)^n+2a_{2n-3},\\ a_{2n+1}=2a_{2n-2}, \quad n\ge 3, \end{cases} \qquad or \qquad \begin{cases} a_0=1\\ a_1=0\\ a_2=-1\\ a_3=2\\ a_4=1\\ a_5=-2\\ \ldots, \end{cases}$$

and

$$\phi(x)=h_0(x,0)-h_2(x,0)+2h_3(x,0)+h_4(x,0)-2h_5(x,0)+\cdots.$$

Example 5 Let $\mathscr{T}=2^{\mathscr{N}_0}\cup\{0\}$. Consider the equation

$$\phi(x)=\sum_{n=0}^{\infty}\frac{2^{n+1}-3}{3^n+4^n}h_n(x,0)+\int_0^x h_1(x,2y)\phi(y)\Delta y.$$

Here

$$\sigma(x)=2x,\quad x\in\mathscr{T}.$$

Then the given equation we can rewrite in the following form

$$\phi(x)=\sum_{n=0}^{\infty}\frac{2^{n+1}-3}{3^n+4^n}h_n(x,0)+\int_0^x h_1(x,\sigma(y))\phi(y)\Delta y.$$

Then

$$\phi(x) = \sum_{n=0}^{\infty} a_n h_n(x,0).$$

Hence,

$$\begin{aligned}
\sum_{n=0}^{\infty} a_n h_n(x,0) &= \sum_{n=0}^{\infty} \frac{2^{n+1}-3}{3^n+4^n} h_n(x,0) + \int_0^x h_1(x,\sigma(y)) \sum_{n=0}^{\infty} a_n h_n(y,0)\Delta y \\
&= \sum_{n=0}^{\infty} \frac{2^{n+1}-3}{3^n+4^n} h_n(x,0) + \sum_{n=0}^{\infty} a_n \int_0^x h_1(x,\sigma(y)) h_n(y,0)\Delta y \\
&= \sum_{n=0}^{\infty} \frac{2^{n+1}-3}{3^n+4^n} h_n(x,0) + \sum_{n=0}^{\infty} a_n h_{n+2}(x,0) \\
&= \sum_{n=0}^{\infty} \frac{2^{n+1}-3}{3^n+4^n} h_n(x,0) + \sum_{n=2}^{\infty} a_{n-2} h_n(x,0) \\
&= -\frac{1}{2} h_0(x,0) + \frac{1}{7} h_1(x,0) + \sum_{n=2}^{\infty} \left(a_{n-2} + \frac{2^{n+1}-3}{3^n+4^n} \right) h_n(x,0),
\end{aligned}$$

whereupon

$$\begin{aligned}
a_0 h_0(x,0) + a_1 h_1(x,0) + \sum_{n=2}^{\infty} a_n h_n(x,0) &= -\frac{1}{2} h_0(x,0) + \frac{1}{7} h_1(x,0) \\
&\quad + \sum_{n=2}^{\infty} \left(a_{n-2} + \frac{2^{n+1}-3}{3^n+4^n} \right) h_n(x,0).
\end{aligned}$$

Therefore

$$\begin{cases} a_0 = -\frac{1}{2} \\ a_1 = \frac{1}{7} \\ a_n = a_{n-2} + \frac{2^{n+1}-3}{3^n+4^n}, \quad n \geq 2. \end{cases}$$

Exercise 2 Find a solution $\phi(x) = \sum_{n=0}^{\infty} a_n h_n(x,0)$ of the following equations.

1. $\phi(x) = e_2(x,0) + e_1(x,0) + \int_0^x h_3(x, y+3)\phi(y)\Delta y, \quad \mathcal{T} = 3\mathcal{Z},$
2. $\phi(x) = \cos_2(x,0) + 4\sin_3(x,0) + \int_0^x h_4(x, y+1)\phi(y)\Delta y, \quad \mathcal{T} = \mathcal{Z},$

3. $\phi(x) = e_1(x,0) + 2 + \cosh_1(x,0) + \int_0^x e_3(x, y+1)\phi(y)\Delta y, \quad \mathscr{T} = \mathscr{Z},$
4. $\phi(x) = \cosh_3(x,0) + \sinh_2(x,0) + \int_0^x h_1(x, 2y)\phi(y)\Delta y, \quad \mathscr{T} = 2^{\mathscr{N}_0} \cup \{0\},$
5. $\phi(x) = \cosh_1(x,0) + e_1(x,0) + e_{-i}(x,0) + \int_0^x h_{10}(x, 3y)\phi(y)\Delta y, \quad \mathscr{T} = 3^{\mathscr{N}_0} \cup \{0\},$
6. $\phi(x) = 10x + 7 - 3\cos_2(x,0) + \int_0^x h_4(x, 2y)\phi(y)\Delta y, \quad \mathscr{T} = 2^{\mathscr{N}_0} \cup \{0\},$
7. $\phi(x) = 2e_{-2i}(x,0) + e_i(x,0) + 2 + \int_0^x h_2(x, y+1)\phi(y)\Delta y, \quad \mathscr{T} = \mathscr{Z},$
8. $\phi(x) = \cos_1(x,0) - 3e_4(x,0) - \int_0^x h_2(x, y+1)\phi(y)\Delta y, \quad \mathscr{T} = \mathscr{Z},$
9. $\phi(x) = \cosh_3(x,0) + \sinh_{-2}(x,0) + 2\int_0^x h_8(x, 2y)\phi(y)\Delta y, \quad \mathscr{T} = 2^{\mathscr{N}_0} \cup \{0\}.$

8.2 Generalized Volterra Integral Equations of the First Kind

Here we consider the generalized Volterra integral equation of the first kind

$$f(x) = \int_0^x K(x, y)\phi(y)\Delta y, \tag{8.4}$$

where the kernel $K(x, y)$ and the function $f(x)$ are given real-valued functions. As in the previous section, we will consider the solution $\phi(x)$ to be analytic, where it has derivatives of all orders, and it possesses Taylor series at $x = 0$ of the form

$$\phi(x) = \sum_{n=0}^{\infty} a_n h_n(x, 0), \tag{8.5}$$

where the coefficients a_n will be determined recurrently. Suppose that

$$f(x) = \sum_{n=0}^{\infty} f_n h_n(x, 0). \tag{8.6}$$

Substituting (8.5) and (8.6) into (8.4) gives

$$\sum_{n=0}^{\infty} f_n h_n(x, 0) = \int_0^x K(x, y) \sum_{n=0}^{\infty} a_n h_n(y, 0)\Delta y.$$

We next equate the coefficients of $h_n(x, 0), \quad n \geq 0$, in both sides of the resulting equation to obtain a recurrence relation in $a_n, \quad n \geq 0$.

Example 6 Consider the equation

$$e_1(x,0)-1=\int_0^x \phi(y)\Delta y.$$

Let

$$\phi(x)=\sum_{n=0}^{\infty} a_n h_n(x,0).$$

Then

$$\begin{aligned}\sum_{n=0}^{\infty} h_n(x,0)-1&=\int_0^x \sum_{n=0}^{\infty} a_n h_n(y,0)\Delta y\\&=\sum_{n=0}^{\infty} a_n\int_0^x h_n(y,0)\Delta y\\&=\sum_{n=0}^{\infty} a_n h_{n+1}(x,0),\end{aligned}$$

whereupon

$$\sum_{n=1}^{\infty} h_n(x,0)=\sum_{n=0}^{\infty} a_n h_{n+1}(x,0).$$

Hence,

$$h_1(x,0)+h_2(x,0)+h_3(x,0)+\cdots=a_0h_1(x,0)+a_1h_2(x,0)+a_2h_3(x,0)+\cdots.$$

Therefore $a_n=1$ for all $n\in\mathscr{N}_0$ and

$$\phi(x)=\sum_{n=0}^{\infty} h_n(x,0)=e_1(x,0).$$

Example 7 Let $\mathscr{T}=\mathscr{Z}$. Consider the equation

$$\sum_{n=4}^{\infty} 3^{n+2}h_n(x,0)=\int_0^x h_3(x,y+1)\phi(y)\Delta y.$$

Here $\sigma(x)=x+1,\quad x\in\mathscr{T}$. Then the given equation can be rewritten in the form

$$\sum_{n=4}^{\infty} 3^{n+2}h_n(x,0)=\int_0^x h_3(x,\sigma(y))\phi(y)\Delta y.$$

Let

$$\phi(x) = \sum_{n=0}^{\infty} a_n h_n(x, 0).$$

Then

$$\begin{aligned}\sum_{n=4}^{\infty} 3^{n+2} h_n(x, 0) &= \int_0^x h_3(x, \sigma(y)) \sum_{n=0}^{\infty} h_n(y, 0)\Delta y \\ &= \sum_{n=0}^{\infty} a_n \int_0^x h_3(x, \sigma(y)) h_n(y, 0)\Delta y \\ &= \sum_{n=0}^{\infty} a_n h_{n+4}(x, 0) \\ &= \sum_{n=4}^{\infty} a_{n-4} h_n(x, 0).\end{aligned}$$

Consequently

$$a_{n-4} = 3^{n+2}, \quad n \geq 4,$$

or

$$a_n = 3^{n+6}, \quad n \geq 0.$$

From here,

$$\phi(x) = \sum_{n=0}^{\infty} 3^{n+6} h_n(x, 0).$$

Example 8 Let $\mathscr{T} = 2^{\mathscr{N}_0} \cup \{0\}$. Consider the equation

$$\sum_{n=0}^{\infty} (3^n - 4^n) h_n(x, 0) = 2\int_0^x h_2(x, 2y)\phi(y)\Delta y.$$

Here $\sigma(x) = 2x, \quad x \in \mathscr{T}$. Then the given equation can be rewritten in the following form

$$\sum_{n=0}^{\infty} (3^n - 4^n) h_n(x, 0) = 2\int_0^x h_2(x, \sigma(y))\phi(y)\Delta y.$$

Let

$$\phi(x) = \sum_{n=0}^{\infty} a_n h_n(x, 0). \tag{8.7}$$

Then

$$\sum_{n=0}^{\infty}(3^n - 4^n)h_n(x,0) = 2\int_0^x h_2(x,\sigma(y))\sum_{n=0}^{\infty} a_n h_n(y,0)\Delta y$$
$$= 2\sum_{n=0}^{\infty} a_n \int_0^x h_2(x,\sigma(y))h_n(y,0)\Delta y$$
$$= 2\sum_{n=0}^{\infty} a_n h_{n+3}(x,0).$$

Hence,

$$(3^0 - 4^0)h_0(x,0) = 0, \quad (3^1 - 4^1)h_1(x,0) = 0, \quad (3^2 - 4^2)h_2(x,0) = 0 \quad on \quad \mathscr{T},$$

which is a contradiction.
Therefore the considered equation has not any solution of the form (8.7).

Exercise 3 Find a solution $\phi(x) = \sum_{n=0}^{\infty} a_n h_n(x,0)$ of the following equations.

1. $\sinh_1(x,0) = 2\int_0^x \phi(y)\Delta y, \quad \mathscr{T} = \mathscr{Z},$
2. $\cosh_1(x,0) + e_1(x,0) = \int_0^x h_7(x, y+1)\phi(y)\Delta y, \quad \mathscr{T} = \mathscr{Z},$
3. $\sinh_{-1}(x,0) = \int_0^x \phi(y)\Delta y, \quad \mathscr{T} = 2^{\mathscr{N}_0} \cup \{0\}.$

8.3 Generalized Volterra Integro-Differential Equations of the Second Kind

We will assume that the solution $\phi(x)$ of the generalized Volterra integro-differential equation of the second kind

$$\phi^{\Delta^n}(x) = f(x) + \lambda\int_0^x K(x,y)\phi(y)\Delta y, \quad \phi^{\Delta^k}(0) = \phi_k, \quad 0 \le k \le n-1, \quad (8.8)$$

is in the form

$$\phi(x) = \sum_{n=0}^{\infty} a_n h_n(x,0). \quad (8.9)$$

Here $K : \mathscr{T} \times \mathscr{T} \longmapsto \mathscr{R}$, $f : \mathscr{T} \longmapsto \mathscr{R}$ are given continuous functions, λ is a parameter.

The first few coefficients a_k, $0 \le k \le n-1$, can be determined by using the initial conditions so that

$$\phi(0) = a_0 = \phi_0,$$

$$\phi^{\Delta}(0) = a_1 = \phi_1,$$

$$\ldots$$

$$\phi^{\Delta^{n-1}}(0) = a_{n-1} = \phi_{n-1}.$$

The remaining coefficients a_k of (8.9) will be determined by applying the series solution method to the generalized Volterra integro-differential equations of the second kind. Substituting (8.9) into (8.8) gives

$$\left(\sum_{k=0}^{\infty} a_k h_k(x,0)\right)^{\Delta^n} = u(x) + \lambda \int_0^x K(x,y) \sum_{k=0}^{\infty} a_k h_k(y,0)\Delta y.$$

We next equate the coefficients of $h_n(x,0)$ into both sides of the resulting equation to determine a recurrence relation in a_k, $k \ge 0$. Solving the recurrence relation will lead to a complete determination of the coefficients a_k, $k \ge 0$, where some of these coefficients will be used from the initial conditions. Having determined the coefficients a_n, $n \ge 0$, the series solution follows immediately upon substituting the derived coefficients into (8.9). The exact solution may be obtained if such an exact solution exists. If an exact solution is not obtainable, then the obtained series can be used for numerical purposes.

Example 9 Consider the equation

$$\phi^{\Delta^2}(x) = 1 + \int_0^x \phi(y)\Delta y, \phi(0) = \phi^{\Delta}(0) = 1.$$

We will search a solution in the form

$$\phi(x) = \sum_{n=0}^{\infty} a_n h_n(x,0).$$

Then, using the initial data,

$$\phi(0) = a_0 = 1,$$

$$\phi^{\Delta}(x) = \left(\sum_{n=0}^{\infty} a_n h_n(x,0)\right)^{\Delta}$$

$$= \sum_{n=0}^{\infty} a_n h_n^{\Delta}(x,0)$$

$$= \sum_{n=1}^{\infty} a_n h_{n-1}(x,0),$$

$$\phi^{\Delta}(0) = a_1 = 1,$$

$$\phi^{\Delta^2}(x) = \left(\sum_{n=1}^{\infty} a_n h_{n-1}(x,0)\right)^{\Delta}$$

$$= \sum_{n=1}^{\infty} a_n h_{n-1}^{\Delta}(x,0)$$

$$= \sum_{n=2}^{\infty} a_n h_{n-2}(x,0)$$

$$= \sum_{n=0}^{\infty} a_{n+2} h_n(x,0).$$

We substitute in the given equation and we find

$$\sum_{n=0}^{\infty} a_{n+2} h_n(x,0) = 1 + \int_0^x \sum_{n=0}^{\infty} a_n h_n(y,0)\Delta y$$

$$= 1 + \sum_{n=0}^{\infty} a_n \int_0^x h_n(y,0)\Delta y$$

$$= 1 + \sum_{n=0}^{\infty} a_n h_{n+1}(x,0).$$

Hence,

$$a_2 h_0(x,0) + a_3 h_1(x,0) + a_4 h_2(x,0) + a_5 h_3(x,0) + \cdots$$

$$= h_0(x,0) + a_0 h_1(x,0) + a_1 h_2(x,0) + a_2 h_3(x,0) + \cdots .$$

Therefore

$$\begin{cases} a_2 = 1 \\ a_3 = a_0 \\ a_4 = a_1 \\ a_5 = a_2 \\ \dots \\ a_n = a_{n-3}, \quad n \geq 6. \end{cases}$$

Consequently $a_n = 1$ for all $n \in \mathcal{N}_0$ and

$$\phi(x) = h_0(x,0) + h_1(x,0) + h_2(x,0) + h_3(x,0) + \cdots$$

$$= e_1(x,0).$$

Example 10 Consider the equation

$$\begin{cases} \phi^{\Delta^3}(x) = h_2(x,0) + 2\int_0^x \phi(y)\Delta y \\ \phi(0) = 1, \quad \phi^{\Delta}(0) = -1, \quad \phi^{\Delta^2}(0) = 0. \end{cases}$$

We will search a solution of the form

$$\phi(x) = \sum_{n=0}^{\infty} a_n h_n(x,0).$$

Using the initial conditions, we obtain

$$\phi(0) = a_0 = 1,$$

$$\phi^{\Delta}(x) = \left(\sum_{n=0}^{\infty} a_n h_n(x,0)\right)^{\Delta}$$

$$= \sum_{n=0}^{\infty} a_n h_n^{\Delta}(x,0)$$

$$= \sum_{n=1}^{\infty} a_n h_{n-1}(x,0),$$

$$\phi^{\Delta}(0) = a_1 = -1,$$

$$\phi^{\Delta^2}(x) = \left(\sum_{n=1}^{\infty} a_n h_{n-1}(x,0)\right)^{\Delta}$$
$$= \sum_{n=1}^{\infty} a_n h_{n-1}^{\Delta}(x,0)$$
$$= \sum_{n=2}^{\infty} a_n h_{n-2}(x,0),$$
$$\phi^{\Delta^2}(0) = a_2 = 0.$$

Also,

$$\phi^{\Delta^3}(x) = \left(\sum_{n=2}^{\infty} a_n h_{n-2}(x,0)\right)^{\Delta}$$
$$= \sum_{n=2}^{\infty} a_n h_{n-2}^{\Delta}(x,0)$$
$$= \sum_{n=3}^{\infty} a_n h_{n-3}(x,0).$$

We substitute in the given equation and we find

$$\sum_{n=3}^{\infty} a_n h_{n-3}(x,0) = h_2(x,0) + 2\int_0^x \sum_{n=0}^{\infty} a_n h_n(y,0)\Delta y$$
$$= h_2(x,0) + 2\sum_{n=0}^{\infty} a_n \int_0^x h_n(y,0)\Delta y$$

$$= h_2(x,0) + 2\sum_{n=0}^{\infty} a_n h_{n+1}(x,0).$$

From here,

$$a_3h_0(x,0) + a_4h_1(x,0) + a_5h_2(x,0) + a_6h_3(x,0) + a_7h_4(x,0) + \cdots$$

$$= 2a_0h_1(x,0) + (2a_1+1)h_2(x,0) + 2a_2h_3(x,0) + 2a_3h_4(x,0) + \cdots .$$

Therefore

$$\begin{cases} a_3 = 0 \\ a_4 = 2a_0 \\ a_5 = 2a_1 + 1 \\ a_6 = 2a_2 \\ a_7 = 2a_3 \\ \cdots \\ a_n = 2a_{n-4}, \quad n \geq 3, \end{cases}$$

whereupon

$$\begin{cases} a_3 = 0 \\ a_4 = 2 \\ a_5 = -1 \\ a_6 = 0 \\ a_7 = 0 \\ a_8 = 4 \\ a_9 = 0 \\ \cdots . \end{cases}$$

Hence,

$$\phi(x) = h_0(x,0) - h_1(x,0) + 2h_4(x,0) - h_5(x,0) + 4h_8(x,0) + \cdots .$$

Exercise 4 Find a solution $\phi(x) = \sum_{n=0}^{\infty} a_n h_n(x, 0)$ of the following equations.

1.
$$\begin{cases} \phi^{\Delta^3}(x) - 2\phi(x) = 1 + 2\int_0^x \phi(y)\Delta y \\ \phi(0) = \phi^{\Delta}(0) = \phi^{\Delta^2}(0) = 1, \end{cases}$$

2.
$$\begin{cases} \phi^{\Delta^2}(x) - \phi^{\Delta}(x) - \phi(x) = 3\int_0^x \phi(y)\Delta y \\ \phi(0) = \phi^{\Delta}(0) = 1, \end{cases}$$

3.
$$\begin{cases} \phi^{\Delta^2}(x) + 5\phi(x) = 3\phi^{\Delta}(x) + \int_0^x h_4(x, \sigma(y))\phi(y)\Delta y \\ \phi(0) = \phi^{\Delta}(0) = 2. \end{cases}$$

Chapter 9
Non-linear Generalized Integral Equations

The generalized Volterra integral equation

$$\phi(x) - \lambda \int_a^x K(x, y)\phi(y)\Delta y = u(x)$$

and the generalized Fredholm integral equation

$$\phi(x) - \lambda \int_a^b K(x, y)\phi(y)\Delta y = u(x)$$

with which we were concerned in the previous chapters, are both linear with respect to the unknown function $\phi(x)$.
In this chapter we shall give a brief sketch of several results for non-linear generalized integral equations.

9.1 Non-linear Generalized Volterra Integral Equations

We consider the equation

$$\phi(x) + \int_a^x F(x, s, \phi(s))\Delta s = u(x). \tag{9.1}$$

We make the following assumptions. The function $F(x, y, z)$ is continuous on a domain D defined by

$$|x| \leq b, \quad |y| \leq b, \quad |z| \leq c, \quad b > a,$$

S.G. Georgiev, *Integral Equations on Time Scales*, Atlantis Studies
in Dynamical Systems 5, DOI 10.2991/978-94-6239-228-1_9

and satisfies the Lipschitz condition with respect to z

$$|F(x, y, z_1) - F(x, y, z_2)| \leq L_1|z_1 - z_2|.$$

The function $u(x)$ is continuous for $|x| \leq b$, vanishes for $x = a$ and satisfies the Lipschitz condition

$$|u(x_1) - u(x_2)| \leq L_2|x_1 - x_2|.$$

Let D' be a domain given by

$$|x| \leq a', \quad a' = \min\{b, \frac{b}{L_2 + M}\}, \quad M = \sup_D |F(x, y, z)|.$$

Then we can define on D' the successive approximations

$$\phi_0(x) = u(x)$$

$$\phi_1(x) = u(x) - \int_a^x F(x, s, \phi_0(s))\Delta s$$

$$\phi_2(x) = u(x) - \int_a^x F(x, s, \phi_1(s))\Delta s$$

$$\ldots$$

$$\phi_n(x) = u(x) - \int_a^x F(x, s, \phi_{n-1}(s))\Delta s, \quad n \geq 3.$$

We can prove that $\{\phi_n(x)\}_{n=1}^{\infty}$ is uniformly convergent on D' and the limit

$$\lim_{n\to\infty} \phi_n(x) = \phi(x)$$

is a unique solution of the Eq. (9.1).

9.2 Non-linear Generalized Fredholm Integral Equations

We consider the equation

$$\phi(x) + \lambda \int_a^b F(x, s, \phi(s))\Delta s = u(x) \tag{9.2}$$

in the unknown function $\phi(x)$. We make the following assumptions. The function $F(x, y, z)$ is continuous on a domain D defined by

$$|x| \leq a, \quad |y| \leq b, \quad |z| \leq c$$

and satisfies the Lipschitz condition with respect to z

$$|F(x, y, z_1) - F(x, y, z_2)| \leq L_1|z_1 - z_2|$$

in D.

The function $u(x)$ is continuous for $a \leq x \leq b$ and

$$\sup_{a \leq x \leq b} |u(x)| = u < c.$$

Let

$$|\lambda| \leq \frac{c - u}{M(b - a)}, \quad M = \sup_{D} |F(x, y, z)|.$$

Then, for such λ, we can define the successive approximations

$$\phi_0(x) = u(x)$$

$$\phi_1(x) = u(x) - \lambda \int_a^b F(x, s, \phi_0(s))\Delta s$$

$$\phi_2(x) = u(x) - \lambda \int_a^b F(x, s, \phi_1(s))\Delta s$$

$$\phi_n(x) = u(x) - \lambda \int_a^b F(x, s, \phi_{n-1}(s))\Delta s, \quad n \geq 3.$$

We can prove that the sequence $\{\phi_n(x)\}_{n=0}^{\infty}$ is uniformly convergent on $[a, b]$ and

$$\lim_{n \to \infty} \phi_n(x) = \phi(x)$$

is a unique solution of the Eq. (9.2).

References

1. M. Bocher, *Integral Equations* (Cambridge University Press, London, 1974)
2. M. Bohner, A. Peterson, *Dynamic Equations on Time Scales: an Introduction with Applications* (Birkhäuser, Boston, 2003)
3. M. Bohner, G. Guseinov, Partial differentiation on time scales. Dyn. Equ. Appl. **13**, 351–379 (2004)
4. L.G. Chambers, *Integral Equations. A Short Course.* (International Textbook Company, London, 1976)
5. C. Corduneanu, *Integral Equations and Applications* (Cambridge University Press, Cambridge, 1991)
6. C. Corduneanu, *Integral Equations and Stability of Feedback Systems* (Academic Press, New York, 1973)
7. R. Courant, D. Hilbert, *Methods of Mathematical Physics*, vols. I and II (Wiley, New York, 1953, 1962)
8. L.M. Delves, J.L. Mohamed, *Computational Methods for Integral Equations* (Cambridge Univ. Press, CambridgeNew York, 1985)
9. L.M. Delves, J. Walsh, *Numerical Solution of Integral Equations* (Oxford University Press, London, 1974)
10. S. Georgiev, Volterra Integro-Differential Equations on Time Scales. Int. J. Appl. Comp. Math. Springer, in press
11. G. Gripenberg, S. Londen, O. Staffans, *Volterra Integral and Functional Equations* (Cambridge Univ. Press, Cambridge, New York, 1990)
12. W. Hackbusch, *Integral Equations: Theory and Numerical Treatment* (Verlag, Boston, Birkhäuser, 1995)
13. A.J. Jerri, *Introduction to Integral Equations with Applications* (Wiley, New York, 1999)
14. R. Kanwal, *Linear Integral Equations* (Birkhauser, Boston, 1997)
15. J. Kondo, *Integral Equations, Oxford Applied Mathematics and Computing Science Series* (The Clarendon Press, Oxford University Press, New York, 1991)
16. R. Kress, *Linear Integral Equations* (Springer, Berlin, 1999)
17. W. Lovitt, *Linear Integral Equations* (Dover, New York, 1950)
18. G. Micula, P. Pavel, *Differential and Integral Equations through Practical Problems and Exercises* (Kluwer, Boston, 1992)
19. S. Mikhlin, *Integral Equations* (Pergamon Press, Oxford, 1957)
20. B.L. Moiseiwitsch, *Integral Equations* (Longman, London and New York, 1977)
21. I. Petrovskii, *Integral Equations* (Grylock Press, New York, 1957)
22. A. Pipkin, *A Course on Integral Equations* (Springer, New York, 1991)
23. W. Pogorzelski, *Integral Equations and Their Applications*, vol. 1 (Pergamon Press, Oxford, 1966)

S.G. Georgiev, *Integral Equations on Time Scales*, Atlantis Studies in Dynamical Systems 5, DOI 10.2991/978-94-6239-228-1

24. A.D. Polyanin, A.V. Manzhirov, *Handbook of Integral Equations*, 2nd edn. (Chapman & Hall/CRC, Boca Raton, FL, 2008)
25. D. Porter, D.S.G. Stirling (1990) Integral Equations. A practical treatment, from spectral theory to applications. Cambridge Texts in Applied Mathematics. (Cambridge University Press, Cambridge, 1990)
26. M. Rahman, *Integral Equations and Their Applications* (WITpress Southampton, Boston, 2007)
27. I. Smirnov, *Integral Equations and Partial Differential Equations* (Addison-Wesley, Reading, Massachusetts, 1964)
28. F. Tricomi, *Integral Equations* (Wiley, New York, 1957)
29. A.M. Wazwaz, *A First Course in Integral Equations* (World Scientific, Singapore, 1997)

Index

S.G. Georgiev, *Integral Equations on Time Scales*, Atlantis Studies in Dynamical Systems 5, DOI 10.2991/978-94-6239-228-1

MIX
Papier aus verantwortungsvollen Quellen
Paper from responsible sources
FSC® C105338

If you have any concerns about our products,
you can contact us on
ProductSafety@springernature.com

In case Publisher is established outside the EU,
the EU authorized representative is:
Springer Nature Customer Service Center GmbH
Europaplatz 3, 69115 Heidelberg, Germany

Printed by Libri Plureos GmbH
in Hamburg, Germany